Ecological Studies, Vol. 165

Analysis and Synthesis

Edited by

Ecological Studies

Volumes published since 1997 are listed at the end of this book.

Springer-Verlag Berlin Heidelberg GmbH

C.A. Brigham M.W. Schwartz (Eds.)

Population Viability in Plants

Conservation, Management, and Modeling of Rare Plants

With 47 Figures and 25 Tables

Springer

Dr. Christy A. Brigham
Dr. Mark W. Schwartz

Department of Environmental Science and Policy
University of California, Davis
One Shields Avenue
Davis, California 95616
USA

Cover illustration: The population viability of Florida scrub plants is strongly affected by fire (see Chap. 11). Endangered and threatened species shown include *Dicerandra frutescens* (top left, photo by Thomas Eisner), *Eryngium cuneifolium* (middle, photo by Amarantha Quintana-Morales), *Hypericum cumulicola* (bottom left, photo by Matthew Trager), and *Eriogonum longifolium* var. *gnaphalifolium* (lower right, photo by Jennifer Schafer).

ISSN 0070-8356
ISBN 978-3-642-07869-9 ISBN 978-3-662-09389-4 (eBook)
DOI 10.1007/978-3-662-09389-4

Library of Congress Cataloging-in-Publication Data

Population viability in plants : conservation, management, and modeling of rare plants /
C.A. Brigham, .M.W. Schwartz (eds.).
p. cm. -- (Ecological studies, ISSN 0070-8356 ; vol. 165)
Includes bibliographical references.

1. Plant conservation. 2. Rare plants--Population viability analysis. I. Brigham, C. A. (Christy A.), 1972- II. Schwartz, M. W. (Mark W.), 1958- III. Ecological studies ; v. 165.

QK86.A1P66 2003
581.68--dc21 2003041555

http://www.springer.de

Originally published by Springer-Verlag Berlin Heidelberg New York in 2003
Softcover reprint of the hardcover 1st edition 2003

Production: Friedmut Kröner, 69115 Heidelberg, Germany
Cover design: *design & production* GmbH, Heidelberg
Typesetting: Kröner, 69115 Heidelberg, Germany

31/3150 YK - 5 4 3 2 1 0 - Printed on acid free paper

Preface

The development of conservation biology as a discipline as well as the increasing public awareness of threatened biodiversity both reflect a bias towards animals over plants. For example, in its early formation, large, vertebrate animals dominated the Endangered Species List. Not until recently has the list come to reflect a more accurate distribution of threatened fauna by including more plants and insects than vertebrate animals. Similarly, one of the common tools of conservation biologists, population viability analysis (PVA), was applied exclusively to animals for approximately the first decade of its use.

In 2002, plants made up 59 % of the species of the US Endangered Species Act. There are 743 species listed as either endangered or threatened and over 100 more are proposed or are candidates for listing. Worldwide, according to the IUCN's 2002 red list of threatened plants and animals, plants made up 49 % of the world list of threatened and endangered organisms. Clearly, plants constitute a large proportion of threatened biodiversity. As such, considerable attention and effort have focused on conserving and restoring plant populations. The US Fish and Wildlife Service is mandated by Congress to write recovery plans for endangered species and has currently written plans for 585 of the 786 threatened or endangered US plants. These plans often require some assessment, formal or informal, of population viability as a prerequisite for delisting. In 1998, the IUCN first published a Red Data for plants of the world, listing 12.5 % of all plant species, at risk. A better understanding of the threats to these nearly 34,000 species that are currently threatened with extinction is required. There is currently a great need for tools to assess viability in plants.

The first published PVA for plants was for Furbish's lousewort (*Pedicularis furbishiae*), an endangered riparian herb of the northeastern US. Published in 1990, this paper by Eric Menges marked the beginning of the application of PVA to plants. Because plants differ from animals in many respects, including patterns of growth and reproduction as well as longevity, the problems encountered in making a PVA for plants often differ from those of animal PVAs.

It has been 12 years since the Furbish's lousewort PVA was published. In that time a number of papers and books have been published on PVAs in general and the application of PVA to plants has increased dramatically (see Chap. 1, this Vol.).

In this volume, we examine threats to plants and take an in-depth look at population viability analysis as applied to plant populations. We build on recent advances in the application of PVA to both plant management and conservation (for recent trends in plant PVAs see discussion in Chap. 1, this Vol.). With plants comprising such a large component of threatened biodiversity, it is worthwhile to specifically address population viability analysis with respect to plants. In our continuing struggle to preserve species, a fuller understanding of current threats to plants, how PVA may be applied to plants, the limitations of the PVA approach with respect to plants, and finally, alternatives to strict PVAs that are likely to be useful for plant conservation may further our abilities to conserve plant biodiversity and thus, biodiversity as a whole. It is with this hope that we offer this volume on plant population viability.

The book contains three sections addressing threats to plants, approaches to modeling plant population viability, and specific difficulties of plant life histories in the context of PVA. Overall, the chapters within the book address the how, when, where, and why of population viability analysis with respect to plants. We also include overviews of the factors that currently threaten plants and how and when these factors should be included in PVAs. Finally, we look forward to the future of plant conservation and PVA and ask what the likely progression of plant conservation will be.

Davis, California,
January 2003

Christy A. Brigham
and *Mark W. Schwartz*

Contents

Contributors

Timothy J. Bell

Chicago State University, Department of Biological Sciences,
9501 South King Drive, Chicago, Illinois 60628, USA

Arjen Biere

Department of Plant Population Biology, Netherlands Institute of Ecology,
NIOO-CTO, PO Box 4, 6666 ZG Heteren, The Netherlands

Marlin L. Bowles

The Morton Arboretum, 4100 Illinois Route 53, Lisle,
Illinois 60532-1293, USA

Christy A. Brigham

Department of Environmental Science and Policy, University of
California, Davis, One Shields Avenue, Davis, California 95616, USA

C. Alex Buerkle

Phillips Hall 330, Department of Biology, University of Wisconsin-Eau
Claire, Eau Claire, Wisconsin 54702-4004, USA

Mark A. Burgman

School of Botany, The University of Melbourne, Parkville 3010, Australia

Daniel F. Doak

Department of Environmental Studies, UCSC, Santa Cruz, California
95064, USA

Bret D. Elderd

Department of Environmental Studies, UCSC, Santa Cruz, California
95064, USA

JANE ELITH

School of Botany, The University of Melbourne, Parkville, 3010 Australia

DAVID W. INOUYE

Department of Biology, University of Maryland, College Park, Maryland 20742-4415, USA

MICHAEL J. LEIGH

Department of Biology, University of Maryland, College Park, Maryland 20742, USA

GARRETT LOWE

Department of Biology, University of Maryland, College Park, Maryland 20742, USA

KATHRYN MCEACHERN

Channel Islands Field Station, Channel Islands National Park, 1901 Spinnaker Drive, Ventura, California 93001-4354, USA

ERIC S. MENGES

Archbold Biological Station, PO Box 2057, Lake Placid, Florida 33862, USA

MANUEL A. MORALES

Department of Biology, Williams College, Williamstown, Massachusetts 01267, USA

J. GERARD B. OOSTERMEIJER

Plant Population Ecologist, Institute for Biodiversity and Ecosystem Dynamics, University of Amsterdam, Section Experimental Plant Systematics, Kruislaan 318, 1098 SM Amsterdam, The Netherlands

N. JOOP OUBORG

Department of Ecology, University of Nijmegen, Toernooiveld 1, 6525 ED Nijmegen, The Netherlands

PEDRO F. QUINTANA-ASCENCIO

Archbold Biological Station, PO Box 2057, Lake Placid, Florida 33862, USA

Loren H. Rieseberg

1001 E. 3rd St., Biology Department, Indiana University, Bloomington, Indiana 47405, USA

Mark W. Schwartz

Department of Environmental Science and Policy, University of California, Davis, One Shields Avenue, Davis, California 95616, USA

Priya Shahani

Department of Ecology and Evolutionary Biology, UCSC, Santa Cruz, California 95064, USA

Diane Thomson

Department of Environmental Science and Policy, University of California, Davis, One Shields Avenue, Davis, California 95616, USA

Diana E. Wolf

1001 E. 3rd St., Biology Department, Indiana University, Bloomington, Indiana 47405, USA

I. Threats to Plant Population Viability

1 Why Plant Population Viability Assessment?

M.W. Schwartz and C.A. Brigham

1.1 Introduction

As Conservation Biology has matured into its role as an integral and applied branch of the Ecological Sciences, there are several topics that have fallen clearly under its umbrella. Assessing the likelihood that a population will persist, a population viability analysis (PVA) is one of several critical central questions of Conservation Biology. PVA has been a popular tool in research and management of populations since its inception. Recently, the number of PVAs performed for plants has increased radically (Menges 2000). Although there has been a long-standing interest in population dynamics in plants (e.g., Silvertown 1987), the issues involved in understanding the population dynamics of rare plants has acquired a recent urgency due to the large numbers of rare plants currently on the endangered species list and threatened worldwide.

Using PVA with plants presents a number of unique challenges. Many aspects of plant life histories present unique challenges to modeling viability (e.g., long-lived stages, hidden life stages). Furthermore, PVA requires an understanding of the threats to plant species and the effects of these threats on population dynamics. These data are unavailable for many endangered species.

This volume addresses three main themes in plant population viability. First, what are the threats and issues facing rare plant populations? Second, what are the modeling approaches available to model plant population viability? And third, what are the peculiarities of plant life histories that make plant models different from animal models and how can these life history attributes be addressed in a modeling framework? Throughout the book, contributing authors have sought to combine discussions of plant biology and natural history with modeling approaches and management considerations. This volume should be useful to those considering plant population viability from a purely theoretical context as well as those faced with managing rare plant populations and seeking guidance in modeling population dynamics.

Ecological Studies, Vol. 165
C.A. Brigham, M.W. Schwartz (Eds.)
Population Viability in Plants

One of the key concepts to emerge from the PVA literature in recent years is that PVAs may not be as reliable as we may have initially hoped (e.g. Beissinger and Westphal 1998; Doak and Morris 1999; Fieberg and Ellner 2000). As a consequence, researchers need to have a clearly defined suite of goals and objectives prior to investing the cost of data collection and analysis for a PVA (Beissinger and Westphal 1998). Not all projects will have the resources to conduct an adequate PVA. Not all plants at risk require a PVA before proceeding with either listing or recovery actions. Nonetheless, ecologists and environmental managers would like to support conservation decision-making with quantitative estimates of the likelihood of population persistence. Hence, a common decision process for endangered species management has been the decision whether to invest the resources in order to conduct a formal PVA. As PVA has entered the conservation biologists' toolkit, it has become important to use the tool judiciously with specific goals in mind. With the general objective of providing guidance on when and how to conduct a PVA in mind, we offer this edited volume to aid researchers conducting research in support of plant species conservation.

1.1.1 Book Structure

The book is divided into three sections. The first section outlines threats to plant population viability and evaluates methods of addressing these threats by PVA. The second section discusses broad modeling approaches and how these approaches might be applied to plants. The final section addresses specific problems of modeling plant populations and how these might be dealt with within the framework of a PVA.

In this first chapter we discuss how plants differ from animals and the unique challenges ecologists and conservation biologists face in trying to understand and manage plant populations. This introductory chapter sets the stage for this edited volume by elucidating several reasons for a book dedicated to plant PVAs: differences in life history attributes; unique conservation challenges; and differences in data issues in plants. These overarching themes are carried through the volume and emerge repeatedly in the different chapters presented within the book. Chapter 2 is a broad overview of threats facing plant populations and includes a categorization of threats as well as a brief assessment of the interactions between threats and viability modeling. Chapter 3 takes an in-depth look at four specific threats, genetic effects, competition, pollination, and herbivory, and evaluates the evidence for their impacts on rare species and the necessity of including them in population viability models. Plant disease is an issue that has rarely been addressed in considerations of population viability but is one that is likely to become more important. This issue is addressed in Chapter 4. Chapter 5 addresses the specific issue of hybridization, which has become more important as agricultural crops and weeds interact with wild populations.

In the second section, we specifically address how to model population viability in plants. The chapters in this section tackle problems associated with modeling the complex life histories of plants. The first chapter in this section, Chapter 6, is an overview of modeling approaches that have been applied to plants and highlights the strengths, weaknesses, and data needs of each of these methods. One type of population viability modeling that has seldom been applied to plants is that of diffusion approximation. This approach holds great promise for modeling plant population viability because it requires only count-based data, the type of data most frequently collected for plants (Morris et al. 1999). The potential and limitations of this approach are addressed in Chapter 7 by Eldred et al. Chapter 8 discusses another type of modeling approach for PVA, that of habitat modeling. This approach holds great promise for plants since many features of plant demography are often driven by habitat factors.

The final section of the book tackles problems of plant life histories head-on. In this section, there are chapters on modeling species with extreme longevity (e.g., many trees and other long-lived perennial; Chap. 9), incorporating biotic interactions into PVA (Chap. 10), addressing disturbance and spatial heterogeneity (Chap. 11), and finally, using PVAs to investigate the future of plant restoration projects (Chap. 12).

Understanding persistence in plants is an issue with more relevance now than ever before. Efforts to manage rare plant populations, design reserves, respond to global climate change, preserve biodiversity, and proactively manage populations all rely on an understanding and an ability to predict population dynamics. This volume provides a state-of-the-art look at population viability in plants. By addressing issues in life history, threats to species, and modeling approaches, we supply the tools needed to advance our knowledge and application of PVA in plants.

1.2 Why Plants Differ

The first question we had to address was whether a volume specifically focused on PVAs for plants was needed. After all, a series of PVA books and reviews have appeared over the past several years that provide thorough background and guidance on PVA (e.g., Soulé 1987; Tuljapurkar and Caswell 1997; Beissinger and Westphal 1998; Groom and Pascual 1998; Sjögren-Gulve and Ebenhard 2000; Caswell 2001). Nonetheless, the need for a PVA book specifically on plants remains for several reasons. First, there are fundamental life history attributes unique to plants that make the challenges inherent in PVA slightly different than for animals. These differences are manifest in the types of data that are available with which to assess plant population viability. Second, the conservation challenges for plants appear to be fundamentally focused on different attributes than for animals. This introductory chapter focuses on these two reasons in detail below.

A third reason to focus a book on PVAs for plants emerges from the perception that PVAs are primarily for and about vertebrate animals. Clearly, the literature on PVAs emerged from vertebrate ecologists (Soulé 1987). It took several years before the first viability assessment of a plant population at risk of extinction was published (Menges 1990). The majority of viability studies continues to focus on vertebrates, with most of the PVA literature directed toward vertebrate targets. Nonetheless, the problem of plant species at risk is very large (Stein et al. 2000). In fact, within the U.S. a larger proportion of plants are at risk of extinction than vertebrates (Stein et al. 2000). Plants have increasingly dominated the list of federally protected species within the U.S. (1.1). Thus, there appears to be a need to increase the attention of plant conservation biologists toward PVAs. A volume dedicated to the topic, we hope, will stimulate such research attention.

Another compelling reason to focus a volume on plant PVA is that plants, perhaps more so than animals, require continued development of population level conservation strategies. Conservation is moving steadfastly toward a focus on larger-scale, spatially explicit habitat, ecosystem-level studies (Christensen et al. 1996; Pickett 1997; Soulé and Terbough 1999). This focus on larger scales is driving many of the high-profile conservation actions toward remote regions where defensible large reserves may reside. Utilizing large-scale approaches may reduce the need for conservation biologists to use PVA as a tool for assessing conservation needs. Certainly it is hoped that these large-scale projects will preemptively protect resources before they become endangered. Rare plants, however, are often habitat specialists restricted to particular edaphic conditions and often not well captured by these large-scale projects. As a consequence, plant conservation may require species-by species approaches more frequently than vertebrates. We provide several examples to illustrate our point.

Serpentine chaparral is the single habitat that houses the largest number of rare plants in California (Pavlik et al. 1994). The second richest habitat for rare plants is vernal pools (Pavlik et al. 1994). These are both patchily distributed habitats with very restricted distributions. As such, large tracts containing numerous patches have seldom been the target of conservation plans that try to preemptively protecting a wide array of species. Species within these edaphic anomalies are frequently addressed on an individual basis. For example, the Ione manzanita (*Arctostaphylos myrtifolia*) inhabits a single patch of Ione soils (Gankin 1957). A protected area is specifically set aside for this species. Conservation of this species will require recovery on this site and this site alone.

Another example is found in the association of plants with centers of high population density within the US. Schwartz et al. (2002) found that nearly one

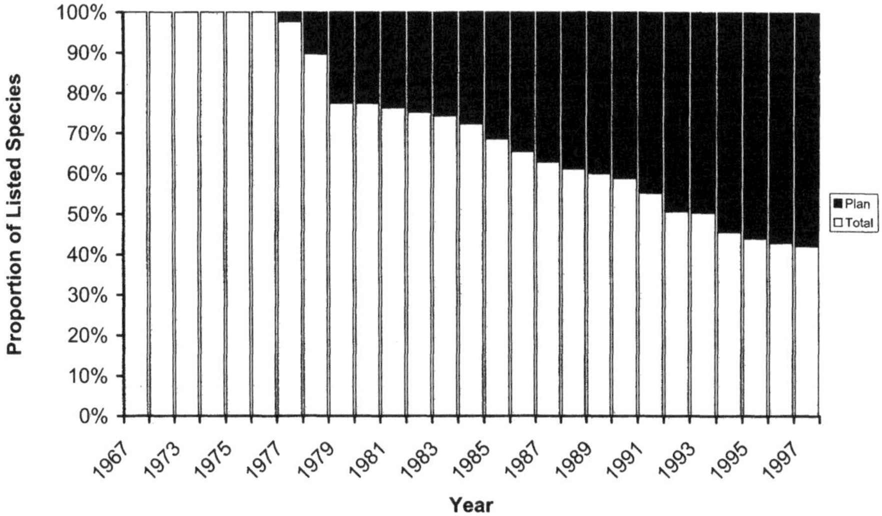

Fig. 1.1. A bar graph showing the increasing percentage of plants (*black*) relative to all other species (*white*) on the US Endangered Species List from the inception of the list (1967) until 1997. As of 1997, 754 of the 1,302 (58 %) taxa listed were plants

quarter of endangered plant occurrences are in the 8.4 % of land that comprises the 40 most populous cities in the US, where 50 % of Americans live. These populations are often already on some kind of protected habitat, and at least for San Francisco, Schwartz et al. found that there is currently little monitoring of population performance. Thus, these authors argue for additional monitoring, as management activities in urban areas could help many populations survive. The majority of these sites surveyed were multiple-use sites that were also managed for public recreation or that were in the public service (i.e., water management, defense).

Despite the many advantages of approaching conservation from an ecosystem standpoint, a single-species approach is likely to continue to play a central role in the conservation of diversity for plants. With numerous PVA projects on-going for high profile vertebrates, one has to think that this will also be the case for vertebrates (e.g., Lahaye et al. 1994; Lamberson et al. 1995; McKelvey 1996; Mills et al. 1996). Given that it is often easy, relative to animals, to estimate population size from year to year, it seems likely that species-level plant conservation may entail extensive use of PVA. Thus, it is timely to present advances in plant PVAs so as to promote the further use and development of these tools to critical conservation problems.

1.3 Life Histories of Plants

The plant kingdom spans an incredible variety of life history attributes, many of which are critically important when considering viability assessment. The simplest distinction between plants and animals is related to distinctiveness of gender expression. Population performance in vertebrates typically relies on estimating fecundity of female individuals (e.g., Mills et al. 1996). In plants, however, species vary in sexual expression from dioecy, the norm for animals, to monoecy. For plants with clear-cut life history attributes, this distinction results in a simple decision: for dioecious plants, population performance can be projected in terms of reproductive females, while the whole population is used for monoecious species. The picture, however, can be considerably more complex. Individuals within populations of a monoecious species may, in actuality, vary in sexual expression from effectively entirely male to effectively entirely female (e.g., Kudo 1993; Pickering and Ash 1993; Kudo and Maeda 1998; Wright and Barrett 1999; Sarkissian et al. 2001). In these cases, determining the unit of study for a population projection can be difficult without first measuring the fitness contribution of individuals based on sexual expression via floral display.

Yet another aspect of sexual reproduction in plants that poses a problem to conservation biologists is hybridization. Hybridization is much more common in plants than in animals (Chap. 5) but the outcome of hybridization among plant species in nature is not clear. Under some conditions hybridization may lead to extinction while in others stable hybrid zones may be formed (Chap. 5). Understanding and predicting the consequences of hybridization may be of critical importance for some endangered plant species.

Plants also pose significant challenges to PVA as a result of clonal growth, which creates two problems for biologists trying to predict future population size. First, clonal growth can obscure the identity of individuals. For example, delineating an individual among beach strawberry (*Fragaria chiloensis*) is impossible without resorting to genetic analysis (Alpert et al. 1993). Given that estimating future population size is generally predicated upon estimating current population size, this poses a significant problem. Even using some proxy for the number of individuals (e.g., number and sizes of tussocks), there remain difficulties in handling clonal growth. In particular, clonal growth, and the budding off of independent ramets, can be a secondary mode of reproduction. Determining if and when the physical connection, and hence the fate, of ramets becomes independent can be very challenging. Thus, matrix models for plant population projection must consider the classification of new ramets as a reproduction (e.g., Montalvo et al. 1997; Damman and Cain 1998; Sharma 2001; Weekley and Race 2001).

The sedentary nature of plants also makes them fundamentally different from animals. The immobility of plants may seem to be simple and obvious to the absurd, but this distinction provides significant advantages to plant PVAs. Assessing survivorship of individuals can be substantially easier in plants than animals. Furthermore, many plant populations have been surveyed through time to estimate population size or density. These data can be used to predict future populations through methods that are numerically much simpler than matrix model PVAs (Caswell 2001; Chap. 7, this Vol.). These simpler methods have strengths and weaknesses but should be considered as possible tools for plants due to the large amount of count-based data on plant species of concern (Chap. 7). There are likely many species for which a PVA could be done now, but for which we have failed to use appropriate methods to create a PVA.

Plants are characterized by indeterminate life-span and indeterminate growth. Plants can even shrink from one year to the next (e.g., Bierzychudek 1999; Schwartz et al. 2000). As a consequence, critical life stages (e.g., seed, seedling, juvenile, adult), not age, are used for population projection. This problem, however, is exacerbated by complex stage transitions, with individuals skipping stages upward and downward (e.g., Schwartz, this Vol.). As a consequence, the typical plant PVA using matrix models results in a matrix that contains many fewer non-zero elements than included within an animal PVA (Fig. 1.2). Additional non-zero elements may make the math harder to do by hand, but with computers this is hardly a constraint. Instead, the additional non-zero elements pose a problem because of error inflation. Additional non-

1) Plant

$$\begin{bmatrix} a_{1,1} \\ a_{2,1} \\ a_{3,1} \\ a_{4,1} \\ a_{5,1} \end{bmatrix} \times \begin{bmatrix} p_{1,1} & p_{1,2} & 0 & p_{1,4} & p_{1,5} \\ p_{2,1} & p_{2,2} & p_{2,3} & 0 & 0 \\ p_{3,1} & p_{3,2} & p_{3,3} & p_{3,4} & 0 \\ 0 & p_{4,2} & p_{4,3} & p_{4,4} & p_{4,5} \\ 0 & 0 & p_{5,3} & p_{5,4} & p_{5,5} \end{bmatrix} = \begin{bmatrix} a_{1,2} \\ a_{2,2} \\ a_{3,2} \\ a_{4,2} \\ a_{5,2} \end{bmatrix}$$

2) Animal

$$\begin{bmatrix} a_{1,1} \\ a_{2,1} \\ a_{3,1} \\ a_{4,1} \\ a_{5,1} \end{bmatrix} \times \begin{bmatrix} 0 & 0 & 0 & p_{1,4} & p_{1,5} \\ p_{2,1} & 0 & 0 & 0 & 0 \\ 0 & p_{3,2} & 0 & 0 & 0 \\ 0 & 0 & p_{4,3} & 0 & 0 \\ 0 & 0 & 0 & p_{5,4} & p_{5,5} \end{bmatrix} = \begin{bmatrix} a_{1,2} \\ a_{2,2} \\ a_{3,2} \\ a_{4,2} \\ a_{5,2} \end{bmatrix}$$

Fig. 1.2. Two hypothetical sets of vectors and matrices for projecting populations. In each, the left vector is a characterization of the number of individuals in each of five age or stage classes within a population at time *t*. Multiplying this vector by the matrix of transition probabilities results in a predicted vector of population sizes within the age or stage classes at time *t*+1. The *top matrix* represents a typical plant population in which individuals may reside in stage classes across time periods, regress, or skip forward stage classes. Thus, for plants many or most of the matrix elements may be non-zero. The *lower equation* describes a typical age class matrix for an animal population in which at each time step all individuals must either move up an age class or die, resulting in numerous zero transition probabilities

zero elements often include some transitions that are rare events, and thus likely to be poorly estimated by demographic data. Increasing life history complexity increases the likelihood of incorrect prediction and of expanded confidence limits around population projections.

Perhaps the most significant attribute that provides major challenges to the biologist in pursuit of a PVA on some plant species is the indistinct fate of propagules. An estimate of average fecundity is not hard for the many vertebrates, nor does fecundity vary much. Moreover, because of site fidelity, fledging rates is often well understood. For animals, the difficulty often lies in assessing adult survivorship. In contrast, individual plants can often produce enormous quantities of seeds, with a very low success rate of these seeds becoming adults. Annual variation in environmental conditions can, likewise, result in orders of magnitude differences in inter-annual seed production from individual plants (e.g., Sork et al. 1993; Koenig et al. 1994; Haase et al. 1995; Hilton and Packham 1997; Herrera et al. 1998; Shibata et al. 1998; Healy et al. 1999; Houle 1999; Koenig and Knops 2000; Selas 2000; Vila and Lloret 2000; Piovesan and Adams 2001). Estimating seed production and accounting for the fates of these seeds, whether they disperse out of the local population or not, whether they succeed when dispersing long distances, how long individuals may remain dormant in a seed bank, and the likelihood of a seed reaching maturity, is often very difficult (Leck et al. 1989; Chap. 7, this Vol.). Similarly, adult dormancy can add complexity to modeling population performance (Shefferson et al. 2001). As a consequence, matrix elements associated with reproduction are often estimated with large error terms. For plants it can be necessary to conduct specific experiments to determine both seed germination rates and seedling survivorship rates. Given that many plants have exceedingly low seedling survivorship rates, this can require an exorbitantly large experiment to do well.

Additionally, plants may rely on other species for persistence (pollination, dispersal, etc.; Chaps. 2, 3, 10) or suffer negative impacts of other species (e.g., herbivory, seed predation; Chaps. 3, 10). While animals may also have strong trophic interactions, these interactions are another aspect of plant populations that need consideration in PVA (Chaps. 3, 10).

1.4 Conclusion: Conservation Challenges for Plants

Destruction and degradation of habitat are the single largest factors associated with the loss of species (Wilcove et al. 1998, 2000). Although both plants and animals succumb to habitat loss, the effects of habitat degradation are likely to be more varied and more directly experienced by plants by virtue of their being sedentary (Chap. 4). Individual plants cannot relocate to better habitat. In addition, what constitutes habitat degradation can be difficult for scientists to assess. Clearly, invasive species have been prominently identified as a driver of habitat degradation (Cronk and Fuller 1995; Mack et al. 2000; Mack and Lonsdale 2001; Chaps. 2, 3, this Vol.). Invasive weeds can displace plants, invasive herbivores can drive populations to low numbers, invasive diseases can increase mortality (Chap. 4). Invasive species may also affect rare species through hybridization (Chap. 5).

It may be much more difficult to discern how and when plant populations are threatened by a multitude of other environmental drivers of global change. For example, increased nitrogen loading is enriching plant communities worldwide (Mickler et al. 2000; Mosier et al. 2001; Eviner et al. 2001). Ecological experiments that have augmented ecosystems with nitrogen universally show that biomass productivity is N-limited and that N-enrichment causes a decrease in plant species diversity (Gross et al. 2000). Thus, scientists ought to expect that N-deposition in an N-limited environment will alter population growth rates of species, with a few species increasing in abundance and many others experiencing a decreased population growth rate. Attributing the decline of a plant population to increased N-deposition is likely to be very difficult. Managing for persistence is likely to be even more difficult. For example, managing fire in excess over historically observed frequencies may be required to volatilize N from tallgrass prairie in order to maintain species diversity. Given the current concern over the use of fire in terms of insect conservation (e.g., Panzer and Schwartz 2000; Swengel 2001), this may not be a realistic option.

Global warming also poses problems relative to detection of population problems. Global warming, however, carries a unique set of options. It has been widely hypothesized that warming scenarios will result in many currently rare and geographically restricted species being restricted to habitats that lie entirely outside the climate envelope in which they now reside (Schwartz 1992). As a consequence, plants with restricted distributions may need to shift northward, or upslope, in order to remain within a zone of favorable climate. The rate at which these populations are likely to respond is predicted to be insufficient to keep pace with climate change (Schwartz 1992, 1993; Dyer 1995; Schwartz et al. 2001), As a consequence, species with restricted ranges that are also climatically limited may experience population persistence problems (Schwartz 1992). A spatial viability model may be required in order to fully understand the consequences of global warming on population persistence probabilities. Nonetheless, stress-induced declines in population growth rate will be very difficult to pin on global warming as the environmental driver. It will be much more likely to observe populations with increased susceptibility to proximate mortality factors (e.g., drought stress, herbivory, disease).

There are, of course, additional drivers of change that may have significant effects on the ability of plants to persist. These include altered disturbance regimes (Chap. 11), the deposition of ozone and other pollutants, as well as grazing and other altered biotic interactions (Chap. 10). Each of these drivers carries particular attributes that require specialized treatment within the context of a PVA in order to assess population status. Despite the development of software tools to assist in viability assessment, individual problems often require unique solutions. It is not enough to know how to input data into software programs that will provide an extinction likelihood projection. Understanding the complexities of individual problems should drive the researcher toward problem formation, which will, in turn, drive the methods best suited to the problem. We hope that this volume provides guidance and examples that will allow the reader to not only discern when to use PVA for a particular plant conservation problem, but also what kind of PVA to use, what other kinds of information to collect in order to make management recommendations, and how to put these together into a cohesive conservation strategy. There is no magic formula. Every problem presents its own difficulties. This book is intended to demonstrate methods and higlight issues by example and is not intended as a step-by-step how-to book (Morris and Doak 2003).

References

Alpert P, Lumaret R, Di Giusto F (1993) Population structure inferred from allozyme analysis in the clonal herb *Fragaria chiloensis* (Rosaceae). Am J Bot 80(9):1002–1006

Beissinger SR, Westphal MI (1998) On the use of demographic models of population viability in endangered species management. J Wildlife Manage 62(3):821–841

Bierzychudek P (1999) Looking backwards: assessing the projections of a transition matrix model. Ecol Appl 9(4):1278–1287

Caswell H (2001) Matrix population models: construction, analysis, and interpretation. Sinauer Associates, Sunderland, MA

Christensen NL, Bartuska AM, Brown JH, et al. (1996) The report of the Ecological Society of America committee on the scientific basis for ecosystem management. Ecol Appl 6(3):665–691

Cronk QCB, Fuller JL (1995) People and plants conservation manuals: plant invaders: the threat to natural ecosystems. Chapman and Hall, London, xiv; p 241

Damman H, Cain ML (1998) Population growth and viability analyses of the clonal woodland herb, *Asarum canadense*. J Ecol 86(1):13–26

Doak DF, Morris W (1999) Detecting population-level consequences of ongoing environmental change without long-term monitoring. Ecology (Washington, DC) 80(5):1537–1551

Dyer JM (1995) Assessment of climatic warming using a model of forest species migration. Ecol Model 79(1–3):199–219

Eviner VT, Chapin FS, Vaughn CE (2001) Nutrient manipulations in terrestrial ecosystems. In: Sala OE, Jackson RB, Mooney HA, Howarth RW (eds) Methods in ecosystem science. Springer, Berlin Heidelberg New York, pp 291–307

Fieberg J, Ellner SP (2000) When is it meaningful to estimate an extinction probability? Ecology (Washington, DC) 81(7):2040–2047

Gankin R (1957) The variation pattern and ecological restrictions of *Arctostaphylos myrtifolia* Parry (Ericaceae): ii, 38 leaves. Thesis, University of California, Davis
Groom MJ, Pascual MA (1998) The analysis of population persistence: an outlook on the practice of viability analysis. In: Fiedler PL, Karieva PM (eds) Conservation biology for the coming decade, 2nd edn. Chapman Hall, New York, pp 4–27
Gross KL, Willig MR, Gough R et al. (2000) Patterns of species density and productivity at different spatial scales in herbaceous plant communities. Oikos 89(3):417–427
Haase P, Pugnaire FI, Incoll LD (1995) Seed production and dispersal in the semi-arid tussock grass *Stipa tenacissima* L. during masting. J Arid Environ 31(1):55–65
Healy WM, Lewis AM, Boose EF (1999) Variation of red oak acorn production. For Ecol Manage 116(1–3):1–11
Herrera CM, Jordano P, Guitian J, Traveset A (1998) Annual variability in seed production by woody plants and the masting concept: reassessment of principles and relationship to pollination and seed dispersal. Am Nat 152(4):576–594
Hilton GM, Packham JR (1997) A sixteen-year record of regional and temporal variation in the fruiting of beech (*Fagus sylvatica* L.) in England (1980–1995). Forestry (Oxford) 70(1):7–16
Houle G (1999) Mast seeding in *Abies balsamea*, *Acer saccharum* and *Betula alleghaniensis* in an old growth, cold temperate forest of north-eastern North America. J Ecol 87(3):413–422
Koenig WD, Knops JMH (2000) Patterns of annual seed production by Northern Hemisphere trees: a global perspective. Am Nat 155(1):59–69
Koenig WD, Mumme RL, Carmen WJ, Stanback MT (1994) Acorn production by oaks in central coastal California: variation within and among years. Ecology (Tempe) 75(1):99–109
Kudo G (1993) Size-dependent resource allocation pattern and gender variation of *Anemone debilis* Fisch. Plant Species Biol 8(1):29–34
Kudo G, Maeda T (1998) Size-dependent variation of phenotypic gender and functional gender of a spring ephemeral, *Anemone debilis* Fisch. Plant Species Biol 13(2–3):69–76
Lahaye WS, Gutierrez RJ, Akcakaya HR (1994) Spotted owl metapopulation dynamics in southern California. J Animal Ecol 63(4):775–785
Lamberson RH, Noon BR, Voss C, Mckelvey KS (1995) Reserve design for territorial species: the effects of patch size and spacing on the viability of the northern spotted owl. Conserv Biol:38–48
Leck MA, Parker VT, Simpson RL (1989) Ecology of soil seed banks. Academic Press, San Diego
Mack RN, Lonsdale WM (2001) Humans as global plant dispersers: getting more than we bargained for. Bioscience 51(2):95–102
Mack RN, Simberloff D, Lonsdale WM, et al. (2000) Biotic invasions: causes, epidemiology, global consequences, and control. Ecol Appl 10(3):689–710
McKelvey R (1996) Viability analysis of endangered species: a decision-theoretic perspective. Ecol Model 92(2–3):193–207
Menges ES (1990) Population viability analysis for an endangered plant. Conserv Biol 4(1):52–62
Menges ES (2000) Population viability analyses in plants: challenges and opportunities. Trends Ecol Evol 15:51–56
Mickler RA, Birdsey RA, Hom J (2000) Ecological studies. Responses of Northern US forests to environmental change. Springer, Berlin Heidelberg New York, pp i-xix; 1–578
Mills LS, Hayes SG, Baldwin C, et al. (1996) Factors leading to different viability predictions for a grizzly bear data set. Conserv Biol 10(3):863–873

Montalvo AM, Conard SG, Conkle MT, Hodgskiss PD (1997) Population structure, genetic diversity, and clone formation in *Quercus chrysolepis* (Fagaceae). Am J Bot 84(11):1553–1564
Morris W, Doak D, Groom M, Kareiva P, Fieberg J, Gerber L, Murphy P, Thomson D (1999) A practical handbook for population viability analysis. The Nature Conservancy Press, New York
Morris WF, Doak DF (2003) Quantitative Conservation Biology: Theory and Practice of Population Viability Analysis. Sinauer & Associates, Sunderland, MA, 480 pp
Mosier AR, Bleken MA, Chaiwanakupt P, et al. (2001) Policy implications of human-accelerated nitrogen cycling. Biogeochemistry (Dordrecht) 52(3):281–320
Panzer R, Schwartz M (2000) Effects of management burning on prairie insect species richness within a system of small, highly fragmented reserves. Biol Conserv 96(3):363–369
Pavlik BM, Skinner MW et al. (1994) California Native Plant Society's inventory of rare and endangered vascular plants of California. California Native Plant Society, Sacramento, CA
Pickering CM, Ash JE (1993) Gender variation in hermaphrodite plants: evidence from five species of alpine *Ranunculus*. Oikos 68(3):539–548
Pickett ST (1997) The ecological basis of conservation: heterogeneity, ecosystems, and biodiversity. Chapman and Hall, New York
Piovesan G, Adams JM (2001) Masting behaviour in beech: linking reproduction and climatic variation. Can J Bot 79(9):1039–1047
Sarkissian TS, Barrett SCH, Harder LD (2001) Gender variation in *Sagittaria latifolia* (Alismataceae): is size all that matters? Ecology (Washington DC) 82(2):360–373
Schwartz MW (1992) Potential effects of global climate change on the biodiversity of plants. For Chron 68:462–471
Schwartz MW (1993) Modeling effects of habitat fragmentation on the ability of trees to respond to climatic warming. Biodiv Conserv 2:51–61
Schwartz MW, Hermann SM, Van Mantgem PJ (2000) Population persistence in Florida torreya: comparing modeled projections of a declining coniferous tree. Conserv Biol 14(4):1023–1033
Schwartz MW, Iverson LR, Prasad AM (2001) Predicting the potential future distribution of four tree species in Ohio using current habitat availability and climatic forcing. Ecosystems 4(6):568–581
Schwartz MW, Jurjavcic NL, O'Brien JM (2002) Conservation's disenfranchised urban poor. Bioscience 52:601–606
Selas V (2000) Seed production of a masting dwarf shrub, *Vaccinium myrtillus*, in relation to previous reproduction and weather. Can J Bot 78(4):423–429
Sharma IK (2001) Understanding clonal diversity patterns through allozyme polymorphism in an endangered and geographically restricted Australian shrub, *Zieria baeuerlenii* and its implications for conservation. Biochem Syst Ecol 29(7):681–695
Shefferson RP, Sandercock BK, Proper J, Beissinger SR (2001) Estimating dormancy and survival of a rare herbaceous perennial using mark-recapture models. Ecology (Washington, DC) 82(1):145–156
Shibata M, Tanaka H, Nakashizuka T (1998) Causes and consequences of mast seed production of four co-occurring *Carpinus* species in Japan. Ecology (Washington, DC) 79(1):54–64
Silvertown J (1987) Introduction to plant population ecology. Longman Scientific and Technical, Essex, 229 pp
Sjögren-Gulve P, Ebenhard T (2000) The use of population viability analyses in conservation planning. Munksgaard International Publishers, Malden, MA, USA

Sork VL, Bramble J et al. (1993) Ecology of mast-fruiting in three species of North American deciduous oaks. Ecology (Washington, DC) 74(2):528–541
Soulé ME (1987) Viable populations for conservation. Cambridge University Press, Cambridge
Soulé ME, Terborgh J (1999) Continental conservation: scientific foundations of regional reserve networks. Island Press, Washington, DC
Stein BA, Kutner LS, Adams JS (2000) Precious heritage: the status of biodiversity in the United States. Oxford University Press, Oxford
Swengel AB (2001) A literature review of insect responses to fire, compared to other conservation managements of open habitat. Biodiv Conserv 10(7):1141–1169
Tuljapurkar S, Caswell H (1997) Structured-population models in marine, terrestrial, and freshwater systems. Chapman and Hall, New York
Vila M, Lloret F (2000) Seed dynamics of the mast seeding tussock grass *Ampelodesmos mauritanica* in Mediterranean shrublands. J Ecol 88(3):479–491
Weekley CW, Race T (2001) The breeding system *of Ziziphus celata* Judd and D.W. Hall (Rhamnaceae), a rare endemic plant of the Lake Wales Ridge, Florida, USA: implications for recovery. Biol Conserv 100(2):207–213
Wilcove DS, Rothstein D, Dubow J et al. (1998) Quantifying threats to imperiled species in the United States. Bioscience 48(8):607–615
Wilcove DS, Rothstein D, Dubow J et al. (2000) Leading threats to biodiversity: what's imperiling US species. In: Stein BA, Kutner LS, Adams JS (eds) Precious heritage: the status of biodiversity in the United States. pp 239–254
Wright SI, Barrett SCH (1999) Size-dependent gender modification in a hermaphroditic perennial herb. Proc R Soc Lond Ser B Biol Sci 266(1416):225–232

2 Threats to Rare Plant Persistence

J.G.B. OOSTERMEIJER

2.1 Introduction

Many plant populations and metapopulations all over the world are being threatened by a variety of factors (Semenova and van der Maarel 2000). In most cases, threats are associated with human actions of some sort (Lande 1998), which implies that we can often take action to remove or at least counter them. However, for many threats it is difficult (1) to understand how they ultimately affect the viability of specific plant populations or metapopulations, (2) to untangle their interaction with other threats, and (3) to come up with effective methods to alleviate them.

These difficulties can be illustrated by one of the most important threats to population viability, habitat fragmentation (Harrison and Bruna 1999). The breaking up of continuous habitat into a series of large to small fragments is generally associated with small population size (Aizen and Feinsinger 1994a; Lande 1998; Oostermeijer et al. 1998b; Kéry et al. 2000; Norderhaug et al. 2000) and restricted exchange of pollen or seeds (Kwak et al. 1998; Sork et al. 1999), which have important demographic consequences for processes on the population and metapopulation level. However, fragmentation also increases edge effects and deterioration of habitat quality (Gascon and Lovejoy 1998; Jules 1998; Debinski and Holt 2000), which may have effects on individual plant growth and flowering. Fragmentation may also alter plant-pathogen and plant-herbivore dynamics (Chap. 4, this Vol.). In addition, the small population size and reduced gene flow may lead to changes in genetic variation through drift and inbreeding (Barrett and Kohn 1991; Young et al. 1996). Few studies in conservation biology have been able to address and integrate all of these parameters in a population viability analysis (PVA) of a single plant species with a fragmented metapopulation. Yet, despite the ongoing discussion on the relative importance of demographic vs. genetic stochasticity for population viability (Lande 1988; Nunney and Campbell 1993; Menges 2000), it is clear that their interactions are the most important (Oostermeijer 2000a). Ecological conditions can be optimal for a certain species, but if the popula-

Ecological Studies, Vol. 165
C.A. Brigham, M.W. Schwartz (Eds.)
Population Viability in Plants

tion is small and the offspring are not able to establish because they are inbred, suitable growing conditions are meaningless. Conversely, high rates of selfing in a naturally outcrossing species are of no significance if the environment provides no safe sites for recruitment of the highly inbred offspring (Oostermeijer 2000a).

However, population viability ultimately is a demographic concept. Populations only go extinct when the average death rate is higher than the average birth rate (Boyce 1992; Menges 2000). This implies that, if we want to determine the contribution of a specific threat to the viability of a specific species, we should attempt to ascertain its effect on these vital rates. When doing this, however, we need to be aware that interactions with other threats may influence our results. For example, if we study the effect of acidification on a small population of a rare plant species, inbreeding may already have lowered the population's tolerance to toxic aluminum levels. Inbreeding may also have reduced the growth rate or competitive ability of seedlings (Oostermeijer et al. 1995b; Kéry et al. 2000), which could result in underestimation of potential recruitment levels. We should be careful with interpreting elasticities of projection matrices derived from data of small, declining populations (Silvertown et al. 1996). In at least one case in The Netherlands, this has led to a discussion of the importance of management in stimulating either recruitment or adult survival in small, remnant populations (Oostermeijer 1995; Ouborg et al. 1995).

Future plant PVAs should try to assess the interacting effects of all the different threats on the demography of a large, viable population. In this chapter, the different types of threats discussed in the conservation biological literature will be classified. This will be followed by a brief discussion of each of these types of threats and its demographic consequences, in order to provide insight on how to handle them in future PVAs. In the last section, the management of the major threats will be discussed.

2.2 Delineating Different Types of Threats

The recent literature documents a wide array of threats to rare plants (Table 2.1). In this chapter, the threats to population viability of plants that were mentioned in the recent literature are classified into three main types: (1) threats imposed by changes in the environment, either by natural or anthropogenic causes, (2) threats resulting from disturbance of crucial interactions with other species, and (3) genetic threats. Threats of the first category also involve direct destruction of plant populations or individuals. The threats of the second and third category are generally associated with small population size, low population density and isolation; hence, they are often a consequence of threats of the first category. We could therefore regard the threats to plant habitats or environmental conditions as primary, and those of categories two and three as secondary threats. In general, there are strong interactions between the different threats. The connecting factor is nearly always human intervention in natural patterns and processes (Lande 1998).

2.2.1 Environmental Threats

2.2.1.1 Climate Change

Climate change is often seen as a major environmental threat to plant populations. Logically, the focus of most of the recent research on the effects of climate change has been on plants that occur at clear ecological boundaries, such as arctic and alpine environments (Inouye and McGuire 1991; Lesica and Steele 1996; Saetersdal and Birks 1997; Stenstrom and Jonsdottir 1997; Stenstrom et al. 1997; Price and Waser 1998; Totland 1999). Interesting manipulative experiments have also been performed in limestone grasslands (Akinola et al. 1998; Sternberg et al. 1999) and on species at the periphery of their range (Woodward 1997; Fox et al. 1999).

Effects of climate change on plants can be direct, affecting growth, flowering and phenology (Lesica and Steele 1996; Price and Waser 1998; Fox et al. 1999; Post and Stenseth 1999; Totland 1999), as well as indirect, e.g., through modification of interactions with herbivores (Fox et al. 1999), pollinators (Stenstrom et al. 1997; Harrison 2000), or effects on microsite availability in the plant community (Sternberg et al. 1999).

In an undisturbed natural landscape subjected to gradual climate change, species would either move their range or adapt to the changing environmental conditions. Below, it will be shown that both processes are severely inhibited for many rare species in the presently fragmented landscape. In this light, it seems quite likely that habitat fragmentation will act as the main bottleneck under (rapid) climate change.

2.2.1.2 Habitat Fragmentation and Degradation

The most frequently cited environmental threats to plant populations are habitat destruction, degradation, and fragmentation, changes in land use, herbicides, eutrophication, and fire suppression (Table 2.1). These threats all directly alter plant habitats (in area, stability, connectivity or quality) to such a strong extent that the viability of many populations is significantly reduced. Most of these specific threats are associated with human actions of some sort, with a few exceptions, such as the natural component of climate change (Zwiers and Weaver 2000) and natural catastrophes like earthquakes and hur-

Table 2.1. An overview of the different threats to populations and metapopulations of rare and endangered plant species described in recent conservation biological literature, classified into three major threat types

Threat	Habitat effect	Demographic effects
Environmental stochasticity		
Climate change	Changed temperature and precipitation, microsite availability, new herbivores	Phenology, growth, fecundity, migration
Habitat destruction	All aspects	All aspects
Erosion, land slides	Substrate structure and availability, microclimate, slope	Survival, recruitment, establishment
Deforestation, urbanization, land clearance, other human actions	Destruction, microclimate, erosion	Survival, growth, fecundity
Habitat degradation	Water, nutrient and microsite availability, vegetation structure, competition	Survival, growth, fecundity
Changes in land use, abandonment	Fragmentation, successional closing of the vegetation, competition	Migration, recruitment, growth, fecundity
Eutrophication, fertilization	Pollution, vegetation structure, competition	Recruitment, establishment, growth, fecundity
Fire suppression	Nutrient and microsite availability, vegetation structure, predation	Seed release, recruitment, establishment, growth, fecundity
Modern agricultural practices	Nutrient and water availability, competition, disturbance intensity and frequency	All aspects
Nitrogen deposition	Nutrient availability, competition, vegetation structure	Seed dormancy, recruitment, growth, fecundity
Pesticides and herbicides	Vegetation structure, competition	Survival, recruitment, growth, fecundity
Tourist activities	Trampling	Survival, recruitment, fecundity
Installing protected reserve	Changed disturbance regime, competition	Recruitment, growth, fecundity
Military training	Trampling, fire, soil disturbance	Survival, recruitment, growth, fecundity
Habitat fragmentation	Isolation, edge effects, small size	Migration, survival, growth, fecundity
Harvesting, overexploitation	Small effect (trampling)	Survival, recruitment, growth, fecundity
Disturbed biotic interactions		
Introduced exotic species	Increased competition, site preemption	Recruitment, growth, fecundity
Demographic swamping	Hybridization, hybrid competition	Survival, growth
Allee effect	Reduced visitation, pollen limitation	Fecundity, recruitment
Declined pollinator abundance	Reduced visitation, pollen limitation	Fecundity, recruitment
Reduced disperser abundance	Impaired seed dispersal	Recruitment, site colonization
Exotic herbivores, release from parasitism	Excessive herbivory	Survival, recruitment, growth, fecundity
Overgrazing	Grazing and trampling damage	Survival, recruitment, growth, fecundity
Successional closing, competition from vegetation	Increased competition, safe site availability	Recruitment, growth
Genetic threats		
Accumulating mutational load	Increased inbreeding depression	Fecundity, survival, growth, recruitment
Inbreeding	Fitness reduction	Fecundity, survival, growth, recruitment
Drift, genetic erosion	Reduced phenotypic plasticity	Impaired demographic response ability
Intraspecific hybridization	Outbreeding depression	Fecundity, survival, growth, recruitment
Interspecific hybridization	Genetic swamping, loss of genetic integrity	Changes in life history traits
Variation in SI alleles	Low fruit or seed set	Fecundity, recruitment

ricanes (Kollmann and Poschlod 1997; Mabry and Korsgren 1998). In temperate regions, most of the natural or semi-natural habitats have already been destroyed, and the leftovers generally are being protected as nature reserves, situated as islands in an often heavily exploited matrix (Eriksson 1996; Lennartsson and Svensson 1996; Berge et al. 1998; Bruun 2000; Norderhaug et al. 2000). This means that, particularly in these regions, we are currently not facing active habitat fragmentation itself, but rather its ecological consequences, such as increased edge effects, small population sizes, isolation and a reduced exchange of individuals and genes (Rathcke and Jules 1993; Kruess and Tscharntke 1994; Young et al. 1996; Gascon and Lovejoy 1998; Lande 1998; Harrison and Bruna 1999). Habitat fragmentation has created many "new rares", i.e., species that were once common but have now been turned into rare species with small, isolated populations (Huenneke 1991). As these species are adapted to living in large, connected populations, they are expected to be threatened more strongly by habitat fragmentation than naturally rare species adapted to highly specific, geographically restricted habitats (Deyrup and Menges 1997; Medail and Verlaque 1997; Tremblay et al. 1998; Walck et al. 1999, see discussion in Chap.3, this Vol.). In this chapter, this latter group of rare species will be referred to as "endemics".

As a consequence of their habitat specificity, geographically restricted endemics generally occur in naturally small and isolated populations (Palacios and Gonzalez-Candelas 1997; Witkowski and Liston 1997; Bosch et al. 1998; Conte et al. 1998; Pandit and Babu 1998) and thus have an intrinsically high risk of extinction due to demographic and environmental stochasticity (Menges 1991a; Lande 1998). These species have probably only managed to survive for long periods of time because their specific habitat has experienced low levels of environmental stochasticity (Medail and Verlaque 1997). Changes in environmental stochasticity resulting from human actions therefore form a very serious threat to the survival of this group of rare species (Allphin and Harper 1997; Medail and Verlaque 1997).

For populations in natural habitats, human suppression of naturally occurring disturbances, such as wildfires (Maina and Howe 2000) and river flooding regimes (Smith et al. 1998), poses a very important threat. For instance, fire is required for the completion of the life cycle of specifically adapted species (Carlson et al. 1993; Bradstock et al. 1998; Menges and Dolan 1998), because it stimulates the release of seeds from capsules and subsequent seed germination (Keith 1997), reduces competition from successionally closed vegetation (Hawkes and Menges 1996; Menges and Hawkes 1998; Lesica 1999), and increases soil nutrient availability (Anderson and Menges 1997). Other species escape fire as buried seeds, but only if ants have transported seeds into their nests (Pierce and Cowling 1991). For the latter species, biotic interactions are vital for long-term viability (see Sect. 2.2).

2.2.1.3 Direct Destruction and Overexploitation of Populations

Some threats do not necessarily concern natural or human-induced changes in plant habitats, but involve direct destruction of plant individuals or populations by human actions (Lande 1998). Unsustainable harvesting of medicinally or commercially interesting species from their natural habitat can lower the viability of local populations to such an extent that they are threatened with extinction (Pinard 1993; Witkowski et al. 1994; Olmsted and Alvarez-Buylla 1995; Maze and Bond 1996; Pandit and Babu 1998; Wolf and Konings 2001). An example of such an extinction is silphium, a still unidentified North-African *Ferula* species (Apiaceae) that was extremely popular with ancient Greeks and Romans as a spice and aphrodisiac (Koerper and Kolls 1999) and was one of the most important Cyrenaic export products (Forbes 1965). It is highly likely that this species became extinct as a consequence of overharvesting (Forbes 1965). If no conservation measures are taken, a similar fate is likely for two overexploited palm species of dry tropical forest in Mexico (Olmsted and Alvarez-Buylla 1995). Commercial picking of the blooms of *Banksia hookeriana* reduces fecundity to such an extent that the regeneration of populations after fire is imperiled (Witkowski et al. 1994). Simulation models are a very useful tool to simulate the effects of different intensities and rates of harvesting and restocking and to evaluate the consequences on population viability (Olmsted and Alvarez-Buylla 1995; Bernal 1998; Sutherland 2001). The genetic consequences of reductions in population size or density by harvesting have never been incorporated into such models. Yet, a recent study on *Pinus contorta* shows that harvesting significantly reduced levels of genetic diversity, compromising the ability of the populations to adapt to changing environmental conditions or to counter pathogens (Rajora et al. 2000).

2.2.2 Disturbed Biotic Interactions

2.2.2.1 Pollen Limitation

Rarity and small population size are associated with a disturbance of relationships with other organisms. This is perhaps not as obvious in naturally rare species, which may have been able to evolve close relations with stable pollinator guilds (Deyrup and Menges 1997; Shaw and Burns 1997; Giblin and Hamilton 1999; Tepedino et al. 1999; Chap. 3, this Vol.). Nevertheless, evidence for pollen limitation of seed and fruit set has also been found for endemic species (Talavera et al. 1996; Bosch et al. 1998; Brown and Kephart 1999; Kaye 1999; Kephart et al. 1999; Robertson et al. 1999), although it has rarely been established whether this posed a demographic problem (Bond 1994; Kearns et al. 1998; Oostermeijer 2000a). Given the long individual life-span of many endemic species, a high annual seed production is probably not needed to maintain a stable population.

In the "new rares", disturbed interactions with pollinators have been studied most frequently in relation to habitat fragmentation. As pollinator diversity and abundance is also reduced in fragmented habitats, pollination service of many plants may be reduced (Rathcke and Jules 1993; Bond 1994). Generally, the Allee effect (Allee et al. 1949) seems to be a very important phenomenon in small plant populations, especially if these have a low density. Pollinators may regard small populations as inferior or unreliable food sources, which leads to low visitation rates. Low densities can result in longer residence times of visitors on individual inflorescences. In self-compatible species, this leads to increased geitonogamy (De Jong et al. 1993). In self-incompatible plants, the probability of receiving compatible pollen is reduced whilst the risk of stigma clogging with self-pollen increases (Ehlers 1999).

2.2.2.2 Dispersal Limitation

Besides leading to pollen limitation, disturbed biotic interactions may also result in dispersal limitation if they involve seed-dispersing organisms (Richardson et al. 2000; Chap. 10, this Vol.). Specific examples of this type of threat are scarce, however, probably because one-to-one relationships between plants and animal dispersal vectors are rare (Bond 1994; Richardson et al. 2000). It is to be expected, though, that the impaired migration of various birds and mammals in fragmented landscapes will also have negative consequences for the seed dispersal of many plants (Santos et al. 1999; Yao et al. 1999; Ortiz-Pulido et al. 2000; Robinson and Handel 2000). On a local scale, it has been hypothesized that the displacement of seed-dispersing ants by an alien ant species will ultimately lead to the extinction of the myrmecochorous plants that depend on dispersal to rodent-free safe-sites for germination (Bond and Slingsby 1984). The displacement of the native by the alien ant is associated with human disturbance (Richardson et al. 1996), which means that species in undisturbed habitat are still safe from this threat. This is a clear example of a synergistic interaction among different threats (Richardson et al. 1996).

2.2.2.3 Interactions with Exotic Species

The introduction of exotic species is a widely cited threat to native plant populations and can also be placed under "disturbed biotic interactions". In particular, strongly competitive alien species may start to dominate local plant communities and crowd out native species (Pysek and Pysek 1995; Lesica and

Shelly 1996; Carlsen et al. 2000; reviewed in Chap. 3, this Vol.), often starting with the extinction of small populations of generally not very competitive, light-demanding threatened species (Wester 1994; Stohlgren et al. 1999; Maina and Howe 2000). Additionally, herbivory by introduced mammals, insects or molluscs may cause levels of damage that increase mortality and decrease fecundity of rare plants to such an extent that extinction may follow (Shaw and Burns 1997; Stiling et al. 2000; Chap. 3, this Vol.). Some biologists also see the high levels of damage by native (sometimes even specialist) herbivores as a serious threat to population viability of rare plants (Stanforth et al. 1997; Bevill et al. 1999). In habitat fragments, colonizing herbivores may be released from parasitization, leading to uncontrolled population explosions that may threaten the rare plants inhabiting these fragments (Kruess and Tscharntke 1994).

Alien pathogens, such as *Phytophthora cinnamomi* in Western Australia, may also have large effects on rare plant species. This root-pathogen may seem to have only a direct impact on a relatively small scale, but may trigger a snowball effect on the entire ecosystem through the death of trees and shrubs and the increase in herb cover (Richardson et al. 1996).

An even less obvious threat caused by the introduction of non-native species is the possible hybridization with related native species (Levin et al. 1996; Rhymer and Simberloff 1996; Chap. 5, this Vol.). Especially if hybridization occurs frequently, it will involve introgression, which reduces the genetic integrity of the native species (Krahulcova et al. 1996; Anttila et al. 1998; Bartsch et al. 1999; Huxel 1999). Because small populations tend to behave as sinks, they are especially prone to hybrid gene flow (Ellstrand 1992). In rare species, in which only a few of such sink populations remain, this "genetic swamping" can ultimately cause extinction (Levin et al. 1996; Anttila et al. 1998).

2.2.2.4 Climate Change

Climate change can have strong effects on the mutualistic interactions among species. Especially when the relationships are very close, such as those between figs and fig-wasps, for example, increased frequencies of drought years can easily lead to the extinction of the specialized pollinator, and ultimately also of the plant (Harrison 2000). Furthermore, the figs also function as a key resource for many frugivores, so that the extinction of a specific fig-wasp population may have large, cascading effects on the ecosystem level (Harrison 2000).

2.2.2.5 Grazing and Trampling

Disturbed biotic interactions may also occur on the community level. These threats involve overgrazing or trampling by large herbivores, leading to increased mortality rates or decreased fecundity (Hurtrez-Bousses 1996; Johnson et al. 1999; Francisco-Ortega et al. 2000). However, reducing the grazing pressure will often result in succession and closing of more open vegetation types, causing species of open habitats to decline (Gibson and Brown 1991; Proulx and Mazumder 1998; Losvik 1999; Partel et al. 1999; Baskin and Baskin 2000). In the semi-natural habitats that cover large areas of Europe, managing succession is a very important conservation tool, because the structure of the vegetation has a decisive effect on the demography of many plant and animal species (Bobbink and Willems 1993; Oostermeijer et al. 1994a; McLaughlin and Mineau 1995; Kollmann and Poschlod 1997; Schlapfer et al. 1998; Hulme et al. 1999; Losvik 1999; Bokdam and Gleichman 2000; Esselink et al. 2000; Lennartsson and Oostermeijer 2001).

2.2.3 Genetic Threats

2.2.3.1 Drift and Inbreeding

In small and isolated populations, drift and inbreeding may significantly alter the genetic structure (Barrett and Kohn 1991; Ellstrand and Elam 1993; Young et al. 1996; Chap. 3, this Vol.). Especially for predominantly or obligately outcrossing species with normally large populations, these genetic threats may negatively affect demographic parameters and lower population viability (Newman and Pilson 1997; Oostermeijer 2000b). An increasing number of studies of genetic variation in rare plant species has shown that the proportion of polymorphic loci and the number of alleles per locus is significantly reduced in small populations (Van Treuren et al. 1991; Prober and Brown 1994; Raijmann et al. 1994; Fischer and Matthies 1998b; Persson et al. 1998; Young et al. 1999; Buza et al. 2000; Gaudeul et al. 2000; Luijten et al. 2000). These observations are best explained by genetic drift. In contrast, the observed heterozygosity is much less frequently reduced, which suggests that inbreeding has not had such clear effects on small populations as drift (Oostermeijer et al. 1996a; Oostermeijer 2000b). A possible reason for this is that many of the plants that have been studied are long-lived perennials. Thus, it is highly likely that only a few generations have passed since the populations became fragmented (Van Treuren et al. 1991, 1993; Young et al. 1996). In fact, the European species that have been studied are often formerly common species from semi-natural grass- or heathlands, which have become rare as a consequence of the large-scale intensification of agricultural practices following World War II. For perennials with a life span of 30–60 years, few generations have passed since the genetic bottleneck (Oostermeijer et al. 1996a; Oostermeijer 2000b). In addition, changes in habitat quality and vegetation dynamics (abandonment, eutrophication, scrub encroachment, etc.) have drastically lowered the demographic turnover rates of these populations, so that they often consist of old adult individuals that probably became established as seedlings around the 1940s or 1950s (Lesica and Allendorf 1992; Oostermeijer et al. 1994a; Raijmann et al. 1994; Luijten et al. 2000).

2.2.3.2 Inbreeding Effects

Although drift may be generally more significant, this does not mean that inbreeding depression is not a problem for remnant populations. In species with a mixed mating system, small populations experience significant reductions in seed quality, which is expressed in lowered germinability, small seedling size and higher mortality (Menges 1991b; Oostermeijer et al. 1994b; Newman and Pilson 1997; Fischer and Matthies 1998a, b; Morgan 1999; Kéry et al. 2000; Luijten et al. 2000; Chap. 3, this Vol.). In part, this can be ascribed to a lower investment in offspring by environmentally stressed maternal plants. When maternal effects have worn off in later stages of the life cycle, the lower heterozygosity associated with higher levels of selfing starts to have detrimental effects on performance [number of flower stalks produced, probability of flowering, number of flowers, number of ovules per flower, etc. (Oostermeijer et al. 1994b)]. Generally, this inbreeding depression is due to the increased expression of mildly deleterious recessive mutations and is detected through comparisons of inbred (often selfed) with outcrossed offspring. However, in small and isolated populations, many of the deleterious alleles may have become fixed. In that case, we expect only small differences between in- and outbred offspring, because both groups show the same effects of population inbreeding (Keller and Waller 2002). Inbreeding effects in such populations can only be detected by performing among-population crosses to reveal heterosis (Keller and Waller 2002). Indeed, heterosis has been found in several rare plant species (Van Treuren et al. 1993; Ouborg and Van Treuren 1994; Oostermeijer et al. 1995a; Luijten 2001; Luijten et al. 2002).

2.2.3.3 Loss of S-alleles

The negative demographic consequences of inbreeding are better documented than those of drift. In self-incompatible species, fruit or seed set tends to be reduced in small populations, suggesting that drift has eliminated significant numbers of S-alleles and decreased the availability of compatible mates (Les et al. 1991; Byers and Meagher 1992; De Mauro 1993; Ågren 1996;

Negron-Ortiz 1998; Vekemans et al. 1998; Kéry et al. 2000; Luijten et al. 2000). In threatened species, it is clear that this disruption of the normal breeding system compromises the regeneration of small populations (De Mauro 1993; Luijten et al. 2000). Theoretically, it can be expected that (long-lasting) bottlenecks would lead to the breakdown of the SI-system in small populations of rare plant species (Reinartz and Les 1994; Glemin et al. 2001). However, in two species that show variation for self-incompatibility, self-compatible individuals were found mostly in large populations (Lipow et al. 1999; Luijten et al. 2000), so that the opportunities for selection for self-compatible plants were strongly reduced. The few remaining populations of the endemic *Centaurea corymbosa* in France still maintain effective self-incompatibility (Colas et al. 1997). Hence, there is – to date – no evidence that small, fragmented populations are evolving self-compatibility to overcome reduced mate availability (see also Karron 1987).

2.2.3.4 Reduced Adaptability

Loss of variation caused by drift will lower the ability of a population to adapt to changes in the environment. However, this long-term threat to viability has hardly been demonstrated for rare or threatened plant species. Fischer et al. (1997) showed that there was significant variation among families in their response to elevated CO_2 in the rare *Gentianella germanica*, with some showing a positive, and most a negative response. This analysis shows that in genetically variable populations of this species adaptation to higher CO_2 levels would be possible. Using random amplified polymorphic DNA (RAPD), the same authors also demonstrated that small populations of this species were genetically depauperate (Fischer and Matthies 1998b). If this loss of variation included the loci involved in the response to changes in ambient CO_2, adaptation could be hampered. In support of this hypothesis, Kéry et al. (2000) showed that families sampled from large populations of *Primula veris* were responding more strongly to a combined fertilization-competition treatment than families from small populations. Likewise, in the clonal buttercup *Ranunculus reptans*, offspring of small populations produced fewer daughter rosettes and flowers than those from large populations and showed a more negative response to competition (Fischer et al. 2000).

2.2.3.5 Genetic Variation in Endemic Plants

Although mainly expected for "new rares", reduced genetic variation and high genetic differentiation associated with small population size and isolation have also been reported for naturally rare species (Palacios and Gonzalez-Candelas 1997; Coates and Hamley 1999; Dolan et al. 1999; Godt and Hamrick

1999; Matolweni et al. 2000). Differences in genetic variation between rare and widespread species survived correction for phylogenetic dependence, whereas differences in within- and between-population structuring did not (Gitzendanner and Soltis 2000). Nevertheless, the long-term persistence of genetically depauperate endemic species suggests that inbreeding and drift do not necessarily have significant effects on demographic performance. Under the (supposed) low environmental stochasticity in the habitat of endemic species (Allphin and Harper 1997), the high genetic variation needed to cope with changes may not be necessary. In addition, the negative consequences of inbreeding are likely to be less pronounced in stable environments (Dudash 1990; Barrett and Kohn 1991; Koelewijn 1998), and purging of deleterious recessive mutations (Hedrick 1994; Frankham 1995) may occur through strong selection on developing seeds or recruiting seedlings (Ledig et al. 1997; Kephart et al. 1999). Despite the considerable attention that endemics have received in the Conservation Biology literature, field studies testing the degree of late-acting inbreeding depression (i.e., later than the seedling stage) in these species are not available.

2.3 Managing Different Types of Threats

2.3.1 Managing Threats to Endemic Species

As such a large proportion of our rare and threatened species are endemics with a restricted distribution and very high habitat specificity, it is a very important question how we should deal with them in conservation programs. Because endemic species have intrinsically high extinction probabilities, extinction by natural causes would, in my opinion, be acceptable and should not necessarily be prevented. However, the boundaries between natural and human-driven extinction have become much more vague now that even a natural phenomenon like climate change seems no longer free of human disturbance (Zwiers and Weaver 2000).

Richardson, Cowling, and Lamont (1996) described how different threats can have synergistic effects on the extinction of endemic species of South-African Cape fynbos and Australian kwongan, two structurally and functionally similar shrubland communities on infertile sandy soils and in a Mediterranean climate. Invasive exotic species, such as pines, seem to be a key threat to these ecosystems, because they trigger many adverse effects, which include changes in the response of the system to fire and increased interception of rainfall (Richardson et al. 1996).

The best strategy for conservation of endemic species seems to be to keep their natural habitat as free as possible from any form of abnormal (human-related) disturbance, such as grazing by livestock, suppression of wildfires, tourism, water extraction, atmospheric deposition of nitrogen and, perhaps most importantly, the invasion of exotic species (Gillespie 1997; Mack et al. 2000). However, ecosystems rich in endemics, such as the fynbos and kwongan mentioned above, are not restricted to nature reserves and the threats are therefore quite difficult to manage effectively. In the fynbos ecosystem, translocation of populations from fragments on private land into protected areas has therefore been suggested, but this brings several questions regarding the materials and methods to be employed (Milton et al. 1999). In the Mediterranean regions of Europe, so-called micro-reserves have been proposed to provide some level of legal protection to the often very specific local habitats of endemic species.

If the quality of the natural habitat in nature reserves can be maintained and environmental stochasticity can be kept as low as possible, genetic variation will probably not pose any problems, unless there is clear evidence that the population has gone through a recent bottleneck and experienced significant drift and/or inbreeding depression. It has sometimes been suggested that interpopulation crosses should be performed to conserve or increase levels of genetic variation in an endemic species (Maki 1999). This management option will be discussed below, in the section on managing genetic threats.

On a landscape scale, there may be sites that appear suitable for an endemic species which are not occupied. In such sites, experimental introduction of seeds can be considered to find out whether there is dispersal limitation within the metapopulation. Such dispersal limitation was demonstrated for *Amphianthus pusillus* (Hilton and Boyd 1996) and *Centaurea corymbosa* (Colas et al. 1997). Decreasing the fraction of suitable, but non-occupied sites in this way will theoretically lower the extinction probability of a regional metapopulation (Eriksson and Kiviniemi 1999).

Although many endemics are restricted to a specific habitat type, Kluse and Doak (1999) recently reported that the endemic species *Chorizanthe pungens* var. *hartwegiana* actually performed better outside its current habitat. Because the apparently more favorable habitat was adjacent to the current one, dispersal limitation seems an unlikely explanation. Apparently, there are still unknown factors that determine the realized habitat of this species. Such factors should be understood before any attempts are made to translocate the species in order to increase the number of occupied sites or improve the level of protection (Milton et al. 1999).

2.3.2 Managing Environmental Threats

Changes in the environment of plant populations involve some of the most significant threats to their existence (Table 2.1). It is impossible to discuss all of these threats individually in this chapter. This section will therefore focus mainly on the very important effects of human-caused habitat fragmentation and habitat degradation. The former includes increased edge effects and reduced connectivity, and the latter comprises eutrophication through atmospheric deposition and agricultural fertilization, ground water manipulations, fire-suppression and effects of various forms of habitat management.

2.3.2.1 Climate Change

The main problem with the possible threats resulting from climate change is that they are nearly impossible to manage. In an undisturbed natural landscape, species would either move their range or adapt to the changing environmental conditions. The problem is that in the presently fragmented landscape, both processes are severely inhibited for many rare species. At present, populations in fragmented habitats already suffer from reduced exchange of individuals or genes and low genetic variation, especially when small and isolated (Young et al. 1996). In this light, it seems quite likely that habitat fragmentation will act as a bottleneck under (rapid) climate change.

2.3.2.2 Habitat Fragmentation

Habitat fragmentation is presently a major threat to the viability of plant populations all around the world. As noted above, fragmentation involves smaller population sizes, reduced migration of genes and individuals and increased edge effects. The small population sizes increase extinction risks through demographic and environmental stochasticity (Menges 1992; Barrett and Husband 1997; Newman and Pilson 1997; Lande 1998; Pfab and Witkowski 2000), the Allee-effect (Widén 1993; Groom 1998; Oostermeijer et al. 1998a; 2000, Stephens and Sutherland 1999) and genetic drift and inbreeding (Oostermeijer et al. 1994b; Fischer and Matthies 1997, 1998a; Newman and Pilson 1997; Eisto et al. 2000; Luijten et al. 2000; Oostermeijer 2000a). Impaired gene flow among fragmented populations has generally been inferred from high levels of genetic differentiation among populations (Van Treuren et al. 1991; Young et al. 1993, 1999; Prober and Brown 1994; Raijmann et al. 1994; Schneller and Holderegger 1996; Young er al. 1996; Fischer and Matthies 1998b; Luijten et al. 2000; Schmidt and Jensen 2000). Probably, there is a close relationship between gene flow levels and the breeding system and pollination modes of species (Berge et al. 1998; Weidema et al. 2000). Newman and Tallmon (2001) recently provided evidence that moderate levels of gene flow significantly reduced the effects of inbreeding depression in experimental populations of *Brassica campestris*.

The most frequently cited key factor to counter the isolation between populations in fragmented habitats is improving or restoring the connectivity by means of the construction of corridors (Beier and Noss 1998). The degree of connectivity and the presence of corridors are indeed often positively related to species richness and the movements of individuals (Beier and Noss 1998; Debinski and Holt 2000). The migration of plant individuals along corridors has hardly been documented, with some exceptions (Fritz and Merriam 1993; Burel 1996; Johansson et al. 1996; Van Dorp et al. 1997; Brunet and von Oheimb 1998; McCollin et al. 2000). Dispersal via the plant's own mechanism will probably occur only if the habitat in the corridor is suitable for establishment (Fritz and Merriam 1993; McCollin et al. 2000) and is predicted to be very slow (<1 m/year, (Van Dorp et al. 1997; Brunet and von Oheimb 1998). However, it can be expected that increased movement of birds and mammals along corridors will also enhance seed transport (Fischer et al. 1996; Kiviniemi 1996; Brunet and von Oheimb 1998; Kiviniemi and Telenius 1998; Oliveira and Ferrari 2000; Ortiz-Pulido et al. 2000). While the destruction of connective habitat strips or stepping-stone patches in a landscape should be avoided (Beier and Noss 1998), when planning to increase connectivity between habitat fragments, it must be realized that: (1) especially the rare species of conservation interest often do not or cannot use the corridors (Collinge 2000), and (2) exotic species or pathogens may invade sensitive small fragments at higher rates and affect local communities negatively (Greenberg et al. 1997; Planty-Tabacchi 1997). In a functioning metapopulation network, particularly small populations tend to behave as sinks and may thus experience a considerable immigration of genes (Ellstrand 1992). In case of conspecific gene flow, this is generally not a problem, but when there is a risk of introgression from a closely related common species, one should be aware of this possibility (Levin et al. 1996; Rhymer and Simberloff 1996; Nagy 1997; Keller et al. 2000).

2.3.2.3 Edge Effects

Edge effects have hardly been studied in fragmented plant populations. The best example is the significantly reduced recruitment of *Trillium ovatum* populations that grow close to the edges of forest fragments in Oregon (Jules 1998). In addition, important negative edge effects on mutualistic relationships were observed in fragmented Argentinian chaco forest (Aizen and Feinsinger 1994a, b). In contrast to continuous forest, several plant species suffered from pollen limitation in the fragments because there the more effective native pollinators were displaced by Africanized honey bees, which transferred much less conspecific pollen to the stigmas (Aizen and Feinsinger 1994a, b). Removing non-native honeybees from fragmented ecosystems may prove necessary to reduce this edge effect.

Where natural habitat fragments are surrounded by a matrix of intensively used agricultural land, as is the case in large parts of Europe, eutrophication through the influx of fertilizers is a main threat to the nutrient balance of the soil at the edges of the reserves (Heil and Diemont 1983; Jansen et al. 1996; Bakker and Berendse 1999; Carroll et al. 1999). In small fragments, such edge effects may be present throughout the entire area. Similar eutrophication effects have been observed where heathland reserves are adjacent to a road (Angold 1997). Eutrophication can disturb competitive relationships between plant species (Smith et al. 1999), in particular between herbs and grasses (Alonso and Hartley 1998). Increased mowing frequencies or other types of nutrient-removing management (e.g., sod-cutting, cattle grazing) may be called for in such situations (Jansen et al. 1996; Bokdam and Gleichman 2000).

The edges of wetland or wet heathland reserves often suffer from the downward seepage of base-rich but nutrient-poor ground water to the surrounding agricultural land where the water table is kept low (Van Wirdum 1993; Van Walsum and Joosten 1994; Runhaar et al. 1996). In The Netherlands, wetland-reserve managers are often forced to compensate for this water loss by letting in water from nutrient-rich "alien" sources, such as rivers and canals (Jansen et al. 1996; Runhaar et al. 1996). This, of course, shifts the problem from water loss to eutrophication. Effects of the latter can be successfully reduced by means of phosphate removal (van Loosdrecht et al. 1997).

In general, increasing the size of reserves and creating buffer zones around them are the most effective methods to counter edge effects (De Vries et al. 1994; Beltman et al. 1996; Larson et al. 2000).

2.3.2.4 Habitat Degradation

The edge effects from external causes, described in the previous section, may dominate habitat quality in the entire reserve if it is too small. In addition to these sources of habitat degradation, there are several other causes, generally associated with human actions. In contrast to natural habitats, where the effects of human actions are often kept to a minimum, semi-natural habitats (e.g., grasslands) need some form of often human-related disturbance. Livestock grazing, mowing and haymaking, as well as sod-cutting are all traditional, small-scale agricultural activities that have prevented vegetation succession so that grassland or heathland communities could establish and persist (Poschlod and Bonn 1998; Bartolome et al. 2000; Verdu et al. 2000). From the perspective of the species characteristic for those grasslands, the cessation of these human activities (i.e., abandonment) is a serious threat to their population viability (Bobbink and Willems 1993; Lennartsson and Svensson 1996; Eriksson and Eriksson 1997; Jensen 1998; Bakker and Berendse 1999; Losvik 1999). In general, the main cause for the decline of many species is that the structure of the vegetation becomes so dense that open patches are no longer available as safe sites for germination and recruitment (Oostermeijer et al. 1994a; Krenova and Leps 1996; Eriksson 1997). Hence, if the high biodiversity of these semi-natural grasslands is to be conserved, farmers or reserve managers have to be involved in their management and should be informed about the optimal management method(s) (Willems 1983; Bobbink and Willems 1993; Norderhaug et al. 2000; Lennartsson and Oostermeijer 2001).

The prevention of wildfires has similar effects on populations of fire-adapted species in that biomass has the chance to build up and seeds will not germinate in the dense litter layer (Hawkes and Menges 1995, 1996; Menges and Kohfeldt 1995; Quintana-Ascencio and Menges 2000). In addition, canopy-stored seeds need to be released by fire, and germination of seeds often needs to be stimulated by smoke and/or heat (Keith 1997). In *Silene regia*, fire suppression was considered to be the primary threat (relative to habitat fragmentation and genetic erosion) to the viability of the remaining populations (Menges and Dolan 1998).

Prescribed burning seems to be the best management tool for threatened fire-adapted species, especially in fragmented habitats in which wildfires cannot be left uncontrolled (Carlson et al. 1993; Gordon 1996; Bradstock et al. 1998; Lesica 1999; Menges et al. 1999; Pendergrass et al. 1999). However, Pendergrass et al. (1999) warned that monitoring the effects of burning treatments on a short time-scale may lead to unrealistically high expectations concerning increases in the population viability of threatened plants. Long-term monitoring in combination with demographic projection models can solve this problem.

Grazing, burning and mowing, along with other management tools used to prevent succession can still create problems for small populations, because they may increase environmental stochasticity considerably. This is especially true when the persons executing the management are not aware of the presence of sensitive populations, or if the disturbance is poorly timed with respect to the species' life cycle (Oostermeijer et al. 1998b, 2002). In addition, high-intensity disturbances also threaten the local fauna (Sterling et al. 1992; Volkl et al. 1993; Schwartz 1994).

2.3.3 Managing Disturbed Biotic Interactions

Section 2.2 described how disturbed interactions among plant individuals, among different plant species, and between plants and animals can lead to reduced population viability. As these interactions are often complex and delicate, managing them for the purpose of the conservation of the involved organisms is an extremely difficult task.

2.3.3.1 Managing Mutualistic Interactions

Next to general measures, such as the conservation of habitats and pollinators, Kearns et al. (1998) have put forward more specific measures to ameliorate disturbed plant-pollinator interactions, although the focus was on crops rather than wild plants. In general, wild insect pollinator populations can, and should, be promoted by decreasing or altering the use of pesticides on crops. It will, however, be difficult to predict the demographic effects of such measures on wild plant populations.

In the absence of specific pollinators, or when scarce pollinator service is limiting recruitment rates, hand pollination is a solution to maintain viable plant populations for the short-term. This can be considered for highly threatened species, for example, during ecological restoration efforts, when local plant and insect communities are often so strongly disturbed that mutualistic and facilitative interactions do not function normally. In successional species, only the first few seedlings cohorts survive to the adult flowering stage. High fecundity in the initial stages of succession may be important if the population size is to be increased. Artificial pollination will additionally increase the outcrossing rate, lowering the inbreeding depression of those first cohorts. Reintroduction of specific pollinators is also an option to reduce pollination limitation. In the case of self-incompatible plant species, introduction of seeds or plants from other populations will increase mate availability. That option will be discussed in the next section, as it mainly concerns genetic management.

Exotic pollinators have sometimes been introduced to take over the role of extinct pollinators (Kearns et al. 1998). Although economically perhaps interesting (i.e. for crops), this is not an ecologically desirable solution, considering the enormous problems that can be caused by exotic species (Hingston and McQuillan 1998). Attempts to remove alien pollinators from the ecosystem have rarely been successful (Kearns et al. 1998), so introductions tend to be irreversible. The introduction of domestic honey bees is also considered risky, as they may compete with and even displace wild bee species (Sugden et al. 1996). Evidence for this is available mainly for the Africanized honey bee in America (Pedro and Decamargo 1991; Aizen and Feinsinger 1994a, b; Frankie et al. 1998), although its effects are often confounded with habitat fragmentation. In Australia, introduced honey bees may have negative effects on the reproduction of some bird-pollinated plants, but positive effects on others (i.e. *Banksia ornata*) which suffer from a shortage of their natural pollinators (Paton 2000). To date, there is little clear evidence for negative effects of the honey bee in Europe (Huryn 1997; Steffan-Dewenter and Tscharntke 2000). Until conclusive evidence is available, a restricted use of bee stands in nature reserves seems wise (Steffan-Dewenter and Tscharntke 2000).

The long-distance dispersal of plants is largely stochastic (Bullock and Clarke 2000), and therefore extremely difficult to study and to predict (Bullock and Clarke 2000; Cain et al. 2000). Manipulation of seed dispersal through the interaction with dispersal vectors seems hardly feasible, except perhaps by creating corridors that promote the migration of ants, birds, and mammals (see Sect. 3.2.2). In particular for animal-dispersed species, it has been suggested to plant perch-trees and/or shrubs in restoration sites to obtain a more "natural" immigration pattern (Robinson and Handel 2000).

In ecologically restored habitats, the dispersal problem can be alleviated for the short term by artificially sowing or planting target species. Although the success of planting adults will most likely be higher (Helenurm 1998; Van Groenendael et al. 1998; Drayton and Primack 2000), my personal preference is sowing, firstly because it resembles natural colonization, secondly because it is easier to introduce high amounts of genetic variation, repeatedly if necessary, and thirdly because the successful germination and establishment of seedlings is a useful "bioassay" to test the demographic suitability of the (restored) site for the target species (Turnbull et al. 2000). In an experiment with *Arnica montana* in The Netherlands, we found that, in contrast to sown seeds, planted seedlings escaped inbreeding depression. Apparently, the initial selection steps that lead to local adaptation had been skipped. Survival of the sown seeds was much lower than that of the planted seedlings, but had the advantage that the less-fit, inbred offspring was immediately eliminated and could never contribute to any future generations (Luijten 2001; Luijten et al. 2002).

2.3.3.2 Managing Herbivores and Pathogens

Herbivores, both natural and alien, can have a statistically significant impact on the viability of plant populations, especially when population size is small (Stanforth et al. 1997; Bevill et al. 1999). Although it may be one of the determinants of a plant's distribution (Bruelheide and Scheidel 1999), herbivory is probably seldom a primary factor in the decline of threatened plants and increased sensitivity to a native herbivore will usually be a by-product of habitat fragmentation or degradation (Pfab and Witkowski 1999; Lesica and Atthowe 2000). Therefore, but also because of the detrimental effects on many other organisms in the same community, using pesticides to reduce the effect of insect herbivory (Bevill et al. 1999) is not an ecologically desirable management method to increase the population viability of threatened plants (Lesica and Atthowe 2000), even in cases in which the target herbivore is not native (Mack et al. 2000). Long-term, ecosystem-wide restoration efforts to enhance the regulating action of natural predators and parasitoids seems a better approach to the control of (invasive) herbivores than trying to eradicate individual species (Mack et al. 2000). Moreover, the problems with alien herbivores and pathogens show again that we have to be extremely careful with introducing alien organisms. Prevention is by far easier and cheaper than post-entry control (Mack et al. 2000).

In the case of large herbivores, such as deer or (in semi-natural grass- or heathlands) cattle or horses, it may sometimes be necessary to temporarily fence off small populations from grazing or browsing. While some species potentially benefit from grazing and trampling in terms of an increased availability of open microsites for recruitment, the flowers or fruits are often completely removed in small populations, impairing the desired response (Oostermeijer et al. 1992). Other species reportedly suffer from direct damage and increased pathogen infection after trampling by cattle (Knowles and Witkowski 2000). As the small population size is often the main problem, the exclosures should be kept in place until monitoring indicates that recruitment is taking place and population size increases. In some cases it may be necessary to protect nature reserves from grazing altogether, but this depends on the historical or traditional context of grazing in the ecosystem (McLaughlin and Mineau 1995; Verdu et al. 2000).

2.3.3.3 Managing Exotic Plant Species

As we have seen before, exotic plant species frequently have strong negative effects on native plants (Pavlik and Manning 1993; Lesica and Shelly 1996; Carlsen et al. 2000). Especially when native populations are small and fragmented, the risk of local extinction as a result of these interactions is considerable (Wester 1994; Duncan and Young 2000; Maina and Howe 2000). As stated above, preventing the introduction of alien species is the most effective method to avoid problems with native plants, as eradication is only rarely successful (Simberloff 1997; Mack et al. 2000). Elimination is only possible when monitoring is effective and the invading populations are detected early on, when they are still small (Simberloff 1997). This is, however, rarely the case. Currently, in The Netherlands, there is a slow but steady invasion of the alien giant hogweed (*Heracleum mantegazzianum*), whose invasiveness and adverse effects are well-known (Pysek and Pysek 1995; Hitchmough and Woudstra 1999; Wadsworth et al. 2000). Nevertheless, hardly any coordinated action is being taken to eradicate the species while it is still possible.

Eradication of alien invasive plants can be attempted with herbicides, by mechanical removal or by introducing natural enemies (Mack et al. 2000). Herbicides such as glyphosate (Roundup) bear the risk of affecting non-target native plants and animals, although studies done so far have not provided much evidence to support this (Gardner and Grue 1996; Simenstad et al. 1996; Marrs and Frost 1997; Lindgren et al. 1999). In spite of these possible side effects, herbicides have been used to "improve" the ecological restoration of Great Plains grasslands by removal of exotic weeds (Masters et al. 1996). Introduced natural enemies may also affect non-target organisms. For example, the Eurasian weevil *Rhynocyllus conicus* was introduced to North America to control the invasive thistle *Carduus nutans*, but currently attacks native thistles as well (Louda et al. 1997). Mechanical removal of weeds is a safer method, but is very labor-intensive and will therefore hardly be effective in later stages of invasions, when the number of individuals and populations is already too large. However, reduction of exotic plants to acceptable levels is sometimes possible with long-term persistent efforts (Randall et al. 1997).

2.3.4 Managing Genetic Threats

2.3.4.1 Managing Hybridization and Introgression

An additional genetic risk of exotic species is a cryptic form of extinction that occurs through hybridization of rare plants with common close relatives (Rhymer and Simberloff 1996; Huxel 1999; Soltis and Gitzendanner 1999). This risk is clearly highest when the close relative is an non-indigenous species with which gene flow was previously impossible (Ayres et al. 1999). Changing environmental conditions such as human disturbance of natural habitats can also bring related species into contact (Francisco-Ortega et al. 2000; Runyeon-Lager and Prentice 2000). Gene flow from (genetically modified) crops into wild species is another problem that will affect the genetic integrity of wild plants (Linder et al. 1998; Bartsch et al. 1999). Again, mainly small populations of the rare species will suffer, as gene flow from the larger populations of the common species will have a much larger effect on their gene pool (Anttila et al. 1998). Management of the hybridization problem is similar to that mentioned for exotic species above. By improving awareness among ecologists and policy-makers with respect to the possibility of hybridization, risky situations can be avoided. Where hybridization risks already exist, any suspected introgression can be detected best with the help of DNA-markers (Ayres et al. 1999; Soltis and Gitzendanner 1999).

2.3.4.2 Managing Risks of Drift, Inbreeding and Outbreeding

Small plant populations often have low levels of genetic variation as a consequence of drift. In addition, inbreeding reduces heterozygosity, resulting in inbreeding depression. In self-incompatible species, drift reduces the number of S-alleles. Likewise, it may alter the morph or sex ratio in species with a heterostylous or dioecious breeding system (Percy and Cronk 1997; Kéry et al. 2000) to such an extent that successful sexual reproduction becomes a problem.

It is clear that first of all we should avoid (further) reductions in population size. This can be done by maintaining or restoring habitat quality and preventing and mitigating habitat fragmentation. However, we are now already facing genetic erosion problems in many rare species which are declining so rapidly that something should be done in the short term (Frankham 1999).

Stimulating genetic exchange, either by means of artificial crossings or by creating corridors, is often mentioned as a method to increase genetic variation within isolated, genetically depauperate populations (i.e., to counter the effects of drift). This method does not seem appropriate for endemic habitat specialists, for which isolation and the resulting low genetic variability and high genetic differentiation are quite likely entirely natural phenomena. Artificial outbreeding would lead to an immediate destruction of the characteristic population genetic structure for these species. Yet, outcrossing among endemic populations has been suggested as a management option (Maki 1999; Kang et al. 2000).

Also for fragmented populations of the "new rares", the question whether we should attempt to artificially restore low levels of genetic variation has still not been answered satisfactorily. Results of empirical studies on the effects of outbreeding among populations of plants are contradictory, which means that it is very difficult to give sound management advice (Reinartz 1995; Frankham 1999; Dudash and Fenster 2000). To a large extent, this is caused by the fact that many studies (often performed with rare species, by conservation biologists!) (1) were too short to consider F_2–F_3 generations, which would be necessary to detect the breakdown of coadapted gene complexes, and (2) were performed in an experimental garden or glasshouse, instead of under field conditions, so that the causes of outbreeding depression could not be tested properly (e.g., Oostermeijer et al. (1995a), Fischer and Matthies (1997), Byers (1998), Hardner et al. (1998)). The few extensive studies that have been done suggest that outbreeding either significantly reduces population fitness (Waser and Price 1994; Keller et al. 2000; Waser et al. 2000) or has variable, minor effects (Fenster and Galloway 2000). These effects are outweighed by the benefits of initial heterosis (Fenster and Galloway 2000) or by the overall increased genetic variation on which selection can act (Moritz 1999). The strongest negative effects were found for crosses over very long distances (100–1,000 km), which suggests that these should at least be avoided (Fenster and Galloway 2000; Keller et al. 2000).

The magnitudes of outbreeding depression that have been demonstrated thus far (Dudash and Fenster 2000; Waser et al. 2000) will probably not dramatically affect the demographic performance of reintroduced populations. Hence, interpopulation crosses should be considered as a useful method to increase genetic variation of small, isolated populations of formerly common species. For endemic species, in which every population often constitutes an important part of the total genetic variation, I suggest that conservation efforts focus primarily on maintaining the viability of remaining populations by habitat protection and, if necessary, management. Introductions of endemics in non-occupied patches of suitable habitat should also be considered (Colas et al. 1997).

In general, reintroductions or reinforcements should use material from as many populations as possible, avoiding the very distant populations (>100 km, Dudash and Fenster 2000). Considerable variation in establishment success among different source populations has been observed (Helenurm 1998; Luijten 2001). Hence, using single populations as a source for reintroduction, for instance those that ecologically resemble the original population as much as possible, is likely to lower the probability of success.

High genetic variation among transplants is important for the success of artificially introduced populations (Newman and Pilson 1997; Kéry 2000; Williams 2001). This can be achieved by sampling multiple families from multiple populations or from large, genetically variable populations. If the number of introduced seeds is high and introduction of seeds is continued for a number of consecutive years, this sampling scheme will lead to high amounts of genetic variation upon which locally specific selection may act to create a new, locally best adapted population. If the number of seeds available for the introduction is limited, however, planting adult or subadult plants pregrown in a glasshouse is to be preferred above sowing seeds (Van Groenendael et al. 1998). Introduction should only be considered on locations for which there is a large likelihood that the abiotic and biotic conditions are suitable for the target species (Drayton and Primack 2000).

In vitro techniques that involve clonal propagation of plant material taken from remnant natural populations (Amo-Marco and Lledo 1996; Benson et al. 2000; Cuenca and Amo-Marco 2000) have the drawback that they do not increase or restore genetic variation. It may be argued that these clonal "offspring" represent the genotypes which are best adapted to a specific location, but this is by no means sure if we are dealing with perennial plants in a changed or changing environment. It should be checked whether the remaining population shows regular demographic turnover or is made up of persisting adult individuals which established decades ago under totally different habitat conditions (Oostermeijer et al. 1994a). In the latter case, increasing the number of individuals by adding clones probably has limited value for population viability. Partly, this problem can be solved by in vitro germination and propagation of seeds.

For self-incompatible, heterostylous or dioecious species, reinforcement or reintroduction by the guidelines suggested above should also restore the variation in S-alleles (Byers and Meagher 1992; De Mauro 1993; Reinartz and Les 1994; Ågren 1996; Kéry et al. 2000; Luijten et al. 2000), style morphs (Ågren 1996; Eckert et al. 1996; Kéry 2000) and the normal sex-ratio (Percy and Cronk 1997), improving the reproductive interactions among plants in the population.

It speaks for itself that detailed monitoring of establishment, growth, flowering, seed production and the development of genetic variation should be an important part of any introduction program (Gordon 1996). Only then can we gain knowledge on this increasingly important conservation measure, and make sure that the correct follow-up actions are taken, should these be necessary (Pfab and Witkowski 2000).

2.4 Dealing with Threats in Plant PVAs

The best method to assess the importance of specific threats for population viability is to try to quantify their demographic consequences (Menges 2000). To this end, demographic monitoring of a range of populations subjected to different threats, or monitoring in combination with experiments or management measures to include or exclude the threat is essential. Although not always integrated into an "official" PVA, these demographic studies have been performed on a number of rare plant species and have yielded very useful information (Nault and Gagnon 1993; Pavlik and Manning 1993; Bastrenta et al. 1995; Lesica 1995; Olmsted and Alvarez-Buylla 1995; Lesica and Shelly 1996; Oostermeijer et al. 1996b; Ratsirarson et al. 1996; Knox 1997; Maschinski et al. 1997; Bernal 1998; Gross et al. 1998; Jules 1998; Kirkman et al. 1998; Menges and Dolan 1998; Quintana-Ascencio et al. 1998; Willems and Melser 1998; Lesica 1999; Knowles and Witkowski 2000; Oostermeijer 2000a; Pfab and Witkowski 2000; Lennartsson and Oostermeijer 2001; Chap. 11, this Vol.).

However, also for species in unmanaged habitats, such as endemics, causes for decline can be revealed by demographic studies, in particular if environmental variables (potential threats) are measured along with demographic censuses (Lesica and Shelly 1995; Thomas 1996; Allphin and Harper 1997; Byers and Meagher 1997; Damman and Cain 1998; Stanley et al. 1998; Valverde and Silvertown 1998; Waite and Farrell 1998; Kluse and Doak 1999; Eriksson and Eriksson 2000).

As can be seen from the studies cited, the maximum amount of information from demographic data can be obtained when matrix projection models are used (Caswell 1989, 2000). These enable very powerful statistical comparisons among sites, years and management methods (Oostermeijer et al. 1996b; Damman and Cain 1998; Valverde and Silvertown 1998; Lennartsson and Oostermeijer 2001), as well as the identification of the most critical stages in the life cycle (Ratsirarson et al. 1996; Floyd and Ranker 1998; Nantel and Gagnon 1999; Heppell et al. 2000). In addition, they provide the opportunity to simulate the demographic effects of different rates of harvesting, and of different frequencies, intensities and timing of natural or anthropogenic disturbances (Bastrenta et al. 1995; Ehrlen 1995; Olmsted and Alvarez-Buylla 1995; Damman and Cain 1998; Menges and Hawkes 1998; Caswell 2000; Heppell et al. 2000; Chap. 11, this Vol.). Careful interpretation of the results of demographic models is necessary though, as these have important implications for management decisions (Silvertown et al. 1996; Caswell 2000).

It is clear that obtaining sufficient demographic data for the construction of matrix projection models takes a long time, although the annual investment of time does not need to be that large. Moreover, information with this degree of detail cannot be obtained for many threatened species, so we should focus on species with different life histories to be able to generalize (Silvertown et al. 1993). Until now, the focus has probably been too much on long-lived perennials. A good system of life history or functional types (Semenova and van der Maarel 2000) is essential if we want to be able to generalize about the responses of different species to threats.

If time is limited, but an assessment of demographic viability is still required, other methods may be considered (Menges and Gordon 1996; Chaps. 7 and 9, this Vol.). The approach of studying the life-stage structure of populations in relation to habitat conditions, originally advocated by the Russian plant population biologist Rabotnov (1969, 1985), proved very useful to evaluate effects of different management methods on populations of the rare *Gentiana pneumonanthe* (Oostermeijer et al. 1994a, 1996c) and *Salvia pratensis* (Hegland et al. 2001). In *G. pneumonanthe*, the different life-stage structures observed in the field were supported very well by detailed demographic data (Oostermeijer et al. 1996b). Although this method is limited to perennial species with recognizable, discrete life stages (e.g., seedlings, juveniles, vegetative adults and reproductive adults), it is also very suitable for medium-intensity monitoring programs (Menges and Gordon 1996).

Despite the huge number of papers on genetic variation in rare plants, genetic threats have hardly been integrated into demographic studies (Menges 2000). The few studies that have been done demonstrate that inbreeding depression and/or low levels of genetic variation may have significant negative effects on population viability (Newman and Pilson 1997; Menges and Dolan 1998; Kéry et al. 2000; Oostermeijer 2000a; Newman and Tallmon 2001). Now that excellent molecular markers are available, it is high time to allocate more research effort on this still largely unexplored topic (Menges 2000).

2.5 Conclusions

Plant populations are threatened by a large variety of factors. Direct destruction of habitats is the most drastic and generally leads to direct local extinction. It can be included into a PVA model as a catastrophe, with large effect but a low probability of occurrence (Lande 1998). Management of such threats is more a political than an ecological issue. However, increasing the awareness of the huge consequences of these human-caused catastrophes among politicians is a major task for ecologists as well (Reinartz 1995).

Other threats have either direct effects on individual plants and hence on population dynamics (i.e., harvesting, picking, trampling) or act through the consequences of habitat fragmentation and degradation on population size, connectivity and the demographic responses to reduced habitat quality. Through increased edge effects, fragmentation also lowers the habitat quality to a considerable degree.

Despite the multitude and complexity of these kinds of threats, a positive conclusion may be that there are lots of possibilities to reduce their effect on plant populations by means of the appropriate management methods. Some of these have to be implemented on the landscape level, for example, by enlarging fragments, increasing connectivity, creating buffer zones, and restoring the original hydrological situation, etc. (see e.g., Van Wirdum 1993; Van Walsum and Joosten 1994; Berge et al. 1998; Higgins et al. 1999), whereas others act on the local scale, for example, by manipulating mowing and grazing, protection from picking or trampling, and removal of exotic weeds (see e.g. Harvey and Meredith 1981; Wester 1994; Bowles et al. 1998; Menges and Dolan 1998; Kettle et al. 2000; Lennartsson and Oostermeijer 2001). This means that the conservation and restoration of wild plant populations needs the concerted action of reserve managers and (conservation) ecologists from different subdisciplines, such as community ecology, restoration ecology, population ecology and ecological genetics (Huenneke 1995). The latter group is increasingly important, because habitat fragmentation has already reduced the genetic variation of many populations and has increased problems of inbreeding depression of plant performance. This means that even though the habitat conditions may have been restored to an optimal situation for many characteristic species, these no longer have the strength to respond to the restored conditions. In an increasing number of cases, introduction, translocation or reinforcement of populations may be necessary (Reinartz 1995). As these measures have very strong genetic and evolutionary implications for threatened species, they should be carefully planned and should only be conducted under supervision of specialists (Reinartz 1995).

At the end of this chapter, it should be emphasized that, although the preceding discussion has mainly focused on individual plant species, conservation should always take place in the context of the key processes of their ecosystem. The consequences of specific management actions for other (native) species in the system should always be taken into account.

Acknowledgements. I thank Hans den Nijs, Albertine Ellis-Adam, and Sheila Luijten for literature contributions and helpful discussions on the broad subject of threats to plant life. Christy Brigham, Mark Schwartz, and two anonymous referees also provided valuable comments on earlier versions of this chapter.

References

Ågren J (1996) Population size, pollinator limitation, and seed set in the self-incompatible *Lythrum salicaria*. Ecology 77:1779–1790

Aizen MA, Feinsinger P (1994a) Forest fragmentation, pollination, and plant reproduction in a chaco dry forest, Argentina. Ecology 75:330–351

Aizen MA, Feinsinger P (1994b) Habitat fragmentation, native insect pollinators, and feral honey-bees in Argentine chaco serrano. Ecol Appl 4:378–392

Akinola MO, Thompson K, Buckland SM (1998) Soil seed bank of an upland calcareous grassland after 6 years of climate and management manipulations. J Appl Ecol 35:544–552

Allee WC, Emersen AE, Park O, Park T, Schmidt KP (1949) Principles of animal ecology. Saunders, Philadelphia

Allphin L, Harper KT (1997) Demography and life history characteristics of the rare Kachina daisy (*Erigeron kachinensis*, Asteraceae). Am Midl Nat 138:109–120

Alonso I, Hartley SE (1998) Effects of nutrient supply, light availability and herbivory on the growth of heather and three competing grass species. Plant Ecol 137:203–212

Amo-Marco JB, Lledo MD (1996) In vitro propagation of *Salix tarraconensis* Pau ex Font Quer, an endemic and threatened plant. In Vitro Cell Dev-Pl 32:42–46

Anderson RC, Menges ES (1997) Effects of fire on sandhill herbs: nutrients, mycorrhizae, and biomass allocation. Am J Bot 84:938–948

Angold PG (1997) The impact of a road upon adjacent heathland vegetation: effects on plant species composition. J Appl Ecol 34:409–417

Anttila CK, Daehler CC, Rank NE, Strong DR (1998) Greater male fitness of a rare invader (*Spartina alterniflora*, Poaceae) threatens a common native (*Spartina foliosa*) with hybridization. Am J Bot 85:1597–1601

Ayres DR, Garcia-Rossi D, Davis HG, Strong DR (1999) Extent and degree of hybridization between exotic (*Spartina alterniflora*) and native (*S. foliosa*) cordgrass (Poaceae) in California, USA determined by random amplified polymorphic DNA (RAPDs). Mol Ecol 8:1179–1186

Bakker JP, Berendse F (1999) Constraints in the restoration of ecological diversity in grassland and heathland communities. Trends Ecol Evol 14:63–68

Barrett SCH, Husband BC (1997) Ecology and genetics of ephemeral plant populations: *Eichhornia paniculata* (Pontederiaceae) in northeast Brazil. J Hered 88:277–284

Barrett SCH, Kohn JR (1991) Genetic and evolutionary consequences of small population size in plants: implications for conservation. In: Falk DA, Holsinger KE (eds) Genetics and conservation of rare plants. Oxford University Press, New York, pp 3–30

Bartolome J, Franch J, Plaixats J, Seligman NG (2000) Grazing alone is not enough to maintain landscape diversity in the Montseny Biosphere Reserve. Agric Ecosyst Environ 77:267–273

Bartsch D, Lehnen M, Clegg J, Pohl-Orf M, Schuphan I, Ellstrand NC (1999) Impact of gene flow from cultivated beet on genetic diversity of wild sea beet populations. Mol Ecol 8:1733–1741

Baskin CC, Baskin JM (2000) Seed germination ecology of *Lesquerella lyrata* Rollins (Brassicaceae), a federally threatened winter annual. Nat Area J 20:159–165

Bastrenta B, Lebreton JD, Thompson JD (1995) Predicting demographic change in response to herbivory: a model of the effects of grazing and annual variation on the population-dynamics of *Anthyllis vulneraria*. J Ecol 83:603–611

Beier P, Noss RF (1998) Do habitat corridors provide connectivity ? Conserv Biol 12:1241–1252

Beltman B, Van den Broek T, Van Maanen K, Vaneveld K (1996) Measures to develop a rich-fen wetland landscape with a full range of successional stages. Ecol Eng 7:299–313
Benson EE, Danaher JE, Pimbley IM, Anderson CT, Wake JE, Daley S, Adams LK (2000) In vitro micropropagation of *Primula scotica*: a rare Scottish plant. Biodiv Conserv 9:711–726
Berge G, Nordal I, Hestmark G (1998) The effect of breeding systems and pollination vectors on the genetic variation of small plant populations within an agricultural landscape. Oikos 81:17–29
Bernal R (1998) Demography of the vegetable ivory palm *Phytelephas seemannii* in Colombia, and the impact of seed harvesting. J Appl Ecol 35:64–74
Bevill RL, Louda SM, Stanforth LM (1999) Protection from natural enemies in managing rare plant species. Conserv Biol 13:1323–1331
Bobbink R, Willems JH (1993) Restoration management of abandoned chalk grassland in the Netherlands. Biodiv Conserv 2:616–626
Bokdam J, Gleichman JM (2000) Effects of grazing by free-ranging cattle on vegetation dynamics in a continental north-west European heathland. J Appl Ecol 37:415–431
Bond WJ (1994) Do mutualisms matter - assessing the impact of pollinator and disperser disruption on plant extinction. Philos Trans R Soc Lond Ser B Biol Sci 344:83–90
Bond WJ, Slingsby P (1984) Collapse of an ant-plant mutualism: the Argentine ant (*Iridomyrmex humilis*) and myrmecochorous Proteaceae. Ecology 65:1031–1037
Bosch M, Simon J, Molero J, Blanche C (1998) Reproductive biology, genetic variation and conservation of the rare endemic dysploid *Delphinium bolosii* (Ranunculaceae). Biol Conserv 86:57–66
Bowles M, McBride J, Betz R (1998) Management and restoration ecology of the federal threatened Mead's milkweed, *Asclepias meadii* (Asclepiadaceae). Ann Missouri Bot Garden 85:110–125
Boyce MS (1992) Population viability analysis. Annu Rev Ecol Syst 23:481–506
Bradstock RA, Bedward M, Kenny BJ, Scott J (1998) Spatially-explicit simulation of the effect of prescribed burning on fire regimes and plant extinctions in shrublands typical of south-eastern Australia. Biol Conserv 86:83–95
Brown E, Kephart S (1999) Variability in pollen load: implications for reproduction and seedling vigor in a rare plant, *Silene douglasii* var. *oraria*. Int J Plant Sci 160:1145–1152
Bruelheide H, Scheidel U (1999) Slug herbivory as a limiting factor for the geographical range of *Arnica montana*. J Ecol 87:839–848
Brunet J, von Oheimb G (1998) Migration of vascular plants to secondary woodlands in southern Sweden. J Ecol 86:429–438
Bruun HH (2000) Patterns of species richness in dry grassland patches in an agricultural landscape. Ecography 23:641–650
Bullock JM, Clarke RT (2000) Long distance seed dispersal by wind: measuring and modelling the tail of the curve. Oecologia 124:506–521
Burel F (1996) Hedgerows and their role in agricultural landscapes. Crit Rev Plant Sci 15:169–190
Buza L, Young A, Thrall P (2000) Genetic erosion, inbreeding and reduced fitness in fragmented populations of the endangered tetraploid pea *Swainsona recta*. Biol Conserv 93:177–186
Byers DL (1998) Effect of cross proximity on progeny fitness in a rare and a common species of *Eupatorium* (Asteraceae). Am J Bot 85:644–653
Byers DL, Meagher TR (1992) Mate availability in small populations of plant species with homomorphic sporophytic self-incompatibility. Heredity 68:353–359

Byers DL, Meagher TR (1997) A comparison of demographic characteristics in a rare and a common species of *Eupatorium*. Ecol Appl 7:519–530

Cain ML, Milligan BG, Strand AE (2000) Long-distance seed dispersal in plant populations. Am J Bot 87:1217–1227

Carlsen TM, Menke JW, Pavlik BM (2000) Reducing competitive suppression of a rare annual forb by restoring native California perennial grasslands. Restor Ecol 8:18–29

Carlson PC, Tanner GW, Wood JM, Humphrey SR (1993) Fire in Key deer habitat improves browse, prevents succession, and preserves endemic herbs. J Wildlife Manage 57:914–928

Carroll JA, Caporn SJM, Cawley L, Read DJ, Lee JA (1999) The effect of increased deposition of atmospheric nitrogen on *Calluna vulgaris* in upland Britain. New Phytol 141:423–431

Caswell H (1989) Matrix population models: construction, analysis and interpretation. Sinauer Associates Inc, Sunderland, MA

Caswell H (2000) Prospective and retrospective perturbation analyses: their roles in conservation biology. Ecology 81:619–627

Coates DJ, Hamley VL (1999) Genetic divergence and the mating system in the endangered and geographically restricted species, *Lambertia orbifolia* Gardner (Proteaceae). Heredity 83:418–427

Colas B, Olivieri I, Riba M (1997) *Centaurea corymbosa*, a cliff-dwelling species tottering on the brink of extinction : a demographic and genetic study. Proc Natl Acad Sci USA 94:3471–3476

Collinge SK (2000) Effects of grassland fragmentation on insect species loss, colonization, and movement patterns. Ecology 81:2211–2226

Conte L, Troia A, Cristofolini G (1998) Genetic diversity in *Cytisus aeolicus* Guss. (Leguminosae), a rare endemite of the Italian flora. Plant Biosyst 132:239–249

Cuenca S, Amo-Marco JB (2000) In vitro propagation of two Spanish endemic species of *Salvia* through bud proliferation. In Vitro Cell Dev-Pl 36:225–229

Damman H, Cain ML (1998) Population growth and viability analyses of the clonal woodland herb, *Asarum canadense*. J Ecol 86:13–26

Debinski DM, Holt RD (2000) A survey and overview of habitat fragmentation experiments. Conserv Biol 14:342–355

De Jong TJ, Waser NM, Klinkhamer PGL (1993) Geitonogamy – the neglected side of selfing. Trends Ecol Evol 8:321–325

De Mauro MM (1993) Relationship of breeding system to rarity in the lakeside daisy (*Hymenoxys acaulis* var. *glabra*). Conserv Biol 7:542–550

De Vries W, Klijn JA, Kros J (1994) Simulation of the long-term impact of atmospheric deposition on dune ecosystems in The Netherlands. J Appl Ecol 31:59–73

Deyrup M, Menges ES (1997) Pollination ecology of the rare scrub mint *Dicerandra frutescens* (Lamiaceae). Fl Sci 60:143–157

Dolan RW, Yahr R, Menges ES, Halfhill MD (1999) Conservation implications of genetic variation in three rare species endemic to Florida rosemary scrub. Am J Bot 86:1556–1562

Drayton B, Primack RB (2000) Rates of success in the reintroduction by four methods of several perennial plant species in eastern Massachusetts. Rhodora 102:299–331

Dudash MR (1990) Relative fitness of selfed and outcrossed progeny in a self-compatible, protandrous species, *Sabatia angularis* L. (Gentianaceae); a comparison in three environments. Evolution 44:1129–1139

Dudash MR, Fenster CB (2000) Inbreeding and outbreeding depression in fragmented populations. In: Young AG, Clarke GM (eds) Genetics, demography and viability of fragmented populations. Cambridge University Press, Cambridge, pp 35–53

Duncan RP, Young JR (2000) Determinants of plant extinction and rarity 145 years after European settlement of Auckland, New Zealand. Ecology 81:3048–3061

Eckert CG, Manicacci D, Barrett SCH (1996) Frequency-dependent selection on morph ratios in tristylous *Lythrum salicaria* (Lythraceae). Heredity 77:581–588

Ehlers BK (1999) Variation in fruit set within and among natural populations of the self-incompatible herb *Centaurea scabiosa* (Asteraceae). Nord J Bot 19:653–663

Ehrlen J (1995) Demography of the perennial herb *Lathyrus vernus*. 2. Herbivory and population-dynamics. J Ecol 83:297–308

Eisto AK, Kuitunen M, Lammi A, Saari V, Suhonen J, Syrjasuo S, Tikka PM (2000) Population persistence and offspring fitness in the rare bellflower *Campanula cervicaria* in relation to population size and habitat quality. Conserv Biol 14:1413–1421

Ellstrand NC (1992) Gene flow by pollen: implications for plant conservation genetics. Oikos 63:77–86

Ellstrand NC, Elam DR (1993) Population genetic consequences of small population size: implications for plant conservation. Annu Rev Ecol Syst 24:217–242

Eriksson A, Eriksson O (1997) Seedling recruitment in semi-natural pastures: the effects of disturbance, seed size, phenology and seed bank. Nord J Bot 17:469–482

Eriksson A, Eriksson O (2000) Population dynamics of the perennial *Plantago media* in semi-natural grasslands. J Veg Sci 11:245–252

Eriksson O (1996) Population ecology and conservation: some theoretical considerations with examples from the Nordic flora. Acta Univ Upsaliensis Symbolae Bot Upsalienses 31:159–167

Eriksson O (1997) Colonization dynamics and relative abundance of three plant species (*Antennaria dioica, Hieracium pilosella* and *Hypochoeris maculata*) in dry semi-natural grasslands. Ecography 20:559–568

Eriksson O, Kiviniemi K (1999) Site occupancy, recruitment and extinction thresholds in grassland plants: an experimental study. Biol Conserv 87:319–325

Esselink P, Zijlstra W, Dijkema KS, van Diggelen R (2000) The effects of decreased management on plant-species distribution patterns in a salt marsh nature reserve in the Wadden Sea. Biol Conserv 93:61–76

Fenster CB, Galloway LF (2000) Inbreeding and outbreeding depression in natural populations of *Chamaecrista fasciculata* (Fabaceae). Conserv Biol 14:1406–1412

Fischer M, Matthies D (1997) Mating structure and inbreeding and outbreeding depression in the rare plant *Gentianella germanica* (Gentianaceae). Am J Bot 84:1685–1692

Fischer M, Matthies D (1998a) Effects of population size on performance in the rare plant *Gentianella germanica*. J Ecol 86:195–204

Fischer M, Matthies D (1998b) RAPD variation in relation to population size and plant fitness in the rare *Gentianella germanica* (Gentianaceae). Am J Bot 86:811–819

Fischer M, Matthies D, Schmid B (1997) Responses of rare calcareous grassland plants to elevated CO_2: a field experiment with *Gentianella germanica* and *Gentiana cruciata*. J Ecol 85:681–691

Fischer M, van Kleunen M, Schmid B (2000) Genetic Allee effects on performance, plasticity and developmental stability in a clonal plant. Ecol Lett 3:530–539

Fischer SF, Poschlod P, Beinlich B (1996) Experimental studies on the dispersal of plants and animals on sheep in calcareous grasslands. J Appl Ecol 33:1206–1222

Floyd SK, Ranker TA (1998) Analysis of a transition matrix model for *Gaura neomexicana* ssp. *coloradensis* (Onagraceae) reveals spatial and temporal demographic variability. Int J Plant Sci 159:853–863

Forbes WA (1965) De antieke keuken (The antique kitchen). Van Dishoeck, Bussum

Fox LR, Ribeiro SP, Brown VK, Masters GJ, Clarke IP (1999) Direct and indirect effects of climate change on St John's wort, *Hypericum perforatum* L. (Hypericaceae). Oecologia 120:113–122

Francisco-Ortega J, Santos-Guerra A, Kim SC, Crawford DJ (2000) Plant genetic diversity in the Canary Islands: a conservation perspective. Am J Bot 87:909–919
Frankham R (1995) Conservation genetics. Annu Rev Genet 29:305–327
Frankham R (1999) Quantitative genetics in conservation biology. Genet Res 74:237–244
Frankie GW, Thorp RW, Newstrom-Lloyd LE, Rizzardi MA, Barthell JF, Griswold TL, Kim JY, Kappagoda S (1998) Monitoring solitary bees in modified wildland habitats: implications for bee ecology and conservation. Environ Entomol 27:1137–1148
Fritz R, Merriam G (1993) Fence row habitats for plants moving between farmland forests. Biol Conserv 64:141–148
Gardner SC, Grue CE (1996) Effects of Rodeo® and Garlon® 3A on nontarget wetland species in central Washington. Environ Toxicol Chem 15:441–451
Gascon C, Lovejoy TE (1998) Ecological impacts of forest fragmentation in central Amazonia. Zool Anal Complex Syst 101:273–280
Gaudeul M, Taberlet P, Till-Bottraud I (2000) Genetic diversity in an endangered alpine plant, *Eryngium alpinum* L. (Apiaceae), inferred from amplified fragment length polymorphism markers. Mol Ecol 9:1625–1637
Giblin DE, Hamilton CW (1999) The relationship of reproductive biology to the rarity of endemic *Aster curtus* (Asteraceae). Can J Bot 77:140–149
Gibson CWD, Brown VK (1991) The effects of grazing on local colonization and extinction during early succession. J Veg Sci 2:291–300
Gillespie RG (1997) Range contraction and extinction vulnerability: what is natural? Mem Mus Victoria 56:401–409
Gitzendanner MA, Soltis PS (2000) Patterns of genetic variation in rare and widespread plant congeners. Am J Bot 87:783–792
Glemin S, Bataillon T, Ronfort J, Mignot A, Olivieri I (2001) Inbreeding depression in small populations of self-incompatible plants. Genetics 159:1217–1229
Godt MJW, Hamrick JL (1999) Population genetic analysis of *Elliottia racemosa* (Ericaceae), a rare Georgia shrub. Mol Ecol 8:75–82
Gordon DR (1996) Experimental translocation of the endangered shrub Apalachicola rosemary *Conradina glabra* to the Apalachicola Bluffs and Ravines Preserve, Florida. Biol Conserv 77:19–26
Greenberg CH, Crownover SH, Gordon DR (1997) Roadside soils: a corridor for invasion of xeric scrub by nonindigenous plants. Nat Area J 17:99–109
Groom MJ (1998) Allee effects limit population viability of an annual plant. Am Nat 151:487–496
Gross K, Lockwood JR, Frost CC, Morris MF (1998) Modeling controlled burning and trampling reduction for conservation of *Hudsonia montana*. Conserv Biol 12:1291–1301
Hardner CM, Potts BM, Gore PL (1998) The relationship between cross success and spatial proximity of *Eucalyptus globulus* ssp. *globulus* parents. Evolution 52:614–618
Harrison RD (2000) Repercussions of El Niño: drought causes extinction and the breakdown of mutualism in Borneo. Proc R Soc Lond Ser B Biol Sci 267:911–915
Harrison S, Bruna E (1999) Habitat fragmentation and large-scale conservation: what do we know for sure? Ecography 22:225–232
Harvey HJ, Meredith TC (1981) Ecological studies of *Peucedanum palustre* and their implications for conservation management at wicken Fen. In: Synge H (ed) The biological aspects of rare plant conservation. Wiley, Chichester, pp 365–378
Hawkes CV, Menges ES (1995) Density and seed production of a Florida endemic, *Polygonella basiramia*, in relation to time since fire and open sand. Am Midl Nat 133:138–148
Hawkes CV, Menges ES (1996) The relationship between open space and fire for species in a xeric Florida shrubland. Bull Torrey Bot Club 123:81–92

Hedrick PW (1994) Purging inbreeding depression and the probability of extinction: full-sib mating. Heredity 73:363–372

Hegland SJ, Van Leeuwen M, Oostermeijer JGB (2001) Population structure of *Salvia pratensis* in relation to vegetation and management of Dutch dry floodplain grasslands. J Appl Ecol 38:1277–1289

Heil GW, Diemont WH (1983) Raised nutrient levels change heathland into grassland. Vegetatio 53:113–120

Helenurm K (1998) Outplanting and differential source population success in *Lupinus guadalupensis*. Conserv Biol 12:118–127

Heppell SS, Caswell H, Crowder LB (2000) Life histories and elasticity patterns: perturbation analysis for species with minimal demographic data. Ecology 81:654–665

Higgins SI, Richardson DM, Cowling RM, Trinder-Smith TH (1999) Predicting the landscape-scale distribution of alien plants and their threat to plant diversity. Conserv Biol 13:303–313

Hilton JL, Boyd RS (1996) Microhabitat requirements and seed/microsite limitation of the rare granite outcrop endemic *Amphianthus pusillus* (Scrophulariaceae). Bull Torrey Bot Club 123:189–196

Hingston AB, McQuillan PB (1998) Does the recently introduced bumblebee *Bombus terrestris* (Apidae) threaten Australian ecosystems? Aust J Ecol 23:539–549

Hitchmough J, Woudstra J (1999) The ecology of exotic herbaceous perennials grown in managed, native grassy vegetation in urban landscapes. Landscape Urban Plan 45:107–121

Huenneke LF (1991) Ecological implications of genetic variation in plant populations. In: Falk DA, Holsinger KE (eds) Genetics and conservation of rare plants. Oxford University Press, New York, pp 31–44

Huenneke LF (1995) Involving academic scientists in conservation research – perspectives of a plant ecologist. Ecol Appl 5:209–214

Hulme PD, Pakeman RJ, Torvell L, Fisher JM, Gordon IJ (1999) The effects of controlled sheep grazing on the dynamics of upland *Agrostis-Festuca* grassland. J Appl Ecol 36:886–900

Hurtrez-Bousses S (1996) Genetic differentiation among natural populations of the rare Corsican endemic *Brassica insularis* Moris: implications for conservation guidelines. Biol Conserv 76:25–30

Huryn VMB (1997) Ecological impacts of introduced honey bees. Q Rev Biol 72:275–297

Huxel GR (1999) Rapid displacement of native species by invasive species: effects of hybridization. Biol Conserv 89:143–152

Inouye DW, McGuire AD (1991) Effects of snowpack on timing and abundance of flowering in *Delphinium nelsonii* (Ranunculaceae) – implications for climate change. Am J Bot 78:997–1001

Jansen AJM, De Graaf MCC, Roelofs JGM (1996) The restoration of species-rich heathland communities in the Netherlands. Vegetatio 126:73–88

Jensen K (1998) Species composition of soil seed bank and seed rain of abandoned wet meadows and their relation to aboveground vegetation. Flora 193:345–359

Johansson ME, Nilsson C, Nilsson E (1996) Do rivers function as corridors for plant dispersal? J Veg Sci 7:593–598

Johnson CF, Cowling RM, Phillipson PB (1999) The flora of the Addo Elephant National Park, South Africa: are threatened species vulnerable to elephant damage? Biodiv Conserv 8:1447–1456

Jules ES (1998) Habitat fragmentation and demographic change for a common plant: *Trillium* in old-growth forest. Ecology 79:1645–1656

Kang U, Chang CS, Kim YS (2000) Genetic structure and conservation considerations of rare endemic *Abeliophyllum distichum* Nakai (Oleaceae) in Korea. J Plant Res 113: 127–138

Karron JD (1987) A comparison of levels of genetic polymorphism and self-incompatibility in geographically restricted and widespread plant congeners. Evol Ecol 1:47–58
Kaye TN (1999) From flowering to dispersal: reproductive ecology of an endemic plant, *Astragalus australis* var. *olympicus* (Fabaceae). Am J Bot 86:1248–1256
Kearns C, Inouye D, Waser N (1998) Endangered mutualisms: the conservation of plant-pollinator interactions. Annu Rev Ecol Syst 29:83–112
Keith DA (1997) Combined effects of heat shock, smoke and darkness on germination of *Epacris stuartii* Stapf., an endangered fire-prone Australian shrub. Oecologia 112:340–344
Keller LF, Waller DM (2002) Inbreeding effects in wild populations. Trends Ecol Evol 17:230–241
Keller M, Kollmann J, Edwards PJ (2000) Genetic introgression from distant provenances reduces fitness in local weed populations. J Appl Ecol 37:647–659
Kephart S, Brown E, Hall J (1999) Inbreeding depression and partial selfing: evolutionary implications of mixed-mating in a coastal endemic, *Silene douglasii* var. *oraria* (Caryophyllaceae). Heredity 82:543–554
Kéry M, Matthies D, Spillmann HH (2000) Reduced fecundity and offspring performance in small populations of the declining grassland plants *Primula veris* and *Gentiana lutea*. J Ecol 88:17–30
Kettle WD, Alexander HM, Pittman GL (2000) An 11-year ecological study of a rare prairie perennial (*Asclepias meadii*): implications for monitoring and management. Am Midl Nat 144:66–77
Kirkman LK, Drew MB, Edwards D (1998) Effects of experimental fire regimes on the population dynamics of *Schwalbea americana* L. Plant Ecol 137:115–137
Kiviniemi K (1996) A study of adhesive seed dispersal of three species under natural conditions. Acta Bot Neerl 45:73–83
Kiviniemi K, Telenius A (1998) Experiments on adhesive dispersal by wood mouse: seed shadows and dispersal distances of 13 plant species from cultivated areas in southern Sweden. Ecography 21:108–116
Kluse J, Doak DF (1999) Demographic performance of a rare California endemic, *Chorizanthe pungens* var. *hartwegiana* (Polygonaceae). Am Midl Nat 142:244–256
Knowles L, Witkowski ETF (2000) Conservation biology of the succulent shrub, *Euphorbia barnardii*, a serpentine endemic of the Northern Province, South Africa. Aust Ecol 25:241–252
Knox JS (1997) A nine year demographic study of *Helenium virginicum* (Asteraceae), a narrow endemic seasonal wetland plant. J Torrey Bot Soc 124:236–245
Koelewijn HP (1998) Effects of different levels of inbreeding on progeny fitness in *Plantago coronopus*. Evolution 52:692–702
Koerper H, Kolls AL (1999) The silphium motif adorning ancient Libyan coinage: marketing a medicinal plant. Econ Bot 53:133–143
Kollmann J, Poschlod P (1997) Population processes at the grassland-scrub interface. Phytocoenologia 27:235–256
Krahulcova A, Krahulec F, Kirschner J (1996) Introgressive hybridization between a native and an introduced species: *Viola lutea* subsp. *sudetica* versus *V. tricolor*. Folia Geobot Phytotx 31:219-&
Krenova Z, Leps J (1996) Regeneration of a *Gentiana pneumonanthe* population in an oligotrophic wet meadow. J Veg Sci 7:107–112
Kruess A, Tscharntke T (1994) Habitat fragmentation, species loss, and biological control. Science 264:1581–1584
Kwak MM, Velterop O, Van Andel J (1998) Pollen and gene flow in fragmented habitats. Appl Veg Sci 1:37–54
Lande R (1988) Genetics and demography in biological conservation. Science 241:1455–1460

Lande R (1998) Anthropogenic, ecological and genetic factors in extinction and conservation. Res Popul Ecol 40:259–269

Larson AC, Gentry LE, David MB, Cooke RA, Kovacic DA (2000) The role of seepage in constructed wetlands receiving agricultural tile drainage. Ecol Eng 15:91–104

Ledig FT, Jacob-Cervantes V, Hodgskiss PD, Eguiluz-Piedra T (1997) Recent evolution and divergence among populations of a rare Mexican endemic, chihuahua spruce, following Holocene climatic warming. Evolution 51:1815–1827

Lennartsson T, Oostermeijer JGB (2001) Demographic variation and population viability in *Gentianella campestris*: effects of grassland management and environmental stochasticity. J Ecol 89:451–463

Lennartsson T, Svensson R (1996) Patterns in the decline of three species of *Gentianella* (Gentianaceae) in Sweden, illustrating the deterioration of semi-natural grasslands. Acta Univ Upsaliensis Symbolae Bot Upsaliensis 31:170–180

Les DH, Reinartz JA, Esselman EJ (1991) Genetic consequences of rarity in *Aster furcatus* (Asteraceae), a threatened self-incompatible plant. Evolution 45:1641–1650

Lesica P (1995) Demography of *Astragalus scaphoides* and effects of herbivory on population growth. Great Basin Nat 55:142–150

Lesica P (1999) Effects of fire on the demography of the endangered, geophytic herb *Silene spaldingii* (Caryophyllaceae). Am J Bot 86:996–1002

Lesica P, Allendorf FW (1992) Are small populations of plants worth preserving ? Conserv Biol 6:135–139

Lesica P, Atthowe HE (2000) Should we use pesticides to conserve rare plants ? Conserv Biol 14:1549–1550

Lesica P, Shelly JS (1995) Effects of reproductive mode on demography and life-history in *Arabis fecunda* (Brassicaceae). Am J Bot 82:752–762

Lesica P, Shelly JS (1996) Competitive effects of *Centaurea maculosa* on the population dynamics of *Arabis fecunda*. Bull Torrey Bot Club 123:111–121

Lesica P, Steele BM (1996) A method for monitoring long-term population trends: an example using rare arctic-alpine plants. Ecol Appl 6:879–887

Levin DA, Francisco-Ortega J, Jansen RK (1996) Hybridization and the extinction of rare plant species. Conserv Biol 10:10–16

Linder CR, Taha I, Seiler GJ, Snow AA, Rieseberg LH (1998) Long-term introgression of crop genes into wild sunflower populations. Theor Appl Genet 96:339–347

Lindgren CJ, Gabor TS, Murkin HR (1999) Compatibility of glyphosate with *Galerucella calmariensis*; a biological control agent for purple loosestrife (*Lythrum salicaria*). J Aquatic Plant Manage 37:44–48

Lipow SR, Broyles SB, Wyatt R (1999) Population differences in self-fertility in the "self-incompatible" milkweed *Asclepias exaltata* (Asclepiadaceae). Am J Bot 86:1114–1120

Losvik MH (1999) Plant species diversity in an old, traditionally managed hay meadow compared to abandoned hay meadows in southwest Norway. Nord J Bot 19:473–487

Louda SM, Kendall D, Connor J, Simberloff D (1997) Ecological effects of an insect introduced for the biological control of weeds. Science 277:1088–1090

Luijten SH (2001) Reproduction and genetics in fragmented plant populations. PhD thesis. University of Amsterdam, Amsterdam, 120 pp

Luijten SH, Dierick A, Oostermeijer JGB, Raijmann LEL, Den Nijs JCM (2000) Population size, genetic variation and reproductive success in the rapidly declining, self-incompatible *Arnica montana* in The Netherlands. Conserv Biol 14:1776–1786

Luijten SH, Kery MMM, Oostermeijer JGB, Den Nijs JCM (2002) Demographic consequences of inbreeding and outbreeding in *Arnica montana*: a field experiment. J Ecol 90: 593–603

Mabry C, Korsgren T (1998) A permanent plot study of vegetation and vegetation-site factors fifty-three years following disturbance in central New England, USA. Ecoscience 5:232–240

Mack RN, Simberloff D, Lonsdale WM, Evans H, Clout M, Bazzaz FA (2000) Biotic invasions: causes, epidemiology, global consequences, and control. Ecol Appl 10:689–710
Maina GG, Howe HF (2000) Inherent rarity in community restoration. Conserv Biol 14:1335–1340
Maki M (1999) Genetic diversity in the threatened insular endemic plant *Aster asa-grayi* (Asteraceae). Plant Syst Evol 217:1–9
Marrs RH, Frost AJ (1997) A microcosm approach to the detection of the effects of herbicide spray drift in plant communities. J Environ Manage 50:369–388
Maschinski J, Frye R, Rutman S (1997) Demography and population viability of an endangered plant species before and after protection from trampling. Conserv Biol 11:990–999
Masters RA, Nissen SJ, Gaussoin RE, Beran DD, Stougaard RN (1996) Imidazolinone herbicides improve restoration of Great Plains grasslands. Weed Technol 10:392–403
Matolweni LO, Balkwill K, McLellan T (2000) Genetic diversity and gene flow in the morphologically variable, rare endemics *Begonia dregei* and *Begonia homonyma* (Begoniaceae). Am J Bot 87:431–439
Maze KE, Bond WJ (1996) Are *Protea* populations seed limited ? Implications for wildflower harvesting in Cape fynbos. Aust J Ecol 21:96–105
McCollin D, Jackson JI, Bunce RGH, Barr CJ, Stuart R (2000) Hedgerows as habitat for woodland plants. J Environ Manage 60:77–90
McLaughlin A, Mineau P (1995) The impact of agricultural practices on biodiversity. Agric Ecosyst Environ 55:201–212
Medail F, Verlaque R (1997) Ecological characteristics and rarity of endemic plants from southeast France and Corsica: Implications for biodiversity conservation. Biol Conserv 80:269–281
Menges ES (1991a) The application of minimum viable population theory to plants. In: Falk DA, Holsinger KE (eds) Genetics and conservation of rare plants. Oxford University Press, New York, pp 450–461
Menges ES (1991b) Seed germination percentage increases with population size in a fragmented prairie species. Conserv Biol 5:158–164
Menges ES (1992) Stochastic modeling of extinction in plant populations. In: Fiedler PL, Jain SK (eds) Conservation biology: the theory and practice of nature conservation, preservation and management. Chapman and Hall, New York, pp 253–276
Menges ES (2000) Population viability analyses in plants: challenges and opportunities. Trends Ecol Evol 15:51–56
Menges ES, Dolan RW (1998) Demographic viability of populations of *Silene regia* in midwestern prairies: relationships with fire management, genetic variation, geographic location, population size and isolation. J Ecol 86:63–78
Menges ES, Gordon DR (1996) Three levels of monitoring intensity for rare plant species. Nat Area J 16:227–237
Menges ES, Hawkes CV (1998) Interactive effects of fire and microhabitat on plants of Florida scrub. Ecol Appl 8:935–946
Menges ES, Kohfeldt N (1995) Life history strategies of Florida scrub plants in relation to fire. Bull Torrey Bot Club 122:282–297
Menges ES, McIntyre PJ, Finer MS, Goss E, Yahr R (1999) Microhabitat of the narrow Florida scrub endemic *Dicerandra christmanii*, with comparisons to its congener *D. frutescens*. J Torrey Bot Soc 126:24–31
Milton SJ, Bond WJ, Du Plessis MA, Gibbs D, Hilton-Taylor C, Linder HP, Raitt L, Wood J, Donaldson JS (1999) A protocol for plant conservation by translocation in threatened lowland Fynbos. Conserv Biol 13:735–743
Morgan JW (1999) Effects of population size on seed production and germinability in an endangered, fragmented grassland plant. Conserv Biol 13:266–273

Moritz C (1999) Conservation units and translocations: strategies for conserving evolutionary processes. Hereditas 130:217–228
Nagy ES (1997) Frequency-dependent seed production and hybridization rates: Implications for gene flow between locally adapted plant populations. Evolution 51:703–714
Nantel P, Gagnon D (1999) Variability in the dynamics of northern peripheral versus southern populations of two clonal plant species, *Helianthus divaricatus* and *Rhus aromatica*. J Ecol 87:748–760
Nault A, Gagnon D (1993) Ramet demography of *Allium tricoccum*, a spring ephemeral, perennial forest herb. J Ecol 81:101–119
Negron-Ortiz V (1998) Reproductive biology of a rare cactus, *Opuntia spinosissima* (Cactaceae), in the Florida Keys: why is seed set very low? Sex Plant Reprod 11:208–212
Newman D, Pilson D (1997) Increased probability of extinction due to decreased genetic effective population size: experimental populations of *Clarkia pulchella*. Evolution 51:354–362
Newman D, Tallmon DA (2001) Experimental evidence for beneficial fitness effects of gene flow in recently isolated populations. Conserv Biol 15:1054–1063
Norderhaug A, Ihse M, Pedersen O (2000) Biotope patterns and abundance of meadow plant species in a Norwegian rural landscape. Landscape Ecol 15:201–218
Nunney L, Campbell KA (1993) Assessing minimum viable population size: demography meets population genetics. Trends Ecol Evol 8:234–239
Oliveira ACM, Ferrari SF (2000) Seed dispersal by black-handed tamarins, *Saguinus midas niger* (Callitrichinae, Primates): implications for the regeneration of degraded forest habitats in eastern Amazonia. J Trop Ecol 16:709–716
Olmsted I, Alvarez-Buylla ER (1995) Sustainable harvesting of tropical trees - demography and matrix models of 2 palm species in Mexico. Ecol Appl 5:484–500
Oostermeijer JGB (1995) De levensvatbaarheid van kleine plantenpopulaties: een reaktie (The viability of small plant populations: a response). De Levende Natuur 96:223–226
Oostermeijer JGB (2000a) Population viability analysis of the rare *Gentiana pneumonanthe*: importance of genetics, demography, and reproductive biology. In: Young AG, Clarke GM (eds) Genetics, demography and viability of fragmented populations. Cambridge University Press, Cambridge, pp 313–334
Oostermeijer JGB (2000b) Is genetic variation important for the viability of wild plant populations? Schriftenreihe Vegetationskunde 32:23–30
Oostermeijer JGB, Den Nijs JCM, Raijmann LEL, Menken SBJ (1992) Population biology and management of the marsh gentian (*Gentiana pneumonanthe* L.), a rare species in The Netherlands. Bot J Linn Soc 108:117–130
Oostermeijer JGB, Van 't Veer R, Den Nijs JCM (1994a) Population structure of the rare, long-lived perennial *Gentiana pneumonanthe* in relation to vegetation and the management in The Netherlands. J Appl Ecol 31:428–438
Oostermeijer JGB, Van Eijck MW, Den Nijs JCM (1994b) Offspring fitness in relation to population size and genetic variation in the rare perennial plant species *Gentiana pneumonanthe* (Gentianaceae). Oecologia 97:289–296
Oostermeijer JGB, Altenburg RGM, Den Nijs JCM (1995a) Effects of outcrossing distance and selfing on fitness components in the rare *Gentiana pneumonanthe* (Gentianaceae). Acta Bot Neerl 44:257–268
Oostermeijer JGB, Van Eijck MW, Van Leeuwen NC, Den Nijs JCM (1995b) Analysis of the relationship between allozyme heterozygosity and fitness in the rare *Gentiana pneumonanthe* L. J Evol Biol 8:739–757
Oostermeijer JGB, Berholz A, Poschlod P (1996a) Genetical aspects of fragmented plant populations: a review. In: Settele J, Margules C, Poschlod P, Henle K (eds) Species survival in fragmented landscapes. Kluwer, Dordrecht, pp 93–101

Oostermeijer JGB, Brugman ML, De Boer ER, Den Nijs JCM (1996b) Temporal and spatial variation in the demography of *Gentiana pneumonanthe*, a rare perennial herb. J Ecol 84:153–166
Oostermeijer JGB, Hvatum H, Den Nijs JCM, Borgen L (1996 c) Genetic variation, plant growth strategy and population structure of the rare, disjunctly distributed *Gentiana pneumonanthe* (Gentianaceae) in Norway. Acta Univ Upsaliensis Symbolae Bot Upsalienses 31:185–203
Oostermeijer JGB, Luijten SH, Krenová ZV, Den Nijs JCM (1998a) Relationships between population and habitat characteristics and reproduction of the rare *Gentiana pneumonanthe* L. Conserv Biol 12:1042–1053
Oostermeijer JGB, Luijten SH, Kwak MM, Boerrigter EJM, Den Nijs JCM (1998b) Rare plants in peril : on the problems of small populations (in Dutch). De Levende Natuur 99:134–141
Oostermeijer JGB, Luijten SH, Petanidou T, Kos M, Ellis-Adam AC, Den Nijs JCM (2000) Pollination in rare plants: is population size important ? Det Norske Videnskaps-akademi I Matematisk Naturvidenskapelige Klasse Skrifter Ny Serie 39:201–213
Oostermeijer JGB, Luijten SH, Ellis-Adam AC, den Nijs JCM (2002) Future prospects for the rare, late-flowering *Gentianella germanica* and *Gentianopsis ciliata* in Dutch nutrient-poor calcareous grasslands. Biol Conserv 104:339–350
Ortiz-Pulido R, Laborde J, Guevara S (2000) Fruit-eating habits of birds in a fragmented landscape: implications for seed dispersal. Biotropica 32:473–488
Ouborg NJ, Van Treuren R (1994) The significance of genetic erosion in the process of extinction. 4. Inbreeding load and heterosis in relation to population size in the mint *Salvia pratensis*. Evolution 48:996–1008
Ouborg NJ, Haeck J, Reinink K, Van Treuren R (1995) Een methode voor het schatten van de levensvatbaarheid van populaties, met Duifkruid als voorbeeld (A method for the assessment of population viability, with *Scabiosa columbaria* as example). De Levende Natuur 96:46–52
Palacios C, Gonzalez-Candelas F (1997) Lack of genetic variability in the rare and endangered *Limonium cavanillesii* (Plumbaginaceae) using RAPD markers. Mol Ecol 6:671–675
Pandit MK, Babu CR (1998) Biology and conservation of *Coptis teeta* Wall. – an endemic and endangered medicinal herb of eastern Himalaya. Environ Conserv 25:262–272
Partel M, Mandla R, Zobel M (1999) Landscape history of a calcareous (alvar) grassland in Hanila, western Estonia, during the last three hundred years. Landscape Ecol 14:187–196
Paton DC (2000) Disruption of bird-plant pollination systems in southern Australia. Conserv Biol 14:1232–1234
Pavlik BM, Manning E (1993) Assessing limitations on the growth of endangered plant populations .1. Experimental demography of *Erysimum capitatum* ssp. *angustatum* and *Oenothera deltoides* ssp. *howellii*. Biol Conserv 65:257–265
Pedro SRD, Decamargo JMF (1991) Interactions on floral resources between the Africanized honey-bee *Apis mellifera* L and the native bee community (Hymenoptera, Apoidea) in a natural cerrado ecosystem in southeast Brazil. Apidologie 22:397–415
Pendergrass KL, Miller PM, Kauffman JB, Kaye TN (1999) The role of prescribed burning in maintenance of an endangered plant species, *Lomatium bradshawii*. Ecol Appl 9:1420–1429
Percy DM, Cronk QCB (1997) Conservation in relation to mating system in *Nesohedyotis arborea* (Rubiaceae), a rare endemic tree from St Helena. Biol Conserv 80:135–145
Persson HA, Lundquist K, Nybom H (1998) RAPD analysis of genetic variation within and among populations of Turk's-cap lily (*Lilium martagon* L.). Hereditas 128:213–220

Pfab MF, Witkowski ETF (1999) Contrasting effects of herbivory on plant size and reproductive performance in two populations of the critically endangered species *Euphorbia clivicola* R.A. Dyer. Plant Ecol 145:317–325

Pfab MF, Witkowski ETF (2000) A simple population viability analysis of the Critically Endangered *Euphorbia clivicola* RA Dyer under four management scenarios. Biol Conserv 96:263–270

Pierce SM, Cowling RM (1991) Dynamics of soil-stored seed banks of 6 shrubs in fire-prone dune fynbos. J Ecol 79:731–747

Pinard M (1993) Impacts of stem harvesting on populations of *Iriartea deltoidea* (Palmae) in an extractive reserve in Acre, Brazil. Biotropica 25:2–14

Planty-Tabacchi AM (1997) Invasions of riparian corridors in Southwestern France by exotic plant species. Bull Fr Peche Piscic 427–439

Poschlod P, Bonn S (1998) Changing dispersal processes in the central European landscape since the last ice age: an explanation for the actual decrease of plant species richness in different habitats? Acta Bot Neerl 47:27–44

Post E, Stenseth NC (1999) Climatic variability, plant phenology, and northern ungulates. Ecology 80:1322–1339

Price MV, Waser NM (1998) Effects of experimental warming on plant reproductive phenology in a subalpine meadow. Ecology 79:1261–1271

Prober SM, Brown AHD (1994) Conservation of the grassy white box woodlands: population genetics and fragmentation of *Eucalyptus albens*. Conserv Biol 8:1003–1013

Proulx M, Mazumder A (1998) Reversal of grazing impact on plant species richness in nutrient-poor vs. nutrient-rich ecosystems. Ecology 79:2581–2592

Pysek P, Pysek A (1995) Invasion by *Heracleum mantegazzianum* in different habitats in the Czech Republic. J Veg Sci 6:711–718

Quintana-Ascencio PF, Menges ES (2000) Competitive abilities of three narrowly endemic plant species in experimental neighborhoods along a fire gradient. Am J Bot 87:690–699

Quintana-Ascencio PF, Dolan RW, Menges ES (1998) *Hypericum cumulicola* demography in unoccupied and occupied Florida scrub patches with different time-since-fire. J Ecol 86:640–651

Rabotnov TA (1969) On coenopopulations of perennial herbaceous plants in natural coenoses. Vegetatio 19:87–95

Rabotnov TA (1985) Dynamics of plant coenotic populations. In: White J (eds) The population structure of vegetation. Handbook of vegetation science. Junk, Dordrecht, pp 121–142

Raijmann LEL, Van Leeuwen NC, Kersten R, Oostermeijer JGB, Den Nijs JCM, Menken SBJ (1994) Genetic variation and outcrossing rate in relation to population size in *Gentiana pneumonanthe* L. Conserv Biol 8:1014–1026

Rajora OP, Rahman MH, Buchert GP, Dancik BP (2000) Microsatellite DNA analysis of genetic effects of harvesting in old-growth eastern white pine (*Pinus strobus*) in Ontario, Canada. Mol Ecol 9:339–348

Randall JM, Lewis RR, Fitter AH, Alexander IJ (1997) Ecological restoration. In: Simberloff D, Schmitz DC, Brown TC (eds) Strangers in paradise. Island, Washington, DC, pp 205–219

Rathcke B, Jules E (1993) Habitat fragmentation and plant-pollinator interactions. Curr Sci India 65:273–276

Ratsirarson J, Silander JA, Richard AF (1996) Conservation and management of a threatened Madagascar palm species, *Neodypsis decaryi*, Jumelle. Conserv Biol 10:40–52

Reinartz JA (1995) Planting state-listed endangered and threatened plants. Conserv Biol 9:771–781

Reinartz JA, Les DH (1994) Bottleneck-induced dissolution of self-incompatibility and breeding system consequences in *Aster furcatus* (Asteraceae). Am J Bot 81:446–455

Rhymer JM, Simberloff D (1996) Extinction by hybridization and introgression. Annu Rev Ecol Syst 27:83–109
Richardson DM, Cowling RM, Lamont BB (1996) Non-linearities, synergisms and plant extinctions in south African fynbos and Australian kwongan. Biodiv Conserv 5:1035–1046
Richardson DM, Allsopp N, D'Antonio CM, Milton SJ, Rejmanek M (2000) Plant invasions – the role of mutualisms. Biol Rev Cambridge Philos Soc 75:65–93
Robertson AW, Kelly D, Ladley JJ, Sparrow AD (1999) Effects of pollinator loss on endemic New Zealand mistletoes (Loranthaceae). Conserv Biol 13:499–508
Robinson GR, Handel SN (2000) Directing spatial patterns of recruitment during an experimental urban woodland reclamation. Ecol Appl 10:174–188
Runhaar J, van Gool CR, Groen CLG (1996) Impact of hydrological changes on nature conservation areas in the Netherlands. Biol Conserv 76:269–276
Runyeon-Lager H, Prentice HC (2000) Morphometric variation in a hybrid zone between the weed, *Silene vulgaris*, and the endemic, *Silene uniflora* ssp *petraea* (Caryophyllaceae), on the Baltic island of Öland. Can J Bot Rev Can Bot 78:1384–1397
Saetersdal M, Birks HJB (1997) A comparative ecological study of Norwegian mountain plants in relation to possible future climatic change. J Biogeogr 24:127–152
Santos T, Telleria JL, Virgos E (1999) Dispersal of Spanish juniper, *Juniperus thurifera*, by birds and mammals in a fragmented landscape. Ecography 22:193–204
Schlapfer M, Zoller H, Korner C (1998) Influences of mowing and grazing on plant species composition in calcareous grassland. Bot Helvetica 108:57–67
Schmidt K, Jensen K (2000) Genetic structure and AFLP variation of remnant populations in the rare plant *Pedicularis palustris* (Scrophulariaceae) and its relation to population size and reproductive components. Am J Bot 87:678–689
Schneller JJ, Holderegger R (1996) Genetic variation in small, isolated fern populations. J Veg Sci 7:113–120
Schwartz MW (1994) Conflicting goals for conserving biodiversity – issues of scale and value. Nat Area J 14:213–216
Semenova GV, van der Maarel E (2000) Plant functional types – a strategic perspective. J Veg Sci 11:917–922
Shaw WB, Burns BR (1997) The ecology and conservation of the endangered endemic shrub, kowhai ngutukaka *Clianthus puniceus* in New Zealand. Biol Conserv 81:233–245
Silvertown JW, Franco M, Pisanty I, Mendoza A (1993) Comparative plant demography – relative importance of life-cycle components to the finite rate of increase in woody and herbaceous perennials. J Ecol 81:465–476
Silvertown JW, Franco M, Menges ES (1996) Interpretation of elasticity matrices as an aid to the management of plant population for conservation. Conserv Biol 10:591–597
Simberloff D (1997) Eradication. In: Simberloff D, Schmitz DC, Brown TC (eds) Strangers in paradise. Island, Washington, DC, pp 221–228
Simenstad CA, Cordell JR, Tear L, Weitkamp LA, Paveglio FL, Kilbride KM, Fresh KL, Grue CE (1996) Use of Rodeo® and X-77® spreader to control smooth cordgrass (*Spartina alterniflora*) in a southwestern Washington estuary 2. Effects on benthic microflora and invertebrates. Environ Toxicol Chem 15:969–978
Smith M, Keevin T, Mettler-McClure P, Barkau R (1998) Effect of the flood of 1993 on *Boltonia decurrens*, a rare floodplain plant. Regul River Res Manage 14:191–202
Smith VH, Tilman GD, Nekola JC (1999) Eutrophication: impacts of excess nutrient inputs on freshwater, marine, and terrestrial ecosystems. Environ Pollut 100:179–196
Soltis PS, Gitzendanner MA (1999) Molecular systematics and the conservation of rare species. Conserv Biol 13:471–483
Sork VL, Nason J, Campbell DR, Fernandez JF (1999) Landscape approaches to historical and contemporary gene flow in plants. Trends Ecol Evol 14:219–224

Stanforth LM, Louda SM, Bevill RL (1997) Insect herbivory on juveniles of a threatened plant, *Cirsium pitcheri*, in relation to plant size, density and distribution. Ecoscience 4:57–66

Stanley RJ, Dickinson KJM, Mark AF (1998) Demography of a pore endemic *Myosotis*: boom and bust in the high-alpine zone of southern New Zealand. Arctic Alpine Res 30:227–240

Steffan-Dewenter I, Tscharntke T (2000) Resource overlap and possible competition between honey bees and wild bees in central Europe. Oecologia 122:288–296

Stenstrom A, Jonsdottir IS (1997) Responses of the clonal sedge, *Carex bigelowii*, to two seasons of simulated climate change. Global Change Biol 3:89–96

Stenstrom M, Gugerli F, Henry GHR (1997) Response of *Saxifraga oppositifolia* L. to simulated climate change at three contrasting latitudes. Global Change Biol 3:44–54

Stephens PA, Sutherland WJ (1999) Consequences of the Allee effect for behaviour, ecology and conservation. Trends Ecol Evol 14:401–405

Sterling PH, Gibson CWD, Brown VK (1992) Leaf miner assemblies – effects of plant succession and grazing management. Ecol Entomol 17:167–178

Sternberg M, Brown VK, Masters GJ, Clarke IP (1999) Plant community dynamics in a calcareous grassland under climate change manipulations. Plant Ecol 143:29–37

Stiling P, Rossi A, Gordon D (2000) The difficulties of single factor thinking in restoration: replanting a rare cactus in the Florida Keys. Biol Conserv 94:327–333

Stohlgren TJ, Binkley D, Chong GW, Kalkhan MA, Schell LD, Bull KA, Otsuki Y, Newman G, Bashkin M, Son Y (1999) Exotic plant species invade hot spots of native plant diversity. Ecol Monogr 69:25–46

Sugden EA, Thorp RW, Buchmann SL (1996) Honey bee native bee competition: focal point for environmental change and apicultural response in Australia. Bee World 77:26–44

Sutherland WJ (2001) Sustainable exploitation: a review of principles and methods. Wildlife Biol 7:131–140

Talavera S, Arista M, Salgueiro FJ (1996) Population size, pollination and breeding system of *Silene stockenii* Chater (Caryophyllaceae), an annual gynodioecious species of southern Spain. Bot Acta 109:333–339

Tepedino VJ, Sipes SD, Griswold TL (1999) The reproductive biology and effective pollinators of the endangered beardtongue *Penstemon penlandii* (Scrophulariaceae). Plant Syst Evol 219:39–54

Thomas LP (1996) Population ecology of a winter annual (*Lesquerella filiformis* Rollins) in a patchy environment. Nat Area J 16:216–226

Totland O (1999) Effects of temperature on performance and phenotypic selection on plant traits in alpine *Ranunculus acris*. Oecologia 120:242–251

Tremblay RL, Zimmerman JK, Lebron L, Bayman P, Sastre I, Axelro F, Alers-Garcia J (1998) Host specificity and low reproductive success in the rare endemic Puerto Rican orchid *Lepanthes caritensis*. Biol Conserv 85:297–304

Turnbull LA, Crawley MJ, Rees M (2000) Are plant populations seed-limited? A review of seed sowing experiments. Oikos 88:225–238

Valverde T, Silvertown J (1998) Variation in the demography of a woodland understory herb (*Primula vulgaris*) along the forest regeneration cycle: projection matrix analysis. J Ecol 86:545–562

Van Dorp D, Schippers P, Van Groenendael JM (1997) Migration rates of grassland plants along corridors in fragmented landscapes assessed with a cellular automation model. Landscape Ecol 12:39–50

Van Groenendael JM, Ouborg NJ, Hendriks RJJ (1998) Criteria for the introduction of plant species. Acta Bot Neerl 47:3–13

Van Loosdrecht MCM, Hooijmans CM, Brdjanovic D, Heijnen JJ (1997) Biological phosphate removal processes. Appl Microbiol Biotechnol 48:289–296

Van Treuren R, Bijlsma R, Van Delden W, Ouborg NJ (1991) The significance of genetic erosion in the process of extinction.1. Genetic differentiation in *Salvia pratensis* and *Scabiosa columbaria* in relation to population size. Heredity 66:181–189
Van Treuren R, Bijlsma R, Ouborg NJ, van Delden W (1993) The significance of genetic erosion in the process of extinction. IV. Inbreeding depression and heterosis effects caused by selfing and outcrossing in *Scabiosa columbaria*. Evolution 47:1669–1680
Van Walsum PEV, Joosten JHJ (1994) Quantification of local ecological effects in regional hydrologic modeling of bog reserves and surrounding agricultural lands. Agric Water Manage 25:45–55
Van Wirdum G (1993) An ecosystems approach to base-rich fresh-water wetlands, with special reference to fenlands. Hydrobiologia 265:129–153
Vekemans X, Schierup MH, Christiansen FB (1998) Mate availability and fecundity selection in multi-allelic self-incompatibility systems in plants. Evolution 52:19–29
Verdu JR, Crespo MB, Galante E (2000) Conservation strategy of a nature reserve in Mediterranean ecosystems: the effects of protection from grazing on biodiversity. Biodiv Conserv 9:1707–1721
Volkl W, Zwolfer H, Romstockvolki M, Schmelzer C (1993) Habitat management in calcareous grasslands – effects on the insect community developing in flower heads of *Cynarea*. J Appl Ecol 30:307–315
Wadsworth RA, Collingham YC, Willis SG, Huntley B, Hulme PE (2000) Simulating the spread and management of alien riparian weeds: are they out of control? J Appl Ecol 37:28–38
Waite S, Farrell L (1998) Population biology of the rare military orchid (*Orchis militaris* L.) at an established site in Suffolk, England. Bot J Linn Soc 126:109–121
Walck JL, Baskin JM, Baskin CC (1999) Relative competitive abilities and growth characteristics of a narrowly endemic and a geographically widespread *Solidago* species (Asteraceae). Am J Bot 86:820–828
Waser NM, Price MV (1994) Crossing-distance effects in *Delphinium nelsonii*: outbreeding and inbreeding depression in progeny fitness. Evolution 48:842–852
Waser NM, Price MV, Shaw RG (2000) Outbreeding depression varies among cohorts of *Ipomopsis aggregata* planted in nature. Evolution 54:485–491
Weidema IR, Magnussen LS, Philipp M (2000) Gene flow and mode of pollination in a dry-grassland species, *Filipendula vulgaris* (Rosaceae). Heredity 84:311–320
Wester L (1994) Weed management and the habitat protection of rare species – a case-study of the endemic Hawaiian fern *Marsilea villosa*. Biol Conserv 68:1–9
Widén B (1993) Demographic and genetic effects on reproduction as related to population size in a rare perennial herb, *Senecio integrifolius* (Asteraceae). Biol J Linn Soc 50:179–195
Willems JH (1983) Species composition and above-ground phytomass in chalk grassland with different management. Vegetatio 52:171–180
Willems JH, Melser C (1998) Population dynamics and life-history of *Coeloglossum viride* (L.) Hartm.: an endangered orchid species in The Netherlands. Bot J Linn Soc 126:83–93
Williams SL (2001) Reduced genetic diversity in eelgrass transplantations affects both population growth and individual fitness. Ecol Appl 11:1472–1488
Witkowski ETF, Liston RJ (1997) Population structure, habitat profile and regeneration of *Haworthia koelmaniorum*, a vulnerable dwarf succulent, endemic to Mpumalanga, South Africa. S Afr J Bot 63:363–370
Witkowski ETF, Lamont BB, Obbens FJ (1994) Commercial picking of *Banksia hookeriana* in the wild reduces subsequent shoot, flower and seed production. J Appl Ecol 31:508–520
Wolf JHD, Konings CJF (2001) Toward the sustainable harvesting of epiphytic bromeliads: a pilot study from the highlands of Chiapas, Mexico. Biol Conserv 101:23–31

Woodward FI (1997) Life at the edge : a 14-year study of a *Verbena officinalis* population's interactions with climate. J Ecol 85:899–906

Yao J, Holt RD, Rich PM, Marshall WS (1999) Woody plant colonization in an experimentally fragmented landscape. Ecography 22:715–728

Young A, Boyle T, Brown T (1996) The population genetic consequences of habitat fragmentation for plants. Trends Ecol Evol 11:413–418

Young AG, Merriam HG, Warwick SI (1993) The effects of forest fragmentation on genetic variation in *Acer saccharum* Marsh. (sugar maple) populations. Heredity 71:277–289

Young AG, Brown AHD, Zich FA (1999) Genetic structure of fragmented populations of the endangered Daisy *Rutidosis leptorrhynchoides*. Conserv Biol 13:256–265

Zwiers FW, Weaver AJ (2000) Climate change: the causes of 20th century warming. Science 290:2081–2082

3 Factors Affecting Persistence in Formerly Common and Historically Rare Plants

C.A. Brigham

3.1 Introduction

3.1.1 General Background

Ecologists and conservation biologists have long-standing intellectual and practical interests in rare species. The large numbers of rare species in the United States and the world have focused research attention on rarity as a phenomenon. In a comprehensive recent review, Fiedler (1995) discussed the history of causes of rarity in plants and considered new findings with regards to the California flora. Her review highlighted the fact that biologists have been thinking about and researching causes of rarity in plants for over 200 years. Prior research has often centered on causes and patterns of rarity among floras, families, or other groups (e.g., Harper 1979; Hedge and Ellstrand 1996; Lokeshad and Vasudeva 1997; Schwartz and Simberloff 1997). General investigations of causes and patterns of rarity in plants also abound (e.g., Griggs 1940; Drury 1980; Kruckeberg and Rabinowitz 1985; Gaston 1994; Fiedler 1995) . Several frameworks for examining rare species have been proposed previously, including Rabinowitz's (Rabinowitz 1981) seven forms of rarity and Caughley's (Caughley 1994) small vs declining population paradigm. Additionally, many investigators have examined the effects of specific factors on rare plant species such as seed predation (Hegazy and Eesa 1991), disturbance (Menges 1992; Kaye 1997), herbivory (Louda and McEachern 1995), and pollination (Menges 1995; Johnson and Bond 1997). Generally, we have many studies of specific species and hypotheses concerning causes of rarity, but no clear picture has emerged of the importance of different factors to rare plant persistence.

Emphasis has recently been placed on conservation and restoration of rare species. These efforts are reflected in the number of population viability analyses of plants published per year, which increased 14-fold between 1973

Ecological Studies, Vol. 165
C.A. Brigham, M.W. Schwartz (Eds.)
Population Viability in Plants

and 1997 (Menges 2000a). Population viability analyses (PVAs) focus on measuring demographic rates and using these measured rates to estimate viability. They seldom include information on species interactions such as pollination, competition, or herbivory, and few PVAs include possible genetic effects on persistence (although see Oostermeijer 2000). Recent reviews of PVAs have identified the need to include more complicated biological information such as species interactions and genetics (e.g., Reed et al. 1998) in order to fully understand and predict species dynamics. Menges (2000a) briefly outlined the possibilities for including interactions in PVAs. In Chapter 10 (this Vol., Morales et al. provide a comprehensive review and approach for incorporating interactions into PVAs. Considering the work involved in constructing a basic PVA it is pertinent to ask whether including information on species interactions and genetics is really necessary for accurate predictions of population viability. In Chap. 2 of this volume, Oostermeijer reviewed threats to population viability of plants and recommended management strategies for ameliorating threats. This chapter will expand on Oostermeijer's analysis by examining four factors in-depth. I ask whether genetics, competition, pollination, and herbivory have strong enough, demonstrable effects on rare plants to necessitate their inclusion in population viability assessments. The null hypothesis in this case is that each factor has insignificant effects on rare plant species persistence and is thus unnecessary in population viability analysis; the alternative hypothesis is that without inclusion of said factors, PVAs will be seriously flawed and of little use in conservation efforts.

Whether a given factor is important in the biology of a rare plant may, in some cases, depend on the type of rare plant examined. If a species has previous experience with the condition of rarity it may show adaptations to this condition. Here, I define two types of rarity in plants and ask whether type of rarity (formerly common and historically rare-defined below) affects a species response to genetic factors, competition, pollination, or herbivory. If different types of rare species show different patterns of response to these factors, then PVAs should take into account the type of species being modeled. Alternatively, if no differences can be seen, then the type of rarity is irrelevant. I consider possible differences between rarity types from the standpoint of mechanisms of persistence. Focusing on persistence instead of causes of rarity may highlight mechanisms that rare species have for maintaining populations over time and provide a framework for generating hypotheses concerning population dynamics of rare plants.

3.1.2 Selection of Factors

This section focuses on genetics, competition, pollination, and herbivory because these factors span the range of frequency of study (from well-studied to barely studied), include both autecological factors and interactions, are frequently cited as causes of rarity, and are generally not included in PVAs (Reed et al. 1998). Additionally, these four factors are commonly considered to have large effects on plant performance. These factors were chosen because they are amenable to formulating hypotheses of differential responses of historically rare versus formerly common species. Clearly, these are not all the factors that may potentially affect rare species (see Chap. 2, this Vol., for a complete review of factors). A review of these four factors is intended to test two hypotheses: (1) These factors have such large effects on plant persistence that it is necessary to include them in PVAs. (2) Historically rare and formerly common species respond differently to each factor.

3.1.3 Types of Rarity in Plants

Two types of rare species are distinguished, species that have recently been reduced to small population size (formerly common), and historically rare species. The first are plants that were formerly more abundant, perhaps even common members of their community, but are now at low abundance. Their populations have been diminished by factors such as habitat destruction, exotic species invasions, and change in disturbance regimes. These species, referred to here as formerly common, can be contrasted with historically rare species (see discussion of old and new rares in Chapter 2, this Vol.). Many floras contain species that have historically been documented as rare, yet still persist (see examples in Kruckeberg and Rabinowitz 1985). Some of these species may be neo-endemics (recently formed species) that may become more common with time (Willis 1922). Similarly, some may be paleoendemics that were more common before shifts in climate. Whatever the cause or potential future, these historically rare species have recent experience with persistence at low abundance (i.e., they have a recent history of rarity), while formerly common species will not have had this experience to the same degree. Formerly common species may have been rare in certain parts of their range, or previously in their history, but they will lack evolutionary pressures of historically rare plants with respect to persistence at low abundance. Although not all rare species can be clearly delineated into these categories (e.g., a species whose abundance fluctuates wildly through time, going through rare and common phases in its recent past), I propose this dichotomy because populations that are historically rare may have certain adaptations to rarity that enable them to persist. For each factor, potential adaptations are discussed and examined by a review of the literature.

Additionally, focusing on historically rare species may advance our understanding of how plant traits influence persistence of a species. Identifying traits that enable persistence at low abundance may help predict which species are likely to continue to decline once their numbers are reduced versus those that may be able to stabilize at low numbers. This information would be helpful in proactive species conservation and prioritizing conservation actions.

Rarity can also be a local, regional, or global phenomenon. A species may be rare in one geographic area but common in others. Conservation efforts often focus on locally rare species. Additionally, some rare species are at low abundance wherever they are found, while others are at high abundance but only in a select few locales (Rabinowitz 1981). The effects of these differences on hypothesized mechanisms of persistence are discussed below.

3.2 Reductions in Genetic Diversity

Genetic diversity in small populations can be reduced by a variety of forces including inbreeding, genetic drift, and bottlenecks (see discussion in Chap. 2, this Vol.). In order for low genetic diversity to concern conservation biologists, two relationships must be demonstrated. First, there must be a genetic consequence of small population size or rarity. This could be demonstrated by small populations having lower diversity than large populations within a species or rare species having lower genetic diversity than common species. Second, this low diversity must be associated with a negative consequence, either reduced fitness of current populations or evidence that this lower diversity will negatively affect the future of the species through reduced ability to adapt to changing conditions. Reductions in genetic diversity may affect fitness through reduced herbivore or disease resistance (Chap. 4, this Vol.), reduced reproductive output, or other fitness effects. Reduced fitness due to decreased genetic variation may occur due to genetic drift or inbreeding. In small populations (less than 100 individuals) allele frequencies may shift radically due to random sampling of the parental generation (genetic drift, reviewed in Ellstrand and Elam 1993). Inbreeding may occur through selfing or biparental inbreeding (Gillespie 1998). Prevention of inbreeding may be achieved through self-incompatibility, differential development of male and female parts, or spatial separation of parts on different individuals (dioecy) in some plant species (Barrrett and Kohn 1991). Genetic effects on rare species persistence can be evaluated by posing four questions: (1) Is there a positive correlation between genetic diversity and population size within a plant species? (2) Do rare plant species show less genetic diversity than common species? (3) Is there evidence for a positive correlation between reduced genetic diversity and reduced fitness in plants? (4) Do the answers to these three questions differ between historically rare and formerly common species?

3.2.1 Is There a Positive Correlation Between Genetic Diversity and Population Size in Plants?

A difficulty in answering this question is that the censused population is often not the same as the effective population (N_{census} vs $N_{e,}$ see discussion in Ellstrand and Elam 1993). Theoretical predictions of genetic effects of population sizes are based on the effective population size, i.e., members of the population that are actually breeding (see Gillespie 1998). In many organisms, including many plants, the effective population size is usually much smaller than the censused population size. There are several implications of the difference between N_e and N_{census}. First, populations that appear large based on census data may in fact have small effective population sizes. Thus, a lack of difference between large and small populations in genetic diversity or fitness may be due to their similarity in effective population size. Second, conversely, one possible mechanism for rare species to maintain genetic diversity and fitness at small population sizes is to involve a large proportion of individuals in mating, either through selfing, increased pollination, or some other mechanism, and thus increase their effective population size. The difference between N_e and N_{census} is an important one; however, it is extremely difficult to measure N_e and thus, for most species we have no idea what the effective population size is. Consequently I this issue will not be addressed here, although the implications of N_e/N_{census} differences will be discussed in the Conclusions.

Censused population size appears to be sometimes, but not always, positively correlated with reduced genetic diversity. Indirect evidence for this positive relationship comes from comparisons of narrow endemics with more widespread species. Hamrick and Godt (1990) evaluated the genetic diversity of narrow endemics versus more widespread species and found that narrow endemics had lower proportions of polymorphic loci, proportions of loci that were heterozygous per individual, and alleles per polymorphic locus than widespread species. Since no measures of actual population size are given for the endemics studied, and given that plant families can differ radically in their base levels of genetic variation, this study provides only an indirect measure of the relationship between population size and genetic diversity. Due to differences in genetic diversity among families, the most appropriate analyses for examination of the effects of population size on genetic diversity are either within-species comparisons or phylogenetically independent contrasts using congeners (e.g., Karron 1987, 1991; Gitzendanner and Soltis 2000).

Ellstrand and Elam (1993), in a review of studies on variation within species, found limited evidence of a reduction in diversity in small populations. Out of ten studies, they found seven that described a positive correlation between population size and genetic diversity within a given species and three that found no relationship. They also reviewed three studies that measured quantitative variation, which may differ from electrophoretically detectable variation, finding that only one showed a positive correlation between population size and quantitative variation.

I found 14 papers, covering 17 rare plant species, published since 1993 that examine the relationship between population size and genetic diversity within a species. Combining the findings of Ellstrand and Elam (1993) and my investigations, 17 studies demonstrate a positive correlation between population size and genetic diversity, nine show no relationship, and one study finds a positive correlation in one part of its range but not in another (*Silene regia*, Dolan 1994; Table 3.1).

Genetic diversity is usually measured with RAPD markers, AFLPs, and allozymes; however, studies have shown that these measures of diversity often do not reflect adaptively significant variation that may be important for persistence (e.g., Black-Samuelsson et al. 1997; Knapp and Rice 1998). A recent review of 71 studies correlating molecular measures (e.g., RAPDs, AFLPs, and allozymes) and quantitative genetic measures (e.g., seed size, height, or leaf shape) found that the mean correlation between the two measures was weak (r=0.217, Reed and Frankham 2001). In addition, the authors found no significant relationship between these two measures for life history traits (r=-0.11). They suggested that processes such as directional selection on quantitative traits, increased variability of polygenic traits, higher mutational input in polygenic traits because of numerous loci, as well as epistatic genetic variation and dominance can all act to decouple molecular and quantitative variation. Thus it is not clear whether studies that rely on molecular markers to assess diversity are measuring genetic diversity of interest.

Additionally, with only 27 studies addressing genetic diversity and population size in rare plants, it is difficult to evaluate whether the genetic methods used affect the detected outcome (but see Ellstrand and Elam 1993). With these caveats, current evidence suggests that population size and genetic diversity are often, but not always, positively correlated. Therefore attempts to determine the genetic diversity of different populations should not be based on population size estimates alone but also must use collected genetic data specific to the species of concern.

3.2.2 Do Rare Plant Species Show Reduced Genetic Diversity in Comparison to More Common Congeners?

Previous authors have argued that rare species should show reduced genetic diversity, either as a cause of rarity (see Watson et al. 1994) or because of processes acting on small populations such as inbreeding, drift, or genetic bottlenecks. Overall, I found that rare species do show significantly reduced variation in comparison to their more common congeners. Several studies show a pattern of reduced genetic diversity in restricted species. Barbel and Selander (1974) tested the hypothesis that ecological habitat breadth was

Table 3.1. Genetic diversity and population size

Species	Reference	Type of rarity	Positive correlation between genetic diversity and within species[a] Population size?	Genetic measure
Acacia anomala	Coates (1988)	Historic	Y	% PL
Aster kantoensis	Maki et al. (1996)	Historic	N	
Calamagrostis cainii	Godt et al. (1996)	Historic	Y	He
Carex misera	Godt et al. (1996)	Historic	N	He
Cordylanthus palmatus	Fleishman et al. (2000)	Formerly common	N	
Eucalyptus albens	Prober and Brown (1994)	Formerly common	Y	%PL, #A/L, He
Eucalyptus caesia	Moran and Hopper (1983)	Historic	Y	% PL
Eucalyptus crucis	Sampson et al. (1989)	Historic, declining	Y	% PL, # A/L, He
Eucalyptus parviflora	Prober et al. (1990)	Historic	N	
Eucalyptus pendens	Moran and Hopper (1987)	Historic	N	
Eucalyptus pulverulenta	Peters et al. (1990)	Historic	Y	% PL, # A/L, He
Gentiana pneumonanthe	Raijmann et al. (1994)	Formerly common	Y	%PL, marginally He, #A/L
Gentianella germanica	Fischer and Matthies (1998)	Formerly common	Y	RAPD DP
Geum radiatum	Godt et al. (1996)	Historic, declining	Y	He
Halocarpus bidwilli	Billington (1991)	Formerly common	Y	%PL, #A/L, He
Lychnis viscaria	Lammi et al. (1999)	Formerly common	Y	Hexp, %PL
Microseris lanceolata	Prober et al. (1998)	Formerly common	N	
Pedicularis palustris	Schmidt and Jensen (2000)	Formerly common	N	
Ranunculus reptans	Fischer et al. (2000)	Formerly common	Y	RAPD DP
Salvia pratensis	van Treuren et al. (1991)	Formerly common	Y	%PL, #A/L
Scabiosa columbaria	van Treuren et al. (1991)	Formerly common	Y	%PL, #A/L
Silene regia	Dolan (1994)	Formerly common	Y in eastern part of range	H
Silene regia	Dolan (1994)	Formerly common	N in western part of range	H

Table 3.1. (*Continued*)

Species	Reference	Type of rarity	Positive correlation between genetic diversity and within species[a] Population size?	Genetic measure
Swainsonia recta	Buza et al. (2000)	Formerly common	Y	#A/L, F
Trichophorum cespitosum	Godt et al. (1996)	Historic	Y	He
Vicia pisiformis	Black-Samuelsson et al. (1997)	Historic, declining	N	
Warea carteri	Evans et al. (2000)	Formerly common	N	
Washingtonia filifera	McClenaghan and Beauchamp (1986)	Historic	Y	%PL

[a] *% PL* Percent polymorphic loci, *#A/L* number of alleles per locus, *He* total genetic diversity, *RAPD DP* diversity of RAPD phenotypes, *Y* yes, *N* no

associated with high genetic diversity by comparing two different pairs of species, a narrow endemic lupine and its broadly distributed congener (*Lupinus subcarnosus* – restricted and *L. texensis* – common) and two *Hymenopappus* species with similarly broad habitat breadths (*H. scabiosaeus* var. *corymbosus* and *H. artemisiaefolius* var. *artemisiaefolius*). They measured allozyme diversity and found, as expected, that the restricted *L. subcarnosus* had significantly less genetic variation than all three of the more broadly distributed species. Furthermore, there was no difference in genetic diversity between the two *Hymenopappus* species.

Recently, Gitzendanner and Soltis (2000) have reviewed the evidence for reduced genetic diversity in rare species by examining the literature for data on genetic variation at population and species levels and measures of population substructuring for pairs of rare and common congeners. They measured population level diversity as percent polymorphic loci per population, mean number of alleles per locus per population, and observed heterozygosity. Genetic diversity of species was measured as percent polymorphic loci per species, mean number of alleles per locus per species, and total genetic diversity. The authors found that rare species have significantly less genetic variation than their common congeners in percentage polymorphic loci, mean number of alleles per locus, and observed heterozygosity, although differences were small. There was no difference between rare and common congeners in total genetic diversity per species, and there was a large positive correlation between rare and common congeners for all genetic measures.

Godt and Hamrick (2001) compared genetic diversity in rare and common congeners of plants in the southeastern United States. They found that rare species were more genetically depauperate in percent polymorphic loci and alleles per polymorphic locus than their more common congeners. However, the range of diversity in these rare plants encompassed nearly the entire range of diversity previously reported for seed plants, suggesting that rare species can have rates of genetic diversity as high as common species. Finally, they also found that rare southeastern plants were more genetically diverse in percent polymorphic loci and alleles per polymorphic locus than other endemics. The authors hypothesized that southeastern plants have maintained higher genetic diversity than other endemics due to their location close to glacial refugia, which may reduce founder effects and increase rates of gene flow during recolonization after glaciation.

In summary, rare plants show a pattern of significantly reduced diversity compared to their more common congeners; however, rare species also show the same range of genetic diversity as their more common counterparts. Thus predicting genetic diversity of an individual species based on its pattern of abundance alone may be misleading or simply inaccurate. Considering the high degree of correlation between diversity in rare and common congeners, sampling more common species may be a good proxy for the diversity of rare species when material of the rare species is unavailable. Finally, although examining genetic diversity in rare and common congeners can give insights into the biology of rare species, it does not directly address the relationship between population size and diversity since many rare species are also locally abundant.

3.2.3 Is There Evidence for a Positive Correlation Between Reduced Genetic Diversity and Reduced Fitness in Plants?

Even if small populations experience reduced genetic diversity, will this reduction in diversity affect fitness? Two arguments have been made for why we should expect a decline in fitness. First, there have been notorious cases of reduced fitness in inbred animals such as cheetahs (O'Brien et al. 1985), and a few studies have detected decreased fitness with decreased heterozygosity in plants (Van Treuren et al. 1993; Oostermeijer et al. 1994; Fischer and Matthies 1998; Buza et al. 2000). Decreases in heterozygosity are presumed to result in a build-up of deleterious alleles that have negative effects on fitness characters, such as plant size, seed size, development stability, and number of seeds produced (genetic load). Reduced genetic diversity may also affect other traits that affect fitness, such as herbivore or disease resistance (Chap. 4, this Vol.). Second, previous authors have argued that reduced genetic diversity will reduce the ability of species to respond to change over evolutionary time (e.g., Ellstrand and Elam 1993; Chap. 2, this Vol.). Unfortunately, this second argument is hard to evaluate because of the large time-scale needed to observe effects of genetic homozygosity on adaptation to changing conditions; therefore it is not evaluated here. My review of the literature for negative fitness consequences of reduced genetic diversity in plants produced equivocal results.

Of 16 studies, covering 13 species, that examined the relationship between genetic diversity and fitness related characters in plants (e.g., seedling survival, fruit production), 6 species showed decreased fitness with decreasing genetic diversity and 4 showed no relationship (Table 3.2). Three species showed mixed results: (1) some fitness characters were affected by genetic diversity but others were not, (2) for a given species, one study showed negative fitness consequences of reduced genetic diversity and another showed no effect (e.g., *Gentiana pneumonanthe,* see Table 3.2). In another study, Menges (1991) found a negative relationship between fitness and population size for the rare plant *Silene regia* but was unable to differentiate between genetic and nongenetic causes.

Assuming that fitness declines with declining genetic diversity is not warranted, based on the evidence that shows it occurs in less than 50% of the cases examined. More studies are needed to establish at what level of genetic diversity plants begin to show negative fitness consequences.

Table 3.2. Genetic diversity and fitness

Species	Reference	Type of rarity	Positive correlation between genetic diversity and fitness?	Genetic measure	Plant traits affected
Arnica montana	Luijten et al. (1996)	Formerly common	Y	Indirect, breeding	Selfed plants had lower seedling weight than outcrossed plants
Aster kantoensis	Inoue et al. (1998)	Formerly common	Y	Indirect, breeding	Selfed plants had fewer flowers/plant than outcrossed plants
Astragalus linifolius	Karron (1989)	Historic	Y	Indirect, breeding	No effects in fecundity measures but selfed plants had lower viability when grown in growth chambers
Astragalus osterhouti	Karron (1989)	Historic	N	Indirect, breeding	
Gentiana pneumonanthe	Oostermeijer et al. (1995)	Formerly common	mixed	Allozymes	Fitness components containing: no. stems, no. flowers/plant, seedling weight, adult weight
Gentiana pneumonanthe	Oostermeijer et al. (1994)	Formerly common	Y	Isozymes	Adult weight and flowering performance
Gentianella germanica	Fischer and Matthies (1998)	Formerly common	Y	RAPDs	Seeds/plant in the field, no. flowers/plant in common garden
Lychnis viscaria	Lammi et al. (1999)	Formerly common	N	Isozymes	

Table 3.2. (*Continued*)

Species	Reference	Type of rarity	Positive correlation between genetic diversity and fitness?	Genetic measure	Plant traits affected
Pedicularis palustris	Schmidt and Jensen (2000)	Formerly common	Mixed	AFLP markers	No. seeds/plant, no. capsules/plant
Salvia pratensis	Ouborg and van Treuren (1995)	Formerly common	N	Allozymes	
Salvia pratensis	Ouborg and van Treuren (1994)	Formerly common	Y	Indirect, breeding	Selfed plants in the greenhouse had smaller mean seed weight, proportion seeds germinating, plant size, regenerative capacity, and survival than outcrossed plants. In the field, selfed plants had lower survival than outcrossed plants.
Salvia pratensis	van Treuren et al. (1991)	Formerly common	Y	Allozymes	Average of growth and reproductive traits in the field
Scabiosa columbaria	van Treuren et al. (1993)	Formerly common	Y	Indirect, breeding	Selfed plants had lower biomass production, root development, adult survival, and seed set than outcrossed plants.
Senecio integrifolius	Widen (1993)	Formerly common	N	Indirect, breeding	
Silene diclinis	Waldmann (1999)	Historic	N	Indirect, breeding	
Swainsonia recta	Buza et al. (2000)	Formerly common	Y	Allozymes	Seed germination

3.2.4 Do Historically Rare and Formerly Common Species Show Similar Patterns of Correlations Between Population Size and Genetic Diversity?

The results of this comparison were equivocal. Although historically rare species less frequently showed a reduction in genetic diversity with reductions in population size than formerly common species, this difference was nonsignificant (Fig. 3.1).

There are two areas in which historically rare and formerly common species may vary. First, the population size at which genetic diversity declines may differ between historically rare and formerly common species. Historically rare species may have mechanisms to maintain genetic diversity at small population sizes. These mechanisms may involve maintenance of outcrossing at small population sizes, maintenance of dispersal between populations, or selection against homozygotes. Additionally, plants with below-ground populations, such as seed banks, bulbs, and rootstocks, may employ these mechanisms to buffer against loss of genetic diversity in aboveground populations. Second, historically rare species may maintain fitness at lower levels of genetic diversity than formerly common species. Historically rare species that have experienced repeated bottlenecks or commonly occurred at low abundance may have purged deleterious alleles and thus be less affected by reduc-

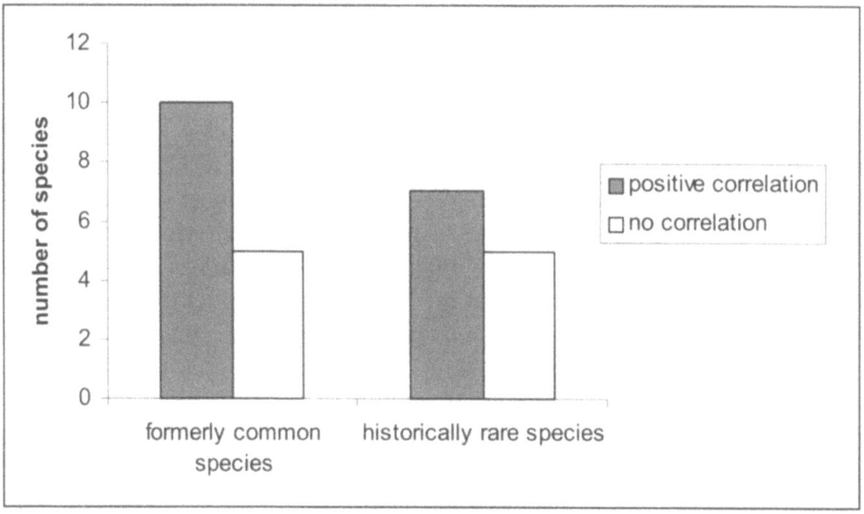

Fig. 3.1. Relationship between measures of genetic diversity and population size grouped by type of rarity. The measure of genetic diversity used varies from study t o study. Data shown include studies cited in Ellstrand and Elam (1993) and all relevant studies found in the literature after 1993. Note that the relationship among historically rare species is more variable than the pattern seen in formerly common species. Data are from Table 3.1

tions in genetic diversity (Lande and Schemske 1985; Ellstrand and Elam 1993; Chap. 2., this Vol.). To examine the possible relationship between type of rarity and population size effects on genetic diversity and fitness, I categorized species from published studies of the relationship between genetic diversity and population size into the categories historically rare or formerly common based on descriptions in the original papers (Table 3.1). Historically rare species showed a slightly different, but nonsignificant, pattern in their relationship between genetic diversity and population size than did formerly common species (Fig. 3.1). A little over half (58 %) of historically rare species (n=12) showed a positive correlation between genetic diversity and population size, while 75 % of formerly common species showed a positive correlation (n=14). While this approach provides some insight into differences between historically rare and formerly common species, it should be noted that not all studies used similar measures of genetic diversity and the species studied varied in their life form. I classified perennial species that are glacial relicts as historically rare although they may be less adapted to rarity than annuals that have been rare for a similar time period because of their much longer generation time (i.e., although an annual species and a perennial species may have been rare for a similar length of time, the annual will have undergone many more generations under small population/rarity selection forces than the perennial).

Among historically rare species that show no relationship between genetic diversity and population size, only *Aster kantoensis* (Maki et al. 1996) and *Vicia pisiformis* (Black-Samuelsson et al. 1997) had low levels of genetic variation regardless of population size. In *Aster kantoensis*, low diversity is ascribed to repeated founder effects. This species colonizes open stream banks until they are closed in by vegetation and thus it repeatedly goes extinct in a given area and must colonize new areas (Maki et al. 1996). *Vicia pisiformis* shows low variation in RAPDs but high differentiation of local populations in morphology (growth and fecundity). Black-Samuelsson et al. (1997) suggested that local populations may be under strong local selection and that these genes may not be detected in analysis of RAPD markers. Three species showed high levels of genetic variation despite small population size (*Acacia anomala*, Coates 1988; *Eucalyptus parvifolia*, Prober et al. 1990; and *Eucalyptus pendens*, Moran and Hopper 1987).

There are 13 studies, covering 11 species, addressing fitness effects of genetic diversity in formerly common plants. Of these 13 studies, 8 showed a positive relationship between genetic diversity and fitness parameters (e.g., Luijten et al. 1996) while 3 showed no relationship (e.g., Lammi et al. 1999) and 2 showed mixed effects (e.g., Oostermeijer et al. 1995; Table 3.2). It remains unclear whether historically rare and formerly common species show similar relationships between genetic diversity and fitness.

Unfortunately, only two studies, covering three historically rare species, have examined effects of reduced genetic diversity on fitness (Table 3.2). Karron (1987) compared levels of inbreeding depression in two narrow endemics and two widespread congeners and found no evidence for inbreeding depression in the fecundity parameters of any of the species. However, there was evidence for inbreeding depression in offspring fitness of inbred crosses in the narrow endemic *Astragalus linifolius*, suggesting it may be on the brink of damaging levels of homozygosity. Waldmann (1999) tested for inbreeding effects in the historically rare endemic *Silene diclinis* and found no significant effects of the crossing treatments on developmental instability of seedlings, a measure of inbreeding depression.

It is difficult to assess whether historically rare plants show fewer negative consequences of small population size than formerly common plants. Historically rare species showed a weaker correlation between genetic diversity and population size than formerly common species, and three of five historically rare species were able to maintain high levels of genetic diversity at small population sizes (Table 3.1). The absence of studies of genetic diversity and fitness in historically rare species makes it difficult to examine whether historically rare species can maintain fitness at low levels of diversity. The small amount of work done to date on historically rare species suggests that they are able to maintain fitness despite severe population restrictions. The relationship between genetic diversity and fitness in historically rare species and rare plants in general needs further study to clarify this issue. Additionally, to fully address questions of genetics and population size, we need more and better estimates of N_e since most predictions of genetic effects of small population size depend on knowing the effective population size.

3.2.5 Implications for PVAs

What does all this mean for PVAs of plants? The evidence is equivocal for both the hypothesis of the importance of genetics in plant persistence and the importance of type of rarity in genetic effects on persistence. There is evidence that population size can affect genetic diversity in plants (Table 3.1; 63 % positive correlations between genetic diversity and population size) and that genetic diversity can affect fitness (Table 3.2). The consistency of the effect, however, is closer to 50 than to 100 %. This evidence supports the argument that inclusion of a genetic analysis in a PVA must be carefully considered. The biology of individual species should determine the factors in a PVA, not broad generalizations of how genetics do or do not matter.

Similarly, clear patterns of differential responses of historically rare and formerly common species did not emerge. There is, however, a suggestion in the data that historically rare species may not experience negative consequences of small population size to the same degree as formerly common species. This point deserves consideration when creating a PVA for a historically rare species. With respect to correlations between genetic diversity and population size, and genetic diversity and fitness, the data on historically rare species are sparse but indicate that a large proportion of historically rare species do not show reductions in diversity or fitness with small population size (Tables 3.1, 3.2).

3.3 Competition

A common argument made with respect to rare species that show patterns of low abundance (as opposed to locally abundant but geographically restricted species) is that this pattern of abundance is due to poor competitive ability (Griggs 1940; Drury 1980; Tanimoto 1996; Tilman 1990a). A number of mechanisms have been proposed to explain persistence of such sparse species, including spatial or temporal escapes from competition (Tilman 1997; Chesson 1996). Tilman (1994), for example, argued for a tradeoff between competitive ability and dispersal ability wherein rare plants are good dispersers but poor competitors. The author posited that some rare plants may be fugitive species, colonizing open habitats and persisting in them until driven out by superior competitors. Similarly, Chesson and Huntly (1997) have proposed a model, termed the "storage effect", wherein coexistence occurs through species maintaining low abundance when conditions favor other species and only becoming abundant when conditions favor their reproduction (temporal escape from competition). The storage effect relies on three premises: (1) temporal fluctuation in resources, (2) covariance between the environment and competition, and (3) buffered population growth through a "storage" life stage (e.g., seeds, bulbs). Both of these theories suggest that rare plant species should show a pattern of competitive avoidance either temporally or spatially and that rare species will prove to be weaker competitors than other members of their community or their more common congeners.

An alternative view to these ideas of competitive escape was proposed by Rabinowitz et al. (1984). They argued that sparse species may be rare for reasons other than competitive ability and these species should be effective competitors in order to persist within their community. If species that are at low abundance within a community have a random or even dispersion (anything but clumped), they will frequently have neighbors that are a different species. In order for these species to persist, they must be able to compete with these more common species. Under this view, we would expect similar competitive abilities between rare and common species.

The theories of Rabinowitz et al. (1984) concerning competitive abilities of rare species rely on a history of coexistence between the rare species and its competitors as well as a history of sparseness. This theory and others do not formally address the potential effects of introduced species where there is no evolutionary history between competitors. Exotic species frequently arrive without native pests or herbivores and can alter resource availability in completely novel ways (e.g., Holmes and Rice 1996; Dyer and Rice 1999). Thus, competition with introduced species may be a cause of decline for many rare species, including both formerly common and historically rare plants. In order to better understand the role of competition in rare plant persistence, the following questions, designed to evaluate competition while keeping previous theories in mind, are posed. First, is there evidence for competition with native species as a cause of rarity in plants? Second, what is the evidence for competition with exotic species playing a role in species declines? Third, will the role of competition differ between historically rare and formerly common species?

3.3.1 Is There Evidence for Competition with Native Species as a Cause of Rarity in Plants?

In greenhouse and field studies of two co-occurring congeners, *Mimulus nudatus* (rare, restricted to serpentine outcrops) and *M. guttatus* (widespread), no major competitive effects of *M. guttatus* on growth or reproduction of *M. nudatus* were found (C. Brigham, unpubl. data). *M. nudatus* is not suppressed as the density of *M. guttatus* increases (Fig. 3.2). Similar patterns were seen when *M. nudatus* was grown in the greenhouse at varying densities with other, more common species in its community (C. Brigham, unpubl. data). Additionally, within habitats where it occurs, competition with other plants has negative effects on *M. nudatus* only in certain sites at certain times in the growing season. Even when competition negatively affects plants, these effects are weaker than abiotic site differences or effects of timing of germination. Although other species have been shown to grow on serpentine to avoid competition with more vigorous plants off serpentine (Kruckeberg 1984), the restriction of *M. nudatus* to serpentine is not simply due to poor competitive ability off serpentine. If grown on non-serpentine soils in the absence of other plants, *M. nudatus* will not produce a functional root system (C. Brigham, pers. observ.; Macnair 1992). Thus, although the habitat restriction of *M. nudatus* is clearly part of the cause for its rarity, this habitat restriction is in turn due, in part, to physiological requirements and not competitive ability.

In reviewing the handful of other studies on competitive abilities of rare plants, I found that the importance of competition depends on the method used to measure it. Competition research falls into two categories: (1) greenhouse trials comparing competitive abilities of rare and common congeners or rare and common species in general (comparison B in Table 3.3), and (2) field-removal experiments testing the effects of neighbors on a measure of fitness such as biomass or seed production (comparison A in Table 3.3). Different research approaches yield different results and conclusions; additionally, only a small number of studies have been conducted, so results must be inter-

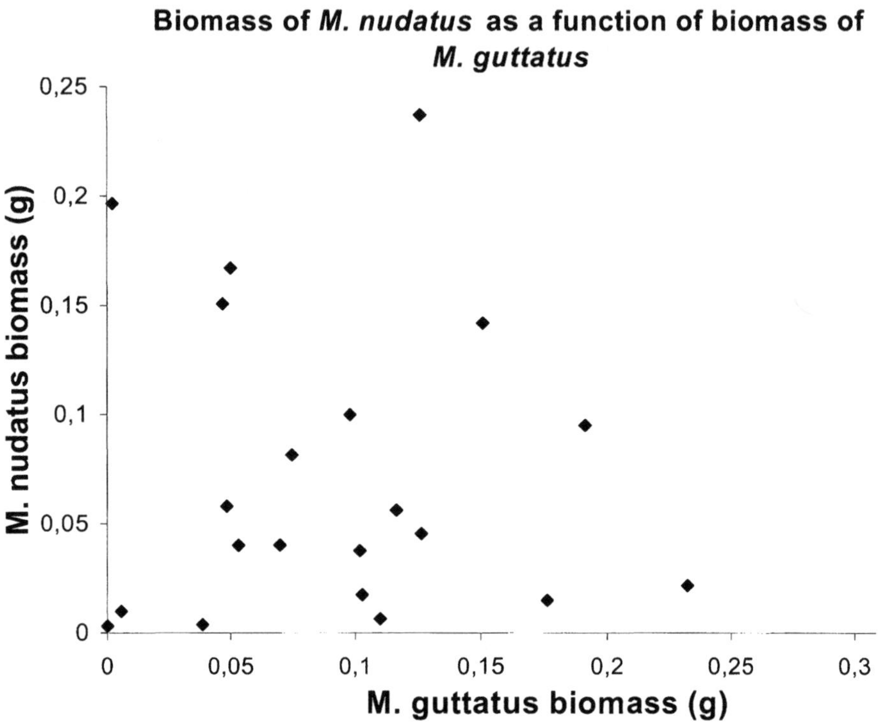

Fig. 3.2. Biomass of *Mimulus nudatus* target plant as a function of *Mimulus guttatus* neighbor biomass when grown in the greenhouse

preted with caution. In greenhouse tests of competitive abilities, rare species were stronger competitors than common species in five out of seven cases (*Agrostis hiemalis, Echinaceae tennesseensis, Festuca paradoxa, Setaria geniculata, Spenopholis obtusata*; Table 3.3). Results of neighbor-removal experiments showed negative effects of competitors on rare species in six out of six cases (*Eryngium cuneifolium, Hypericum cumulicola, Polygonella basiramia,Ssolidago shortii, Eriocaulon kornickianum*, and "uncommon native Mojave annuals"; Table 3.3), but see the example of *Mimulus nudatus* above.

Clearly, more work needs to be done on competition and rare species in order to determine which case is more common: species that are rare because they are poor competitors, or species that are good competitors in order to persist at low abundance. The findings of my research and review serve to refute the simple assertion that rare species are restricted to uncommon habitats or low abundance because they cannot compete with common neighbors or in other habitats. The distribution of a rare species may give clues to its competitive ability. Rare species that are sparse within a community may be good competitors (i.e., able to compete with more common species for colonization sites) while rare species that are locally abundant in few sites may be poor competitors restricted to suitable microhabitats.

Table 3.3. Competition and type of rarity

Species	Reference	Type of rarity	Type of competitor	Reduction in fitness in the field by competitors (A) Or: weaker competitor than more common species in greenhouse tests (B)
Agrostis hiemalis	Rabinowitz et al. (1984)	Historic	Native grasses	N (B)
Cirsium vinaceum	Huenneke and Thomson (1995)	Historic	Exotic herb	Y (B) in greenhouse, N (A) in field
Echinacea tennesseensis	Baskin et al. (1997)	Historic	Native/common congener	N (B)
Eriocaulon kornickianum	Watson et al. (1994)	Formerly common	Natives	Y (A)
Eryngium cuneifolium	Quintana-Ascencio and Menges (2000)	Historic	Native shrubs	Y (A)
Festuca paradoxa	Rabinowitz et al. (1984)	Historic	Native grasses	N (B)
Hypericum cumulicola	Quintana-Ascencio and Menges (2000)	Historic	Native shrubs	Y (A)
Polygonella basiramia	Quintana-Ascencio and Menges (2000)	Historic	Native shrubs	Y(A), biomass, not survival
Setaria geniculata	Rabinowitz et al. (1984)	Historic	Native grasses	N (B)
Solidago shortii	Baskin et al. (1997)	Historic	Native/common congener	Y (B)
Solidago shortii	Walck et al. (1999)	Historic	Exotic herbs	Y (A)
Spenopholis obtusata	Rabinowitz et al. (1984)	Historic	Native grasses	N (B)
Uncommon native Mojave annuals	Brooks (2000)	Unknown	Exotic annual grasses	Y (A)

3.3.2 What Is the Role of Exotic Species in Plant Declines?

Exotic species are frequently cited as a cause for species declines. In a recent review of 181 endangered-species recovery plans, Lawler et al. (unpublished data) found that exotic species were the second most frequently cited threat to endangered species and constituted the primary threat in 22 % of plans examined. Although we have few studies in this area, it is clear from the limited literature that exotic species can have large negative impacts on rare species. In three studies on exotic species effects, exotic plants were found to have negative effects on natives in all three cases (Table 3.3).

Brooks (2000) found that *Bromus madritensis*, *Schismus arabicus*, and *S. barbatus*, three exotic grasses common in the Mojave Desert of California, negatively affected the growth and reproduction of uncommon native annuals. Thinning exotics resulted in an increase in overall native species density and biomass but only one out of forty common species increased significantly; thus the observed increase was due to changes in the uncommon species. The effect on native species was significant in only one of the two years studied, a year of high rainfall in December that resulted in high annual plant productivity. This study demonstrates that exotic species can have negative impacts on rare species but the effects depend on the conditions of a given year.

Walck et al. (1999) studied the effects of competition from two introduced plants on the rare plant *Solidago shortii*. They found that the two exotics studied (*Coronilla varia* and *Festuca arundinacea*) significantly reduced the number of flowering ramets, seedling establishment, and yield of *S. shortii*. Interspecific competition with exotics was significantly greater than intraspecific competition, clearly implicating exotics in the continuing decline of *S. shortii*.

Huenneke and Thomson (1995) examined the potential for interference competition between the endangered thistle *Cirsium vinaceum* and an invasive European teasel, *Dipsacus sylvestris*. They found significant competitive effects of teasel on *C. vinaceum* in the greenhouse. Despite their greenhouse results, they were unable to detect effects of teasel on *C. vinaceum* in the field; however, this may be due to limited replication and confounding factors in the field. Additionally, they found no differences between the habitat characteristics of the two species, and *D. sylvestris* appears to be invading *C. vinaceum* habitats. These authors concluded that there was strong potential for invading teasel to have negative impacts on the rare thistle.

Despite the limited number of studies, competition with exotic plants, when studied, has been found to have strong effects on rare species. Considering the continuing high rates of exotic species spread and invasion, these impacts will most likely only increase and should be better documented in the future.

3.3.3 Competition in Historically Rare and Formerly Common Species

Finally, should we expect the role of competition to differ between historically rare and formerly common species? The majority of competition studies in rare species have been on historically rare species (see Table 3.3). Since data are lacking in this area, I propose several hypotheses to be tested concerning possible historically rare/formerly common differences in competitive ability.

- Historically rare species that are rare because of poor competitive ability should be poorer competitors than formerly common species, assuming the historic abundance of the formerly common species also reflects competitive ability. Furthermore, historically rare species that are restricted to disturbance-originated habitats, such as openings after fire and recently exposed stream banks, are likely restricted to these temporary habitats in part because of poor competitive ability in closed habitats. These species should perform poorly in competition trials.
- Historically rare species that are rare for some reason other than competitive exclusion may in fact be equal or stronger competitors with their neighbors than formerly common species (see examples in Baskin et al. 1997). Formerly common species will have evolved in environments where they were frequently surrounded by conspecifics. Historically rare species that are sparse where they occur may have evolved in environments where they were frequently surrounded by other species. Thus, historically rare species may have been under selective pressure to maintain themselves when surrounded by other species and develop strong competitive abilities.

Unfortunately, there is currently little evidence to test these hypotheses. As mentioned previously, Rabinowitz (1984) found that sparse prairie grasses were strong competitors when surrounded by more common species. *Mimulus nudatus* did not show significant negative effects of competition from neighbors in the greenhouse and in two out of three sites studied in the field (Brigham and Schwartz, unpubl. data).

Studies of competitive abilities of historically rare species show mixed results (see Table 3.3) perhaps in part because of the different causes of rarity in the species involved. More studies need to measure the impact of competition on rare species and compare competitive abilities of historically rare species with neighbors and more common congeners. Only when such studies have accumulated will we be able to evaluate the relative importance of competition in rare species population dynamics.

3.3.4 Implications for PVA

Competition appears to play a role in decreasing plant fitness for many of the species studied, but not all (Table 3.3). Currently most PVAs include measured demographic rates from field-grown plants that most likely include the effects of neighbors on plant persistence. Thus, explicit inclusion of competitive effects may not be necessary for many PVAs. If, however, PVA is being done to evaluate management options, manipulations of competitor densities to determine the importance of competition in persistence are in order.

3.4 Loss of Pollinators at Low Abundance

A number of recent articles suggest that loss of pollinator service may be a problem for small populations of plants (Kearns et al. 1998; Chaps. 2, 10, this Vol.; Spira 2001). A large proportion of flowering plant species relies on insect visitation for reproduction (Chap. 10, this Vol.). As Oostermeijer (Chap. 2, this Vol.) pointed out, small populations of plants may not be attractive resources to pollinators. Here, I review the published evidence to determine whether there is support for the hypothesis that pollinator failure is a problem for rare plants. Furthermore, I consider how historically rare and formerly common species might be expected to differ with regard to pollination and review the evidence for my hypothesized differences.

3.4.1 Are Rare Plants in General Pollinator-Limited?

Existing evidence neither supports nor rejects the hypothesis that pollinator limitation is a major factor in rare-plant demography. Research on pollinator limitation either compares seed production between natural and supplemental pollination or observes pollination and records subsequent seed set. In 14 studies with data on pollinator limitation for 18 species, I found evidence for pollinator limitation in eight species (Table 3.4). Of the remaining species, one was pollen-limited in one part of its range but not another, and nine were not pollen-limited (Table 3.4).

In addition to these studies on rare species, there is some evidence that fragmentation may be disrupting plant-pollinator interactions that may in turn lead to species declines (see reviews by Rathcke and Jules 1993; Bond and Van Wilgen 1996; Kearns et al. 1998; Chap. 10, this Vol.). Disruptions have been seen in common species undergoing fragmentation (see review in Rathcke and Jules 1993), but evidence for pollinator disruption leading to rarity is hard to document and may often be confounded with other factors. For instance, many plant species in Hawaii are declining and may have lost pollinator ser-

Table 3.4. Pollination and type of rarity

Species	Reference	Type of rarity	Pollen-limited?	Notes
Aster curtus	Bigger (1999)	Historic	N	Locally abundant
Aster curtus	Giblin and Hamilton (1999)	Historic	N	
Clianthus puniceus	Shaw and Burns (1997)	Historic	N	
Dedeckera eurekensis	Wiens et al. (1989)	Historic	N	
Delphinium bolosii	Bosch et al. (1998)	Historic	Y	
Dicerandra frutescens	Deyrup and Menges (1997)	Historic	N	
Eupatorium resinosum	Byers (1995)	Historic	Y	
Gentiana cruciata	Petanidou et al. (1995)	Historic	Y	Reproduces clonally
Gentiana pneumonanthe	Petanidou et al. (1995)	Formerly common	N	Relictual large population
Lupinus padre-crowleyi	Taylor (1981)	Historic	N	
Lupinus sulphureus-kincaidii	Kaye and Kirkland (1999)	Formerly common	Y	
Magnolia schiedeana	Dieringer and Espinosa (1994)	Historic	Y/N	One of two populations is pollen-limited. Plants also resprout.
Penstemon penlandii	Tepedino et al. (1999)	Historic	N	
Peraxilla colensoi	Robertson et al. (1999)	Formerly common	Y	
Peraxilla tetrapetala	Robertson et al. (1999)	Formerly common	Y	
Rhodendron championiae	Ng and Corlett (2000)	Historic	N	
Rhodendron moulmainense	Ng and Corlett (2000)	Historic	N	
Rhododendron hongkongense	Ng and Corlett (2000)	Historic	Y	All four species are less pollen-limited than common species
Rhododendron simiarum	Ng and Corlett (2000)	Historic		

vices but are also impacted by exotic species and habitat conversion (U.S. Fish and Wildlife Service, 1998). Australian mistletoes are in decline putatively due to introduced herbivores, but a recent work by Robertson et al. (1999) also implicates declines in native bird faunas resulting in pollinator failure and low seed set. Additionally, disruption of plant-pollinator interactions may be in the early stages, resulting in limited evidence for population declines due to pollinator failure.

3.4.2 How Might We Expect Plant-Pollinator Relationships to Differ for Historically Rare and Formerly Common Species?

Historically rare and formerly common species may differ in ability to maintain pollinator service at small population sizes or in their reliance on pollinator service in general. Historically rare species may have adaptations for maintaining pollinator service while at low density or in small populations. Large floral rewards or developing relationships with specific pollinators may be the means by which rare plants maintain pollinator service. For instance, some species of rare *Eucalyptus* in Australia produce large flowers with large nectar reserves apparently to entice pollinators to travel long distances between plants (Moran and Hopper 1987). Species that are frequently at low abundance may also use generalist pollinators who will respond to the density of flowering neighbors (Chap. 10, this Vol.).

Some rare species may reduce their reliance on pollinator visitation by increasing self-compatibility. For instance, in the rare, bee-pollinated, annual plant *Agalinis skinneriana*, a population with thousands of individuals had a self-compatibility rate of 85 % while a small population with hundreds of individuals had a self-compatibility rate of 99 % (Dieringer 1999). Dieringer hypothesized that increasing self-compatibility has evolved in the smaller population as a mechanism to assure reproduction. The rare plant *Delphinium bolosii* (Bosch et al. 1998) shows a pollination shift in comparison with its common ancestor *D. fissum*. *D. bolosii* is self-compatible while *D. fissum* is self-incompatible. Similarly, the self-incompatible common species *Stephanomeria exigua*, which is found throughout California and Oregon, is thought to have originated from a self-compatible population restricted to a single 150-acre hilltop in southeast Oregon (example in Kruckeberg and Rabinowitz 1985). Thus, in some cases, increasing self-compatibility may reduce reliance on outside pollen (thus maintaining fruit-set when populations are small and isolated) but not on pollinator service per se (many self-compatible species still require pollinator service to set seed, e.g., *D. bolosii*). It is also clear that not all historically rare species will be self-compatible (Fiedler 1987).

In contrast to historically rare species, formerly common species may lack adaptations to small population size and may show sharp declines in pollinator service as populations become small and fragmented. If historically rare species have adaptations to achieve adequate reproduction at small population sizes (e.g., large floral rewards for pollinators, reliance on generalist pollinators, or increased selfing), then they should be less likely to show a pattern of pollen limitation than formerly common species.

Of the 18 species mentioned previously for which pollen limitation has been examined, 14 are historically rare and 4 are formerly common. Historically rare species showed a near 50/50 split between species that were pollen-limited and those that were not (46.1 % pollen-limited, 53.9 % not, Table 3.4, Fig. 3.3), while three of four formerly common species were pollen-limited. For the historically rare species examined, all of the pollen-limited species were perennials and three of the six species could reproduce clonally. While current data are limited, they do show higher rates of maintenance of pollinator service in historically rare versus formerly common species.

Clearly, species dependence on pollination varies (Bond 1994), pollination rates may fluctuate through time, and species that are not currently pollen-limited may be in the future and vice versa. Although more research is needed in this area, the trend is consistent with my hypothesis for historically rare versus formerly common species and maintenance of pollinator function. Loss of pollinator function in formerly common plants appears to be a serious potential problem as populations are reduced.

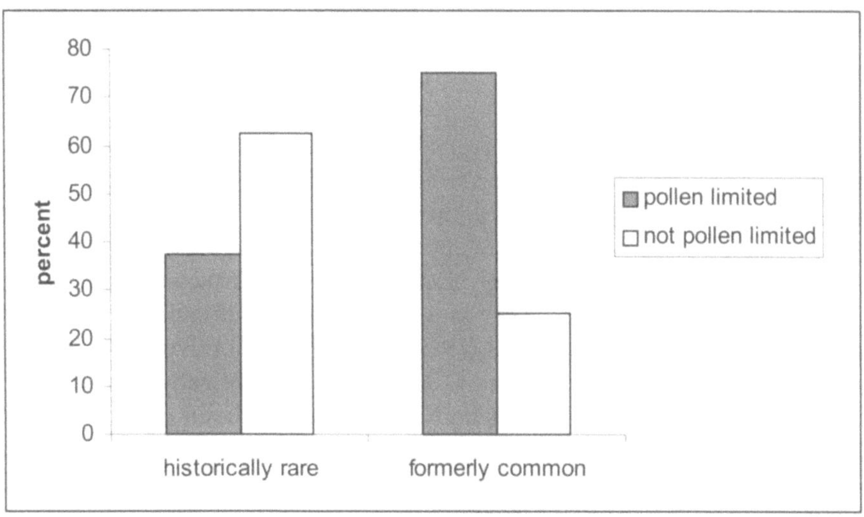

Fig. 3.3. Percentage species that are pollen-limited for historically rare (n=13) and formerly common species (n=4). Data are from Table 3.4

3.4.3 Implications for PVA

For many rare species, especially formerly common ones, disruption of plant-pollinator mutualisms may be a serious issue. Three out of four studies of pollination in formerly common species found that plants were pollen-limited and the one species that was not pollen-limited was in a large, relictual population. The literature provides much evidence that pollinators often respond to population size or plant density (e.g., Agren 1996; Groom 2000; Jennersten 1988), yet most PVAs do not include any measure of pollination or possible change in reproductive rate with declining population size due to pollinator failure (but see Oostermeijer 2000). In a review of PVAs of plants, Menges (2000) found that only 3 studies out of 99 included some aspect of pollination viability. Although too few studies exist to support strong conclusions, it appears that pollinator service will be more of a problem for formerly common species than historically rare ones. Even with the few studies we have in hand, there is a trend in the direction of less pollen-limitation in historically rare species (Table 3.4, and Fig. 3.3). Thus, including pollination may be less vital in PVAs of historically rare species.

3.5 Herbivory and Seed Predation

3.5.1 Review of the Evidence for Effects of Herbivory and Seed Predation on Rare Plants

Few studies have examined herbivory in rare plant species; in fact, for herbivory on vegetative parts, there are only two studies in the literature (but see discussion in Chaps. 4 and 10, this Vol.). Bevill et al. (1999) found that herbivory on juveniles of the threatened thistle *Cirsium pitcheri* significantly reduced survival and seed set over plants that were sprayed with insecticide. They hypothesized that herbivory may be limiting the population at one of its sites. Bruelheide and Scheidel (1999) studied slug herbivory on the rare montane plant *Arnica montana*. *A. montana* is historically restricted to nutrient-poor montane meadows and has recently declined due to nutrient deposition and changes in land use; this species is now endangered. The authors found that slug damage to plants in the field was high and increased with decreased elevation. Using a common garden experiment, they found that slug herbivory prevents plants from persisting at lower elevations when they could otherwise survive in these habitats. In another study, Scheidel and Bruelheide (1999) performed feeding trials with three slug species and found that for all three species *A. montana* was one of the top four preferred foods.

Finally, Fiedler (1987) studied herbivory on rare *Calochortus* lilies (*C. pulchellus, C. obispoensis, and C. tiburonensis*) by pocket gophers (*Thomomys bottae*), blacktail jackrabbits (*Sylvilagus audubonii*), Audubon cottontail, brush rabbit (*S. bachmani*) and mule deer (*Odocoileus lemionus*) in California. She found that two of the three rare species (*C. obispoensis* and *C. tiburonensis*) suffered severe damage to their basal leaves (a primary photosynthetic organ) over 3 years for all size classes studied. Thus, in the few studies that have been conducted, it is clear that herbivory can have a large impact on rare species.

Seed predation as a factor in controlling population dynamics has been studied for a small number of rare species. Hegazy and Eesa (1991) found that 94 % of seeds of the rare legume *Ebenus armitagei* were eaten or damaged by bruchid beetles, *Callosobruchus maculatus*. Predispersal seed predation was also heavy on *Astragalus australis var. olympicus* (over 80 % at one site), a rare legume restricted to the Olympic Mountains in Washington (Kaye and Kirkland 1999). In contrast, seed predation was extremely low on a rare lupine, *Lupinus sulphureus ssp. Kincaidi*, with an average of only 6 % ovule predation (Kaye and Kirkland 1999).

It appears that both herbivory and seed predation can have negative impacts on rare species. These phenomena have not been commonly examined, and thus it is difficult to predict how important they are in reducing persistence of rare species.

3.5.2 Historically Rare vs Formerly Common Species: Expectations

I propose that rare plant species that have few defenses and are locally abundant are likely to experience herbivore attacks, while those that are unpalatable or extremely sparse are more likely to escape herbivore damage.

Field data are lacking concerning general patterns of herbivory on rare species; however, some hypotheses can be generated from the plant-insect literature. For instance, the resource concentration hypothesis (Root 1973) suggests that insects will congregate where resources are concentrated. Thus, if a rare species is at low density, it may not constitute enough of a resource to build up large populations of herbivores and thereby avoid high rates of herbivore damage. Additionally, work with movement rates and residence times of phytophagous insects has shown that emigration rates are higher and residence times shorter for small vs large host-plant patches (e.g., Kareiva 1985; McCauley 1991). Kareiva (1985) found that flea beetles were more likely to emigrate from smaller patches and that single plants were without beetles because of loss through emigration. Additionally, host plants from which beetles emigrated were not readily recolonized. Kareiva (1985) hypothesized that there may be a lower limit to patch sizes that support herbivore populations. McCauley (1991) studied populations of *Tetraopes tetraopthalmus*, a specialist herbivore on milkweed (*Asclepias syriaca*), and found that these beetles have longer residence times in larger patches of milkweed. These studies indicate that rare species may be able to escape herbivory by remaining at low abundance.

Responses of insects to patch size and/or plant density will most likely depend on the species involved and may be quite complicated. Segarra-Carmona and Barbosa (1990) found that a generalist legume borer, *Etiella zinckenella*, preferred low-density patches of soybeans, despite no change in plant growth between high- and low-density patches. This was, however, an agricultural system and density was manipulated, not patch area.

The ability of a plant species to avoid herbivory through low abundance depends on several factors. First, the specificity of the herbivore must be such that other surrounding plants do not also act as hosts. If surrounding plants are also hosts, the area may be able to support high herbivore densities and the rare species may suffer high levels of damage. Bonser and Reader (1995) found that herbivory rates were higher in high-biomass plant neighborhoods than in low-biomass neighborhoods, supporting the resource concentration hypothesis and the idea that plant neighborhoods can affect herbivory rates. Furthermore, to apply these theories to rare plants assumes that the rare species will be at low abundance in its neighborhood, which is not the case for rare species that are locally abundant but restricted geographically.

Alternatively, if a rare species has a specialist herbivore then insects may become more concentrated on plants as they decline in number. If a species is historically rare, we might hypothesize that it would never occur in sufficient numbers to be able to support a specialist herbivore. However, many insects specialize on a genus (for instance, *Cirsium*), in which case an insect might specialize on a more common species and then go on to attack a rare species that co-occurs with it. Bevill et al. (1999) hypothesize this mechanism to explain the high rates of herbivory seen on juveniles of *Cirsium pitcheri*, a rare thistle. Additionally, rare species that are locally abundant may constitute enough of a resource to evolve specialist herbivores or attract generalist herbivores under the resource concentration hypothesis.

How much of a role does abundance play in herbivore selection of plants? Rausher and Feeney (1980) investigated the role of host abundance vs host quality in a butterfly that lays eggs on two different species in the genus *Aristolochia*. The butterfly searches preferentially for one of the two species, and which species it searches for depends on what time it is during the growing season. They found that early in the season, when both host species were equal in quality (as measured by the probability of eggs surviving to adulthood), the butterfly preferentially used the more abundant species. Later in the season, when one of the species has become less hospitable for juveniles, the butterfly switched to the other species and preferentially searched for it, even when it was at low abundance. This study indicates that, if species are of equal quality to the insect, then abundance will play a role in herbivore choice;

however, if plants differ in quality, being at low abundance may not provide a refuge from herbivore attack.

Vasconcelos (1997) studied the foraging behavior of the generalist herbivore *Atta laevigata* (leaf-cutter ants) in a tropical old-field that was undergoing succession in Brazil. Overall, there was a significant positive correlation between leaf-cutter ant attacks and species relative abundance (Vasconcelos 1997). Consumption of even the most preferred species followed patterns of plant abundance: as three preferred species declined in abundance over time (*Cecropia ulei, Cecropia concolor*, and *Casearia* sp.), ant attacks on them declined. Similarly, as a fourth preferred species, *Bellucia imperialis*, increased in abundance, ant attacks on it increased. Additionally, ants did not attack only the six rarest species, out of a total of 43 in the plot, over the 18-month study period. Vasconcelos (1997) hypothesized that they escaped attack not because they were unpalatable but because they occurred at such a low frequency that they were not encountered during the study period. These data suggest that rare species may be able to escape herbivore attack by declining in abundance or maintaining low abundance.

Landa and Rabinowitz (1983) examined the relationship between level of defense (in this case leaf toughness), abundance, and herbivore preference for three common and four sparse prairie grasses. They found a significant positive correlation between toughness and abundance and a non-significant negative correlation between toughness and preference rank by a generalist herbivore (a grasshopper). Additionally, they found that the herbivore tended to prefer sparse grasses to abundant ones with the exception of *Festuca paradoxa*, which was one of the least abundant and least preferred species. Landa and Rabinowitz (1983) also analyzed data from Gangwere et al. (1976) and found that abundance was negatively correlated with preference by grasshoppers for 23 species found in old-fields in Michigan. These studies suggest that less abundant species may be more preferred by herbivores but fail to assess whether this preference would result in damage in the field.

Clearly, factors other than abundance will affect rates of herbivore damage. Plant ecological traits such as individual investment in defense and type of defense of the plant may all affect damage rates (Chaps. 4, 10, this Vol.). Additionally, insect herbivore factors such as mobility, degree of specialization, and feeding mode can all have strong effects on what plants are eaten. In spite of these complications, evidence suggests that plants at low abundance can both escape herbivory (Kareiva 1985) or suffer high rates of herbivory as insects are concentrated on the dwindling resource (Stanforth et al. 1997). More research needs to be done on rates of herbivore damage in rare species with different levels of abundance and palatability in order to investigate relationships between abundance, palatability, and herbivory in rare species. It is currently unclear whether low palatability and low abundance will result in low rates of herbivory, while moderate palatability and local abundance will result in high rates of herbivory in rare species. Additionally, herbivory needs to be evaluated as a negative factor overall for rare species persistence.

Finally, some rare species may actually be rare because they are the preferred food of particular herbivores. Herbivores have been shown to have potentially large consequences on plant distributions (e.g., Harper 1969, 1977; Rausher and Feeney 1980; Parker and Root 1981; Carson and Root 2000). While rarity may provide an escape from herbivory, it is certainly not clear that this is the always the case. Research that examines rates of herbivory on historically rare species is needed to elucidate relationships between rarity and herbivory.

3.5.3 Implications for PVA

Few studies have been conducted on herbivory and seed predation in rare plants; thus it remains difficult to evaluate their importance in rare plant persistence. If rare species can actually escape herbivory through low abundance, then for such locally rare species herbivore dynamics will not play important roles in plant persistence and are not necessary for PVAs. There is evidence, however, that for some species, such as *Cirsium pitcheri* (Stanforth et al. 1997), herbivory can play an important role in plant demography. Currently, few PVAs include any information on herbivory. Menges (2000) found one study out of 99 that included herbivory in its model structure. Individual species biology will have to be taken into account when assessing whether to include an herbivory component in a PVA. Furthermore, including such a dynamic will necessarily complicate persistence models since herbivory rates may be tied in unknown ways to current population sizes (Chap. 10, this Vol.).

3.6 Conclusions

All of the factors examined in this chapter have the potential to reduce rare plant populations. The evidence for each as an important factor in rare plant population dynamics varies. Evidence of negative consequences of reduced genetic diversity is still equivocal. Six of 14 species studied showed no correlation between genetic diversity and fitness. I do not doubt that homozygosity can be damaging, yet it appears that in many of the species studied damaging levels have not been reached. The evidence for negative impacts of competition with other plants on rare species is equally weak. There are few studies in this area and what studies there are show mixed effects. Pollinator failure has a large potential to negatively impact formerly common species, and patterns in the literature support the idea that pollinator service may be disrupted as population sizes decline. Pollinator effects on historically rare species are less clear. Finally, although the potential impacts of herbivory seem large, there is not enough evidence in the literature to say how much of a negative impact herbivory may have on rare plant species.

For comparisons of historically rare and formerly common species, the literature shows intriguing patterns. In genetic studies I found evidence that formerly common species more frequently showed negative consequences of small population size than did historically rare species (Fig. 3.1). Additionally, I found some support for my hypothesis that historically rare species may be better able to maintain pollination than formerly common species (Fig. 3.3). Thus, there is some evidence that historically rare species may have adaptations to rarity, however, these adaptations are not universal. In all of the areas examined, the number of studies used to evaluate my hypotheses was not large. Clearly, as more work is done on the effects of these factors on historically rare and formerly common species, the evidence for the hypothesized differences between these two types of rare species will need to be re-evaluated. Furthermore, other factors that affect species persistence such as temporal variability in population size, seed production, individual biomass, seed dispersal, and hybridization may be examined in a similar manner to test for differences between historically rare and formerly common species.

Identifying patterns of traits that allow for persistence of populations at small numbers, low density, or a limited number of locales (all patterns of occurrence seen in rare species) will help us identify which species are likely to persist if they become rare. Previous authors have examined rare species with the goal of identifying factors that make species prone to rarity and using these "risk factors" to identify species of concern for the future. Here I have taken the opposite tack and tried to hypothesize what traits should aid in persistence and searched for evidence of these traits in historically rare species. Such traits might also be examined in floras that are threatened or declining and used to make predictions about how species should respond to fragmentation or reduction. Such studies would be another test case for the hypothesized relationships between traits and persistence and may also be helpful in making conservation decisions.

3.6.1 Suggestions for Future Directions

This review highlights those areas for which we currently do not have the studies we need to advance our understanding of rare plant dynamics. There are two main issues that need to be addressed with respect to the effects of genetic diversity on rare plant persistence. First, studies in which the effective population size of the species is known or at least estimated are needed. Since theoretical predictions concerning effects of population size on genetics rely on knowing the effective population size and not just the absolute numbers in a population, methods to quantify or estimate which plants are actually breeding in a population need to be developed. Some methods already exist to estimate effective population size but are time consuming and difficult. Second, recent evidence showing a lack of correlation between molecular and quantitative measures of genetic diversity suggest that molecular markers may not be the relevant criteria for examining fitness and persistence effects (Reed and Frankham 2001). It is necessary to measure quantitative genetic variation directly if we want to know whether genetic diversity is declining in a rare species and whether these declines will have fitness effects. Implementing these suggestions will clearly make studies of genetic variation, population size, and fitness much more difficult; however, it seems clear that these steps are necessary to make progress on questions in this area.

There are three areas of competition research that need to be expanded. We need more field and greenhouse experiments comparing the competitive abilities of rare and common congeners in order to assess whether a difference in competitive ability is a potential cause of rarity. In addition, these sorts of comparisons need to be followed by field competition experiments with the rare species. Transplant experiments of rare species into occupied and unoccupied habitats with and without competitors will allow us to determine whether competition from other species is involved in range limitations and to evaluate how important competition is as a limiting factor in occupied habitats. Finally, we need more documentation of the effects of invasive species on rare plants in order to evaluate how much of a threat exotic plants are to rare plant species.

There is currently a large degree of concern with regard to rare plant species and maintenance of pollinator mutualisms. Several recent articles have suggested that failure of mutualisms may pose a serious threat to rare species persistence (Kearns et al. 1998; Chaps. 2, 10, this Vol.; Spira 2001). Despite these concerns, we currently have few studies that actually evaluate the role of pollination in rare species declines. In order to evaluate this claim we need carefully conducted pollinator observations, comparisons of pollination rates in populations of different sizes, and pollen supplementation to determine whether lack of pollination is a limiting factor. Furthermore, the recent suggestion that rare species may benefit from pollinator attraction by neighboring species (Oostermeijer et al. 1998) needs to be explored further through neighborhood manipulation experiments.

The area of herbivore effects on rare species is woefully understudied. We need observations of damage levels on rare plants in the field. To test whether rare plants are rare because of herbivory, we need to conduct herbivore exclusion experiments. In order to examine the relationship between abundance and herbivory, abundance manipulation experiments using rare plants and accounting for factors such as plant investment in defense, history of plant abundance, and historic relationships between the plant involved and insect herbivores should be carried out.

3.6.2 Implications for PVA

All of the factors I have examined have been cited in the literature as attributes that can affect population performance. In each case I asked whether the data show each factor to be consistently important in determining rare plant performance. In no area was this the case. Yes, current PVAs do not often include the areas examined here: genetics, competition, pollination, and herbivory. But in many cases these factors do not seem to have large influence on rare plant dynamics. The problem is, in some cases they do. Differentiating between when each factor is important and should be included in a PVA and when it is not must be left up to careful study of the species biology of the rare species involved. No sweeping generalizations concerning what must or must not be included in a PVA come out of this analysis.

Acknowledgements. Thanks to the Spice Lab – Mark Schwartz, Kelly Lyons, Jason Hoeksema, Felipe Dominguez-Lozano, Ruc Sanchez, and Diane Thomson – for helpful suggestions and discussion. Thanks also to Tiffany Bensen, Kevin Rice, Susan Harrison, and one anonymous reviewer for reviewing this manuscript.

References

Argen J (1996) Population size, pollinator limitation, and seed set in the self-incompatible herb *Lythrum salicaria*. Ecology 77:1779-1790

Barbel GR, Selander RK (1974) Genetic variability in edaphically restricted and widespread plant species. Evolution 28:619–630

Barrett S, Kohn J (1991) Genetic and evolutionary consequences of small population size in plants: implications for conservation. In: Falk D, Holsinger K (eds) Genetics and conservation of rare plants. Oxford University Press, Oxford, pp 3-30

Baskin JM, Snyder KM, Walck JL, Baskin CC (1997) The comparative autecology of endemic, globally-rare, and geographically widespread, common plant species: three case studies. Southwest Nat 42:384–399

Bevill RL, Louda SM, Stanforth LM (1999) Protection from natural enemies in managing rare plant species. Conserv Biol 13:1323–1331

Bigger DS (1999) Consequences of patch size and isolation for a rare plant: pollen limitation and seed predation. Nat Areas J 19:239-244

Billington HL (1991) Effect of population size on genetic variation in a dioecious conifer. Conserv Biol 5:115–119

Black-Samuelsson S, Eriksson G, Gustafsson L, Gustafsson P (1997) RAPD and morphological analysis of the rare plant species *Vicia pisiformis* (Fabaceae). Biol J Linn Soc 61:325–343

Bond WJ (1994) Do mutualisms matter? Assessing the impact of pollinator and disperser disruption on plant extinction. Philos Trans R Soc Lond B: Biol Sci 344:83–90

Bond WJ, Van Wilgen BW (1996) Population and community biology series, 14. Fire and plants. In: Population and community biology series; fire and plants. Chapman and Hall, New York, p viii; 263 p

Bonser SP, Reader RJ (1995) Plant competition and herbivory in relation to vegetation biomass. Ecology 76:2176–2183
Bosch M, Simon J, Molero J, Blanche C (1998) Reproductive biology, genetic variation and conservation of the rare endemic dysploid *Delphinium bolosii* (Ranunculaceae). Biol Conserv 86:57–66
Brooks ML (2000) Competition between alien annual grasses and native annual plants in the Mojave Desert. Am Midl Nat 144:92–108
Bruelheide H, Scheidel U (1999) Slug herbivory as a limiting factor for the geographical range of *Arnica montana*. J Ecol 87:839–848
Buza L, Young A, Thrall P (2000) Genetic erosion, inbreeding and reduced fitness in fragmented populations of the endangered tetraploid pea Swainsona recta. Biol Conserv 93:177–186
Byers D (1995) Pollen quantity and quality as explanations for low seed set in small populations exemplified by *Eupatorium* (Asteraceae). Am J Bot 82:1000–1006
Carson WP, Root RB (2000) Herbivory and plant species coexistence: community regulation by an outbreaking phytophagous insect. Ecol Monogr 70:73–90
Caughley G (1994) Directions in conservation biology. J Anim Ecol 63:215-244
Chesson P, Huntly N (1996) Maintenance of species diversity by temporal environmental variation within the growing season. Bull Ecol Soc Am 77:79
Chesson P, Huntly N (1997) The roles of harsh and fluctuating conditions in the dynamics of ecological communities. Am Nat 150:519–553
Coates DJ (1988) Genetic diversity and population genetic structure in the rare chittering grass wattle, *Acacia anomala*. Aust J Bot 36:273–286
Deyrup M, Menges ES (1997) Pollination ecology of the rare scrub mint *Dicerandra frutescens* (Lamiaceae). Fl Sci 60:143–157
Dieringer G (1999) Reproductive biology of *Agalinis skinneriana* (Scrophulariaceae), a threatened species. J Torrey Bot Soc 126:289-295
Dieringer G, Espinosa-S JE (1994) Reproductive ecology of *Magnolia schiedeana* (Magnoliaceae), a threatened cloud forest tree species in Veracruz, Mexico. Bull Torrey Bot Club 121:154–159
Dolan R (1994) Patterns of isozyme variation in relation to population size, isolation, and phytogeographic history in royal catchfly (*Silene regia*; Caryophyllaceae). Am J Bot 81:965–972
Drury WH (1980) Rare species of plants. Rhodora 82:3–48
Dyer AR, Rice KJ (1999) Effects of competition on resource availability and growth of a California bunchgrass. Ecology 80:2697-2710
Ellstrand NC, Elam DR (1993) Population genetic consequences of small population size: implications for plant conservation. Annu Rev Ecol Syst 24:217–242
Evans ME, Dolan R, Menges ES, Gordon DR (2000) Genetic diversity and reproductive biology in *Warea carteri* (Brassicaceae), a narrowly endemic Florida scrub annual. Am J Bot 87:372–381
Fiedler PL (1987) Life history and population dynamics of rare and common mariposa lilies *Calochortus pursh* Liliaceae. J Ecol 75:977–996
Fiedler PL (1995) Rarity in the California flora: new thoughts on old ideas. Madrono 42:127–141
Fischer M, Matthies D (1998) Effects of population size on performance in the rare plant *Gentianella germanica*. J Ecol 86:195–204
Fischer M, Husi R, Prati D, Peintinger M, van Kleunen M, Schmid B (2000) RAPD variation among and within small and large populations of the rare clonal plant *Ranunculus reptans* (Ranunculaceae). Am J Bot 87:1128–1137
Fleishman E, Launer AE, Switky KR, Yandell U, Heywood J, Murphy DD (2001) Rules and exceptions in conservation genetics: genetic assessment of the endangered plant

Cordylanthus palmatus and its implications for management planning. Biol Conserv 98:45–53
Gangwere SK, Evans FC, Nelson ML (1976) The food habitats and biology of Acrididae in an old-field community in southeastern Michigan. Great Lakes Entomol 9:83–123
Gaston KJ (1994) Rarity. Chapman and Hall, London
Giblin DE, Hamilton CW (1999) The relationship of reproductive biology to the rarity of endemic *Aster curtus* (Asteraceae). Can J Bot 77:140–149
Gillespie JH (1998) Population genetics. Johns Hopkins Press, Baltimore
Gitzendanner MA, Soltis PS (2000) Patterns of genetic variation in rare and widespread plant congeners. Am J Bot 87:783–792
Godt MJW, Hamrick JL (2001) Genetic diversity in rare southeastern plants. Nat Area J 21:61–70
Godt MJW, Johnson BR, Hamrick JL (1996) Genetic diversity and population size in four rare southern Appalachian plant species. Conserv Biol 10:796–805
Griggs FR (1940) The ecology of rare plants. Bull Torrey Bot Soc 67:575–594
Hamrick JL, Godt MJW (1990) Allozyme diversity in plant species. In: Brown AHD, Clegg AL, Kahler AL, Wier BS (eds) Plant population genetics, breeding, and genetic resources. Sinauer, Sunderland, MA, p 449
Groom MJ, Preuninger TE (2000) Population type can influence the magnitude of inbreeding depression in *Clarkia concinna* (Onagraceae). Evol Ecol 14:155–180
Harper JL (1969) The role of predation in vegetational diversity. Brookhaven Symp Biol 22:48–62
Harper JL (1977) Population biology of plants. Academic Press, New York
Harper KT (1979) Some reproductive and life history characteristics of rare plants and implications of management. Great Basin Nat 3:129–137
Hedge SG, Ellstrand NC (1996) Life history differences of rare and common flowering plant species of California and British Isles. Am J Bot 83:67
Hegazy AK, Eesa NM (1991) On the ecology, insect seed-predation, and conservation of a rare and endemic plant species: *Ebenus armitagei* (Leguminosae). Conserv Biol 5:317–324
Holmes TH, Rice KJ (1996) Patterns of growth and soil-water utilization in some exotic annuals and native perennial bunchgrasses of California. Ann Bot 78:233–243
Huenneke LF, Thomson JK (1995) Potential interference between a threatened endemic thistle and an invasive nonnative plant. Conserv Biol 9:416–425
Inoue K, Masuda M, Maki M (1998) Inbreeding depression and outcrossing rate in the endangered autotetraploid plant *Aster kantoensis* (Asteraceae). J Hered 89:559–562
Jennersten O (1988) Pollination in *Dianthus deltoides* (Caryophyllaceae): effects of habitat fragmentation on visitation and seed set. Conserv Biol 2:359–366
Johnson SD, Bond WJ (1997) Evidence for widespread pollen limitation of fruiting success in Cape wildflowers. Oecologia (Berl) 109:530–534
Kareiva P (1985) Finding and losing host plants by *Phyllotreta*: patch size and surrounding habitat. Ecology 66:1809–1816
Karron JD (1987) Breeding systems and inbreeding depression in locally endemic and geographically widespread species of *Astragalus* Fabaceae. Am J Bot 74:653
Karron JD (1989) Breeding systems and levels of inbreeding depression in geographically restricted and widespread species of *Astragalus* (Fabaceae). Am J Bot 76:331–340
Karron JD (1991) Patterns of genetic variation and breeding systems in rare plant species. In: Falk DA, Holsinger KE (eds) Genetics and conservation of rare plants. Oxford University Press, New York
Kaye TN (1997) Effects of timber harvest practices on populations of *Cimicifura elata*, a rare plant of northwestern North American forests. Bull Ecol Soc Am 78:122

Kaye TN, Kirkland M (1999) Effects of timber harvest on *Cimicifuga elata*, a rare plant of western forests. Northwest Sci 73:159-167

Kearns CA, Inouye DW, Waser NM (1998) Endangered mutualisms: the conservation of plant-pollinator interactions. Annu Rev Ecol Syst 29:84–112

Knapp EE, Rice KJ (1998) Comparison of isozymes and quantitative traits for evaluating patterns of genetic variation in purple needlegrass (*Nassella pulchra*). Conserv Biol 12:1031–1041

Kruckeberg AR (1984) California serpentines: flora, vegetation, geology, soils, and management problems. University of California Press, Berkeley

Kruckeberg AR, Rabinowitz D (1985) Biological aspects of endemism in higher plants. Annu Rev Ecol Syst 16:447–479

Lammi A, Siikamaki P, Mustajarvi K (1999) Genetic diversity, population size, and fitness in central and peripheral populations of a rare plant *Lychnis viscaria*. Conserv Biol 13:1069–1078

Landa R, Rabinowitz D (1983) Relative preference of *Arphia sulphurea* (Orthoptera: Acrididae) for sparse and common prairie grasses. Ecology 64:392–395

Lande R, Schemske DW (1985) The evolution of self-fertilization and inbreeding depression in plants: 1. Genetic models. Evolution 39:24–40

Lokeshad R, Vasudeva R (1997) Patterns of life history traits among rare/endangered flora of South India. Curr Sci (Bangalore) 73:171–172

Louda SM, McEachern AK (1995) Insect damage to inflorescences of the threatened dune thistle, *Cirsium pitcheri*. Bull Ecol Soc Am 76:358

Luijten SH, Oostermeijer JGB, Van Leeuwen NC, Den Nijs HCM (1996) Reproductive success and clonal genetic structure of the rare *Arnica montana* (Compositae) in The Netherlands. Plant Syst Evol 201:15–30

Macnair MR (1992) Preliminary studies on the genetics and evolution of the serpentine endemic mimulus-nudatus curran. In: Baker AJM, Proctor J, Reeves RD (eds) The vegetation of ultramafic (serpentine) soils. 1st International Conference On Serpentine Ecology, Davis, California, 19-22 June 1991. Intercept, Ltd, Andover

Maki M, Masuda M, Inoue K (1996) Genetic diversity and hierarchical population structure of a rare autotetraploid plant, *Aster kantoensis* (Asteraceae). Am J Bot 83:296–303

McCauley DE (1991) The effect of host plant patch size variation on the population structure of a specialist herbivore insect, *Tetraopes tetraophthalmus*. Evolution 45:1675–1684

McClenaghan LR, Beauchamp AC (1986) Low genetic differentiation among isolated populations of the California fan palm (*Washingtonia filifera*). Evolution 40:315–322

Menges ES (1992) Habitat preferences and response to disturbance for *Dicerandra-frutescens* a lake wales ridge Florida endemic plant. Bull Torrey Bot Club 119:308–313

Menges ES (1995) Factors limiting fecundity and germination in small populations of *Silene regia* (Caryophyllaceae), a rare hummingbird-pollinated Prairie Forb. Am Midl Nat 133:242–255

Menges E (2000a) Applications of population viability analyses in plant conservation. Ecol Bull 48:73–84

Menges E (2000b) Population viability analyses in plants: challenges and opportunities. Trends Ecol Evol 15:51–56

Moran GF, Hopper SD (1983) Genetic diversity and the insular population structure of the rare granite rock species, *Eucalyptus caesia*. Aus J Bot 31:161–172

Moran GF, Hopper SD (1987) Conservation of genetic resources of rare and widespread eucalypts in remnant vegetation. In: Saunders DA, Arnold GW, Burbridge A, Hopkins AJM (eds) Nature conservation: the role of remnants of native vegetation. Beatty and Sons, Chipping-Norton, Surrey, p 410

Ng S-C, Corlett RT (2000a) Comparative reproductive biology of the six species of Rhododendron (Ericaceae) in Hong Kong, South China. Can J Bot 78:221–229

Ng S-C, Corlett RT (2000b) Genetic variation and structure in six Rhododendron species (Ericaceae) with contrasting local distribution patterns in Hong Kong, China. Molec Ecol 9:959–969

O'Brien S, Roelke ME, Marker L, Newman A, Winkler CA, Meltzer D, Colly L, Evermann JF, Bush M, Wildt DE (1985) Genetic basis for vulnerability in the cheetah Acinonyx-Jubatus-Jubatus. Science 227:1428–1434

Oostermeijer J (2000) Population viability of the rare *Gentiana pneumonanthe*: the importance of genetics, demography, and reproductive biology. In: Young A, Clarke G (eds) Genetics, demography and viability of fragmented populations. Cambridge University Press, Cambridge, pp 313–333

Oostermeijer JGB, Van Eijck MW, Den Nijs JCM (1994) Offspring fitness in relation to population size and genetic variation in the rare perennial plant species *Gentiana pneumonanthe* (Gentianaceae). Oecologia (Berl) 97:289–296

Oostermeijer JGB, Van Eijck MW, Van Leeuwen NC, Den Nijs JCM (1995) Analysis of the relationship between allozyme heterozygosity and fitness in the rare *Gentiana pneumonanthe* L. J Evol Biol 8:739–759

Oostermeijer G, Luijten S, Kwak M, Boerrigter E, Den Nijs H (1998) Rare plants in peril: on the problems of small populations. Levende Natuur 99:134–141

Ouborg NJ, Van Treuren R (1994) The significance of genetic erosion in the process of extinction. IV. Inbreeding load and heterosis in relation to population size in the mint *Salvia pratensis*. Evolution 48:996–1008

Parker MA, Root RB (1981) Insect herbivores limit habitat distribution of a native composite, *Machaeranthera canescens*. Ecology 62:1390–1392

Peters GM, Lonie JS, Moran GF (1990) The breeding system, genetic diversity, and pollen sterility in *Eucalyptus pulverulenta*, a rare species with small disjunct populations. Aust J Bot 38:559–570

Petanidou T, Den Nijs JC, Oostermeijer JGB (1995a) Pollination ecology and constraints on seed set of the rare perennial *Gentiana cruciata* L. in The Netherlands. Acta Botanica Neerlandica 44:55–74

Prober SM, Brown AHD (1994) Conservation of the grassy white box woodlands: Population genetics and fragmentation of *Eucalyptus albens*. Conserv Biol 8:1003–1013

Prober SM, Tompkins C, Moran GF, Bell JC (1990) The conservation genetics of *Eucalyptus paliformis* and *E. parviflora*, two rare species from south-eastern Australia. Aust J Bot 38:79–95

Prober SM, Spindler LH, Brown AHD (1998) Conservation of the grassy white box woodlands: effects of remnant population size on genetic diversity in the allotetraploid herb *Microseris lanceolata*. Conserv Biol 12:1279–1290

Quintana-Ascencio PF, Menges ES (2000) Competitive abilities of three narrowly endemic plant species in experimental neighborhoods along a fire gradient. Am J Bot 87:690–699

Rabinowitz D (1981) Seven forms of rarity. In: Synge H (ed) The biological aspects of rare plant conservation. Wiley, New York, pp 205–217

Rabinowitz D, Rapp JK, Dixon PM (1984) Competitive abilities of sparse grass species: means of persistence or cause of abundance? Ecology 65:1144–1154

Raijmann LEL, Van Leeuwen NC, Kersten R, Oostermeijer JGB, Den Nijs HCM, Menken SBJ (1994) Genetic variation and outcrossing rate in relation to population size in *Gentiana pneumonanthe* L. Conserv Biol 8:1014–1026

Rathcke BJ, Jules ES (1993) Habitat fragmentation and plant-pollinator interactions. Curr Sci (Bangalore) 65:273–277

Rausher MD, Feeney P (1980) Herbivory, plant density, and reproductive success: the effects of *Battus philenor* on *Aristolochia reticulata*. Ecology 61:905–917

Reed DH, Frankham R (2001) How closely correlated are molecular and quantitative measures of genetic variation? A Meta-analysis. Evolution 55:1095–1103

Reed JM, Murphy DD, Brussard PF (1998) Efficacy of population viability analysis. Wildl Soc Bull 26:244–251
Robertson AW, Kelly D, Ladley JJ, Sparrow AD (1999) Effects of pollinator loss on endemic New Zealand mistletoes (Loranthaceae). Conserv Biol 13:499–508
Root RB (1973) Organization of a plant-arthropod association in simple and diverse habitats: the fauna of collards (*Brassica oleracea*). Ecol Monogr 43:95–124
Sampson J, Hopper SD, James SH (1989) The mating system and population genetic structure in a bird-pollinated mallee, *Eucalyptus rhodantha*. Heredity 63:383–393
Scheidel U, Bruelheide H (1999) Selective slug grazing on montane meadow plants. J Ecol 87:828–838
Schmidt K, Jensen K (2000) Genetic structure and AFLP variation of remnant populations in the rare plant *Pedicularis palustris* (Scrophulariaceae) and its relation to population size and reproductive components. Am J Bot 87:678–689
Schwartz MW, Simberloff D (1997) Patterns of rarity in the flora and vertebrate fauna of North America. Bull Ecol Soc Am 78:33
Segarra-Carmona A, Barbosa P (1990) Influence of patch plant density on herbivory levels *by Etiella zinckenella* (Lepidoptera: Pyralidae) on *Glycine max* and *Crotalaria pallida*. Environ Entomol 19:640–647
Shaw WB, Burns BR (1997) The ecology and conservation of the endangered endemic shrub, kowhai ngutukaka *Clianthus puniceus* in New Zealand. Biol Conserv 81:233–245
Spira T (2001) Plant-pollinator interactions: a threatened mutualism with implications for the ecology and management of rare plants. Nat Area J 21:78–88
Stanforth LM, Louda SM, Bevill RL (1997) Insect herbivory on juveniles of a threatened plant, *Cirsium pitcheri*, in relation to plant size, density and distribution. Ecoscience 4:57-66
Tepedino VJ, Sipes SD, Griswold TL (1999) The reproductive biology and effective pollinators of the endangered beardtongue *Penstemon penlandii* (Scrophulariaceae). Plant System Evol 219:39–54
Tilman D (1994) Competition and biodiversity in spatially structured habitats. Ecology 75:2–16
Tilman D (1990a) Constraints and trade-offs: Toward a predictive theory of competition and succession. Oikos 58:3-15
Tilman D (1990b) Mechanisms of plant competition for nutrients the elements of a predictive theory of competition. In: Grace JB, Tilman D (eds.) Perspectives on Plant Competition. Academic Press, San Diego
US Fish and Wildlife Service (1998) Recovery plan for oahu plants. US Fish and Wildlife Service, Portland, Oregon, p 207
van Treuren R, Bijlsma R, Ouborg NJ, van Delden W (1991) The significance of genetic erosion in the process of extinction. IV. Inbreeding depression and heterosis effects caused by selfing and outcrossing in *Scabiosa columbaria*. Evolution 47:1669–1680
Van Treuren R, Bijlsma R, Ouborg NJ, Van Delden W (1993) The significance of genetic erosion in the process of extinction: IV. Inbreeding depression and heterosis effects caused by selfing and outcrossing in *Scabiosa columbaria*. Evolution 47:1669–1680
Vasconcelos HL (1997) Foraging activity of an Amazonian leaf-cutting ant: responses to changes in the availability of woody plants and to previous plant damage. Oecologia (Berl) 112:370–378
Walck JL, Baskin JM, Baskin CC (1999) Effects of competition from introduced plants on establishment, survival, growth and reproduction of the rare plant *Solidago shortii* (Asteraceae). Biol Conserv 88:213–219

Waldmann P (1999) The effect of inbreeding and population hybridization on developmental instability in petals and leaves of the rare plant *Silene diclinis* (Caryophyllaceae). Heredity 83:138–144

Watson LE, Uno GE, McCarty NA, Kornkven AB (1994) Conservation biology of a rare plant species, *Eriocaulon kornickianum* (Eriocaulaceae). Am J Bot 81:980–986

Widen B (1993) Demographic and genetic effects on reproduction as related to population size in a rare, perennial herb, *Senecio integrifolius* (Asteraceae). Biol J Linn Soc 50:179-195

Wiens D, Nickrent DL, Davern CI, Calvin CL, Vivrette NJ (1989) Developmental failure and loss of reproductive capacity in the rare paleoendemic shrub *Dedeckera-eurekensis*. Nature 338:65–67

Willis JC (1922) Age and area: a study in geographical distribution and origin of species. Cambridge University Press, London

4 The Relationship Between Plant-Pathogen and Plant-Herbivore Interactions and Plant Population Persistence in a Fragmented Landscape

N.J. OUBORG and A. BIERE

4.1 Introduction

Contemporary landscapes are characterized worldwide by an increasing degree of fragmentation. Habitat patches are decreasing in size and the distance to neighboring patches is increasing. Populations inhabiting these patches are often small and isolated from con-specific populations. Since the early 1970s it has been recognized that this spatial configuration may influence the viability of remnant populations. Recently, quantitative modeling has been developed to evaluate the long-term viability of populations, an approach summarized under the term "population viability analysis" (PVA, for a review of PVA methods, see Chap. 6, this Vol.).

PVA in plants has been dominated by single species approaches (Menges 2000). All biotic and abiotic influences on the viability of the population are integrated into measures of spatial and temporal variation in demographic parameters, which are then projected into the future (Boyce 1992; Menges 2000). Other organisms are involved in as far as they influence this demographic variation, and are basically treated as inert parts of the environment of the focal population (e.g. Ehrlén 1995).

However, other organisms are not inert aspects of the focal species environment. Rather they are the part of this environment that reacts, responds, and interacts in a dynamical way (Chap. 10, this Vol.). These interactions will often influence the demography of the population we are interested in, and therefore may be important aspects of the viability of this population. Ignoring details of species interactions in current PVAs has the advantage of computational simplicity, which results in more tractable results. However, incorporating species interactions into PVAs may be important for at least two reasons.

First, studying these interactions and the influence of fragmentation on them will help us to understand the mechanisms that underlie at least part of

Ecological Studies, Vol. 165
C.A. Brigham, M.W. Schwartz (Eds.)
Population Viability in Plants

the demographic variation entered into the PVA. Second, when projecting into the future, considering species interactions will allow us to make more reliable predictions of extinction probabilities (Chap. 10, this Vol.).

In this chapter, we will review the concepts, hypotheses and evidence for the role of plant-disease and plant-herbivore interactions within the framework of habitat fragmentation and viability analysis of small, isolated plant populations. (For a discussion of the impacts of other biotic interactions on population viability see Chap. 10, this Vol.)

4.2 Effects of Habitat Fragmentation on Species Interactions

Plants are part of communities and are subjected to interactions with other species belonging to these communities. Because plants are able to fix carbon through photosynthesis they play a central role in any community. Three types of interaction can be distinguished. First, plants are sessile organisms, and therefore must cope with their neighbors; as such, they are subjected to intra- and inter-specific competition for resources (nutrients, light) (Chap. 3, this Vol.). Second, plants may rely for their survival and reproduction on interactions with other species. For instance, animal-pollinated plant species rely for their reproduction on interactions with pollinators (Chaps. 3, 10, this Vol.). These may be generalists or specialists. Plants may also be involved in mutualistic mycorrhizal associations, which help plants to efficiently capture nutrients.

Third, an important category is the interaction of plants with herbivores and pathogens. On the one hand, characteristics of both individual plants and of plant community composition and structure will influence the ecology and evolution of herbivores, pathogens, parasitoids, and predators (Fritz and Simms 1992; Chap. 3, this Vol.). On the other hand, herbivores and pathogens may directly affect growth, reproductive success, seedling establishment, and individual survival, and thus influence the population dynamics of plants (Olff et al. 1999). In as far as these demographic parameters are important for long-term viability, herbivores and pathogens play a role in determining extinction probabilities of plant populations.

In the rest of this chapter we will concentrate on the effects of fragmentation and small, isolated populations on the interaction between plants and herbivores and diseases. We will consecutively present four processes that may influence plant-herbivore and/or plant-pathogen interactions: genetic drift, inbreeding, population thresholds, and multitrophic interactions, and will review the existing empirical evidence for their role.

4.3 The Interaction Between Plants and Diseases and Herbivores

Plants in natural populations are generally challenged by a wide variety of herbivores and pathogens and can be strongly affected by these natural enemies. Pathogens can severely reduce the survival or reproduction of individual plants (e.g. Paul and Ayres 1986; Augspurger 1988; Wennström and Ericson 1990). The expected severity of effects on plant fitness depends on the type of pathogen that is encountered: soil-borne pathogens, foliar pathogens that only cause local lesions, and pathogens that systemically infect their host plants (Wennström 1994; Jarosz and Davelos 1995). Effects on individual hosts may extend to effects on population size, dynamics, and population structure (Burdon 1987), as well as community structure (Dobson and Crawley 1994; Peters and Shaw 1996), for instance by altering the relative competitive abilities of species (e.g. Paul and Ayres 1986; Clay 1990) with subsequent changes in their relative abundance, or by affecting succession rates (Van der Putten et al .1993) and local species diversity (Packer and Clay 2000). Similar impacts can be described for herbivores (Olff et al. 1999). Changes in the size or connectivity of plant populations may therefore not only have direct effects on the viability of plant populations but also indirect effects, if such changes affect the interactions between plants and their natural enemies.

Interactions of plants with their herbivores and pathogens have been investigated in detail both for agricultural and natural systems. Plants have evolved a variety of different mechanisms by which they defend themselves against their natural enemies. Direct defenses of plants include at least three types of mechanisms: avoidance, resistance, and tolerance. Avoidance mechanisms (one of the forms of "passive resistance"; sensu Burdon 1987) reduce the probability of encounters between susceptible host stages and natural enemies. Examples are altered patterns of phenology and reduced apparency of host tissue. For instance, plants that advance or delay the timing of production of their most susceptible tissues relative to the peak abundance of infective stages of an important pathogen or relative to the peak oviposition by an important insect herbivore may show reduced levels of attack (Biere and Honders 1996). Resistance mechanisms (or active resistance sensu Burdon 1987) reduce the probability of successful infection or colonization and growth of natural enemies once they are encountered. Resistance may be based on morphological traits such as spines or trichomes, as well as physiological or biochemical traits, including reduced nutritional quality of host tissue and high constitutive levels of defense chemicals. Finally, tolerance mechanisms (sensu Burdon 1987) allow plants to maintain high survival or reproduction despite a high level of infestation by a herbivore or pathogen.

Of all these mechanisms, the molecular and genetic basis of gene-for-gene (GFG) resistance to pathogens is probably the best documented, and this mechanism has served as the basis for the majority of models of the evolution of host-parasite interactions. The GFG hypothesis, originally formulated by Flor (1956), states that for each gene determining resistance (*R*) in the host there is a corresponding gene for avirulence (*Avr*) in the pathogen with which it specifically interacts. Resistance only occurs when the challenging pathogen carries an *Avr* allele at the avirulence locus and the host carries an *R* allele at the corresponding resistance locus. Any other combination (*R-avr*, *r-Avr*, *r-avr*) will not result in recognition of the pathogen and renders the host susceptible. Since resistance is generally dominant to susceptibility, diploid hosts that are heterozygous at resistance loci are usually resistant to pathogens carrying the *Avr* allele at the corresponding avirulence locus. Within a single pathosystem, there may be many different pairs of *R* loci in the host and corresponding *Avr* loci in the pathogen. Pairs of loci that confer resistance are commonly found to be epistatic to loci that render the host susceptible. In other words, the presence of an *R* and *Avr* allele at a single pair of loci is sufficient to confer resistance.

Although GFG interactions have been shown for interactions between crop species and a range of biotrophic pathogens, as well as for some necrotrophic pathogens, insect herbivores, and nematodes (Thompson and Burdon 1992), it is unclear how widespread such single-gene resistance interactions are in natural populations.

An example of a GFG interaction in a natural system is the interaction between *Linum marginale* and the rust fungus *Melampsora lini*, endemic to Australia. Studies on the dynamics of host-resistance types and pathogen-virulence types in this system have yielded valuable insight into the spatial scale at which such GFG interactions occur. Contrary to expectation, the frequency of different races within local pathogen populations in this system appears to be poorly correlated with the frequency of different resistance types within the corresponding local host populations (Jarosz and Burdon 1991). Stochastic processes during population crashes rather than natural selection within local populations appear to be the main cause of large year-to-year variation in the frequencies of *R* and *Avr* alleles. For instance, after a severe epidemic, host genotypes resistant to the pathogen races that were present at high frequency during the epidemic surprisingly had not increased but decreased in frequency, opposing the view that natural selection in local populations governs GFG coevolution (Burdon and Thompson 1995). The authors suggested that individual populations are mainly influenced by genetic drift, extinction, and gene-flow among populations within the same epidemiological region, and that GFG coevolution is likely to take place at the metapopulation rather than the local population level. These type of studies form the basis for our understanding and predictions of the consequences of decreases in host population size or connectivity for disease susceptibility.

4.4 Habitat Fragmentation and Disease Susceptibility

Our thinking about the interactions between population size, connectivity, and disease incidence has been dominated by the idea that fragmentation will lead to loss of genetic variation and that this will increase disease susceptibility of the population. For instance, Hamilton (1980; Hamilton et al. 1990) formulated an influential hypothesis on the importance of sexual reproduction and genetic recombination as a means to promote genetic variation in the host species in order to be able to keep track of the continuous replenishment of variants in the pathogen, which in most cases has shorter generation times and hence higher evolutionary rates than the host.

An influential paper by O'Brien and Evermann (1988) supported the rather general conviction that small and isolated populations may be highly susceptible to diseases. Much of the evidence in that paper was based on studies on mammals. Exemplary is the well known series of studies by O'Brien and co-workers on the African cheetah (O'Brien et al. 1985, 1987). In African populations very low levels of genetic variation could be detected, both in non-selective marker genes and in functional MHC genes, the major histocompatibility complex genes in vertebrates involved in immune responses. Evidence exists that genetic variation in this MHC system is essential for effective long-term immune responses (e.g. Hughes 1991; Hedrick and Miller 1994; Hill et al. 1997). The cheetah population lacking variation for this gene complex suffered from high incidences of feline peritonitis. O'Brien et al. concluded that the low levels of genetic variation were the consequence of one or two major bottlenecks in the past and resulted in increased susceptibility to diseases. O'Brien and Evermann (1988) presented evidence from other vertebrates basically supporting the same notion. However, this notion and the evidence for it have since been debated. For instance, Nunney and Campbell (1993) argued that the evidence is correlational instead of causal, leaving room for other causal explanations. Indeed, the supposed causal relationship between habitat fragmentation, low MHC diversity, and increased susceptibility to diseases has been criticized on several points (Caro and Laurenson 1994).

Here we will present a number of specific hypotheses on the relationship between reduction in population size and increases in isolation, on the one hand, and disease incidence and herbivore pressure, on the other. The ideas presented concur with various recent papers underlining the importance of incorporating spatial structure into concepts of host-pathogen dynamics and evolution and into epidemiological inferences (Hess 1996; Thrall and Burdon 1997; Holmes 1997; Antonovics et al. 1997).

4.4.1 Disease Incidence and Genetic Drift

One of the most influential ideas affecting our view of the threats associated with habitat fragmentation is that small, isolated populations may be increasingly affected by genetic erosion and inbreeding depression (Soulé and Wilcox 1980; Frankel and Soulé 1981; Ellstrand and Elam 1993; Young et al. 1996; Booy et al. 2000; Young and Clarke 2000). Small populations are expected to experience an increased impact of genetic drift, leading to loss of genetic variants from the population, and inbreeding, leading to an increase in homozygosity. These phenomena, summarized under the term "genetic erosion", will lead to inbreeding depression, i.e. a decrease in the average fitness, in small populations (Chap. 3, this Vol.). This process has been studied in many species over the last decade, and several case studies have shown that inbreeding may be a factor of importance (e.g. *Salvia pratensis* and *Scabiosa columbaria* (Ouborg 1993a; van Treuren 1993); *Gentiana pneumonanthe* (Oostermeijer 1996); *Gentianella germanica* (Fisher and Matthies 1997, 1998).

Small populations are subjected to genetic drift, the random fluctuation of allele frequencies from generation to generation as a consequence of sampling error. The effect of genetic drift that has received the most attention is that these random fluctuations may eventually lead to either loss or fixation of alleles. The rate of this loss is proportional to the effective population size. As these drift effects occur independently in each local population, classical population genetic theory indicates that genetic drift will strongly influence the spatial structure of genetic variation (Wright 1969). For instance, Brakefield (1989) showed that the variance of allele frequencies among (small) populations is more sensitive to genetic drift, and an increase will become evident in an earlier stage, than the variance in allele frequencies of average-sized populations. Indeed evidence was found for such an effect in small populations of the threatened species *Salvia pratensis* and *Scabiosa columbaria* (van Treuren et al. 1991).

In the case of resistance or avoidance genes, selection most likely will override the effects of drift when populations are infected (but see Burdon and Thompson 1995). However, in disease-free situations drift may play a more prominent role even though resistance or avoidance alleles may be selected against, due to what is known as cost of resistance (e.g. Biere and Antonovics 1996). If small, isolated populations are free of pathogens for a substantial number of generations, drift may increase the variance in resistance among them. When attacked by a pathogen this will result in high variance in disease incidence among small populations, and lower variance among large populations. Potentially, some small populations may be more resistant to the pathogen than the average of all populations.

Empirical evidence for this effect of habitat fragmentation on disease resistance and disease incidence is lacking. This may, however, be the result of the lack of attention given to this hypothesis. Although conclusive evidence for the role of drift may be hard to obtain, comparisons of variance among small populations and variance among large populations have yielded valuable insights (e.g. Ouborg et al. 1991; Van Treuren et al. 1991). This approach may be readily extended to variance in disease incidence among populations.

4.4.2 Disease Incidence and Inbreeding

In host populations, inbreeding is a process that will change genotype frequencies and may therefore affect host-pathogen interactions. The effect of inbreeding on disease levels in a host population can be imposed in two different ways. First, inbreeding may affect host responses to pathogen exposure that reduce the level of infection once the pathogen is encountered (cf. active resistance; Burdon 1987). Second, inbreeding may affect those host traits that reduce contact with the pathogen (avoidance, cf. passive resistance; Burdon 1987). Provided that there is genetic variation for the traits involved in active and passive resistance, and there is some degree of genetic dominance for them, inbreeding will result in inbreeding depression in these traits.

Documentation of inbreeding effects on host-pathogen interactions in natural populations is still scanty. The limited data available show equivocal patterns. Significant inbreeding depression was found for resistance to two herbivores in maize (Ajala 1992) and for rust resistance in slack pine (Matheson et al. 1995), but in the seaside daisy *Erigeron glaucus* effects of inbreeding on herbivore resistance depended on the resistance status of the parental plants used to generate the tested progeny families (Strauss and Karban 1994), while in *Datura stramonium* no effect of inbreeding on resistance to two herbivores could be detected (Nunezfaran et al. 1996).

The complicated nature of the influence of inbreeding on host-pathogen interactions was demonstrated in a study of the *Silene alba* (=*S. latifolia*)–*Microbotryum violaceum (=Ustilago violacea)* pathosystem (Ouborg et al. 2000). *M. violaceum* is a fungal pathogen that produces its spores in the anthers of its caryophyllaceous host species, thereby sterilizing the host plant. The spores are transmitted by pollinating insects. Frequent extinction and subsequent recolonization seem to be natural features of the dynamics of this system (Antonovics et al. 1994, 1997). When sites are recolonized, host populations are founded by a very limited number of genotypes. McCauley et al. (1995) estimated the average effective number of colonists to be around four, with only limited mixing of individuals from different populations. Inbreeding will be a common process in local populations of this host species, and the inbreeding level will increase with time.

Ouborg et al. (2000) performed two experiments to estimate the effect of inbreeding in the host on components of resistance against *Microbotryum* infection. In one experiment 65 inbred lines (with inbreeding coefficients of f=0 to f=0.59), originating from eight different populations, were tested for active resistance against fungal infection by artificially inoculating individuals. Overall, a very small, but significant increase in active resistance upon inbreeding was observed. However, the effect of inbreeding on resistance varied significantly among populations and among lines. The variance in inbreeding effects among lines was huge, such that inbreeding resulted in increased resistance in some lines and decreased resistance in others. In a second experiment, 12 inbred lines from one population were screened for passive resistance. It has been shown that flower traits, more specifically flower size and nectar reward (Biere and Antonovics 1996; Shykoff and Bucheli 1995; Shykoff et al. 1997), are important components of avoidance to *Microbotryum* infection, an insect-vectored, florally transmitted disease. Significant inbreeding depression was found for petal size and nectar volume, thereby potentially enhancing avoidance of spore transmission by insects.

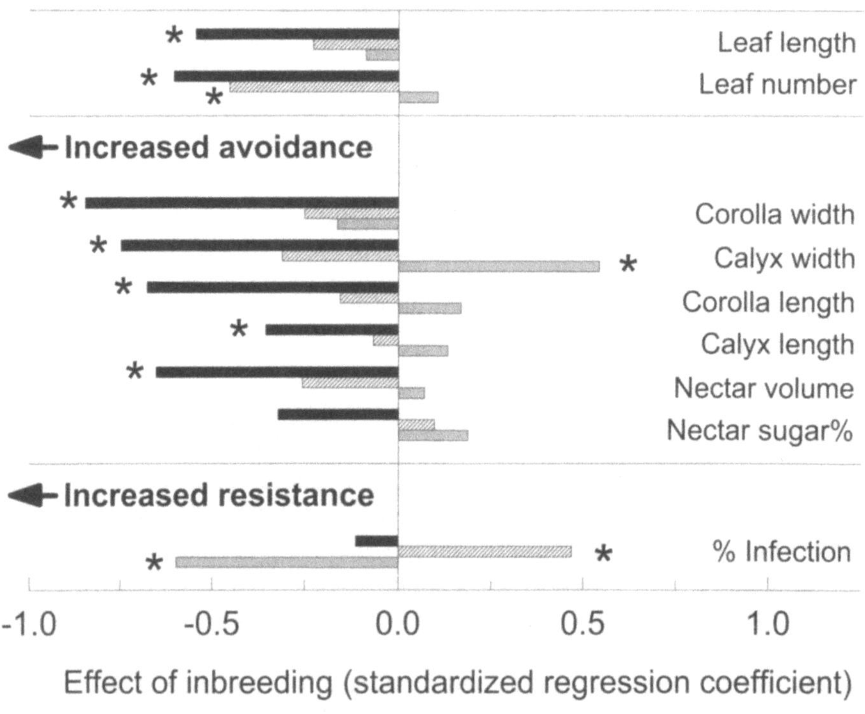

Fig. 4.1. Effects of inbreeding in *Silene alba* on putative avoidance traits and on resistance to the pollinator-transmitted pathogen *Microbotryum violaceum*. Effects of inbreeding are quantified as standardized regression coefficients from linear regressions of trait values on inbreeding coefficients (0, 0.125, 0.25, 0.375, 0.593). Data for female plants from three different inbreeding lines initiated from different mother plants originating from a single population that show strong (*black bars*), weak (*hatched bars*) and no (*gray bars*) inbreeding depression in these traits. *Asterisks* indicate significant effects of inbreeding ($P<0.05$). No correlation exists between avoidance related traits and active resistance. (Ouborg et al. 2000)

However, as with active resistance, inbreeding effects differed significantly among lines. This is illustrated in Fig. 4.1, in which inbreeding effects on avoidance- and resistance-related traits are depicted for three different lines. Moreover, inbreeding effects on active and passive resistance components were uncorrelated, making the net effect of inbreeding on field resistance unpredictable.

If habitat fragmentation is associated with increased levels of inbreeding in local remnant populations, then these results have important implications for the way host-pathogen interactions should be incorporated in PVAs. The results demonstrate that the link between inbreeding, inbreeding depression, and increased susceptibility to diseases may not be uniform and hence will be difficult to incorporate. The effects very much depend on genotypic composition of the population, on inbreeding history and on the covariance structure of the resistance traits involved.

4.4.3 Disease Incidence and Population Dynamics: Thresholds

The interaction between a plant species and its pathogens or herbivores is affected by the demography of both and their mutual influence on each other. This dynamic interaction in turn may be influenced by the increase in interpopulation distance and the decrease in local population size, associated with habitat fragmentation.

Local populations in a fragmented habitat are more isolated from each other. This may hamper disease transmission, as pathogens may be unable to disperse. Thus more isolated populations are less likely to attract diseases (Simberloff 1988). Indeed, positive correlations between disease incidence in local populations and disease status of neighboring populations have been found (Burdon et al. 1995; Antonovics et al. 1997; Ericson et al 1999). If this is generally true, then this sheds new light on management decisions. It is generally assumed that increasing connectivity among populations in a fragmented landscape will increase the viability of both local populations and the metapopulation (e.g. Harrison and Fahrig 1995). However, increasing connectivity might also increase the likelihood of disease transmission, possibly resulting in decreased population viability (Dobson and May 1986; Simberloff 1988; Hobbs 1992; Hess 1994, 1996; Thrall et al. 2000) The net effect of increased connectivity depends on the balance between positive and negative effects; intermediate levels of connectivity may be optimal (Hess 1996).

In a fragmented landscape most remnant populations are small. Population size is an important factor in host-pathogen dynamics. In the most general form, the interaction can be described by:

$$DN/dt=(a-b)N-\alpha I$$

and

$$DI/dt=\beta I(N-I)-(\alpha+b+v)I$$

where N=total population size, I=number of infected individuals, a=per capita birth rate, b=per capita death rate, α=disease induced mortality, β=disease transmission parameter, and v=recovery rate of diseased hosts (Getz and Pickering 1983). When the dynamic properties of this model are explored, it turns out that a threshold host-population size exists below which the disease can not persist in the host population and will be eliminated in time. Below this threshold size, the host population offers too little resources for the pathogen to persist. The threshold value is dependent on the magnitude of disease-induced mortality α, on the natural mortality rate b, on the recovery rate v and on the transmission efficiency β (Anderson and May 1979a,b). Highly pathogenic diseases with low transmission efficiency will experience relatively high host-population size thresholds and thus will less likely persist in small populations.

It has been pointed out that the existence of a threshold can be expected when disease transmission is density-dependent, as in randomly dispersed (cf. wind-dispersed) pathogens; however, when disease transmission is frequency-dependent, as in the case of vector-borne diseases, no threshold is predicted (Getz and Pickering 1983). In those cases, the basic model predicts that pathogens can not stably co-exist with their hosts; the disease either spreads and leads to extinction of both host and pathogen, or the disease is eliminated from the host population. Several modifications have been made to the basic model, including the ability of density dependence to act differentially on healthy and diseased individuals (Thrall et al. 1993), and interactions between pollination and disease transmission (Ingvarsson and Lundberg 1993). These modifications showed that even under frequency-dependent transmission stable co-existence of host and pathogen is possible, under certain conditions. A recent theoretical treatment of plant-parasite interactions (Gubbins et al. 2000) demonstrated that under realistic assumptions a threshold population of susceptible hosts exist below which the parasite is unable to invade the plant population. Furthermore, under stricter conditions, there are also threshold densities for the infected hosts and parasite populations below which invasion does not occur.

Thus, epidemiological theory predicts that for certain types of diseases and under certain circumstances small populations may have a low probability of being diseased. A few studies have investigated the relationship between population size and disease incidence. Jennersten et al. (1983) investigated the interaction between the plant species *Viscaria vulgaris* and the smut fungus *Ustilago violaceae* in populations in southwest Sweden. No *Viscaria* populations smaller than 35 individuals were found to be infected, and there was a positive correlation between population size and infection rate. Carlsson and Elmqvist (1992) found infection of *Silene dioica* in Northern Sweden with

Ustilago violacea to be host-population-size- and host-density dependent. Young populations that became diseased were larger. In two small populations, actual failure of pathogen establishment was recorded, while the pathogen went extinct in another small population. The same was found in American populations of *Silene alba* infected with *Ustilago violacea* (Alexander 1990).

A final example of a threshold effect was found in *Salvia pratensis* populations in the Netherlands. *Salvia* is a threatened species in the Netherlands, with most remaining populations being small and isolated (Ouborg et al. 1991; Ouborg 1993b). The species produces flower stalks with up to 30 whorls of 6 flowers each; each flower potentially produces 4 seeds. Seeds are infected by the gall wasp *Aylax salviae*, which lays eggs in the seed, thereby induces the seed to transform into a gall, and thus effectively reduces seed output. In a survey of 20 populations only populations larger than 300 individuals (5 out of 20 populations) were found to be infected. In populations containing galls, the infection may be as high as 100 % of all individuals, although most individuals will still produce some viable seeds (Ouborg unpubl. data).

These examples show that small populations may be relatively protected against diseases and herbivores. It is, however, unclear whether the observed threshold is indeed a dynamical equilibrium for small populations, or whether absence of the disease is caused by isolation or by age of the population (Jennersten et al. 1983; Carlsson and Elmqvist 1992; Alexander 1990). The phenomenon deserves more attention, including field studies and experimental manipulation to check for causes.

4.4.4 Disease Incidence and Multitrophic Interactions

Studying host-pathogen or host-herbivore systems in isolation (i.e. while disregarding other organisms in the same community) has taught us a great deal about the mechanisms and dynamics of their interactions. However, as they are part of communities, many other organisms may be involved in, and significantly affect, host-pathogen or host-herbivore interactions. Obvious examples are cases in which insect vectors are necessary to transmit disease propagules from one host individual to the next, such as aphid transmission of many viruses, and regulation of herbivores by their predators or parasitoids. It seems very likely that effects of habitat fragmentation on hosts mediated by interactions with their pathogens and herbivores may not be understood without explicitly including such vector relationships or higher trophic interactions if such interactions are strong. Even in interactions involving only a limited number of species, the interrelationships may be complex. For instance, spores of *Microbotryum violaceum* are transmitted between healthy and diseased *Silene alba* individuals mainly by noctuids, especially *Hadena bicruris*. This noctuid serves at least three important roles.

First, it feeds on the nectar of *Silene alba* and is a main, though not exclusive, pollinator of this host (Jürgens et al. 1996). Second, it is an important vector of the anther smut fungus in diseased populations. Third, it is a seed predator specialized on *Silene alba*. After pollination, the noctuid oviposits on the ovaries of flowers of female plants. Emerging larvae eat their way into the young seed capsule and feed on the developing seeds. When these are consumed, they predate the seeds from a few other ripe fruit capsules and pupate in the soil (Brantjes 1976). The noctuid is commonly attacked by two braconid parasitoids, the specialist endoparasitoid *Microplitis tristis* and the generalist ectoparasitoid *Bracon variator*. The complexity of this set of interactions may be representative for many other host-pathogen or host-herbivore systems. Habitat fragmentation may affect any of the components in such a multitrophic web. The net effect on the plant depends on which component of the web is affected most. For instance, if the insect vector is affected more by isolation than the plant, isolated plant populations may be free of disease. On the other hand, if the parasitoid is affected most (see below), the herbivore is no longer under the control of the parasitoid and the plant population may become heavily predated. For predicting the effect of habitat fragmentation on host-pathogen or host-herbivore interactions, it is therefore essential to collect data about the extent of the multitrophic food web, the strength of each interaction in the web, and the effect of habitat fragmentation on each of the members of the food web. We are currently exploring this system to study the effect of habitat connectivity on the incidence of host, pathogen, herbivore and parasitoids in local patches, combined with experimental studies on dispersal and colonization ability of the different components of the system.

In a number of papers, Tscharntke and co-workers demonstrated the relevance of multitrophic interactions in assessing the effects of habitat fragmentation. In manually established islands of red clover (*Trifolium pratense*), isolated patches were colonized by most of the available herbivores but only a few of the available parasitoid species. Herbivores experienced only 19–60 % of the parasitization of non-isolated patches (Kruess and Tscharntke 1994). In another experiment, 32 natural stinging-nettles (*Urtica dioica*) patches of different size and degree of isolation were investigated. Habitat fragmentation reduced species richness, but not all species groups were affected to the same degree and in the same way. Monophagous insects were most affected by the area of the patch, while predatory insects were most affected by the degree of isolation (Zabel and Tscharntke 1998). Finally, in experiments with bush vetch (*Vicia sepium*) plants in pots, overall colonization success of insects decreased with increasing distance. However, parasitoids suffered more from habitat loss and isolation than their phytophagous hosts. The minimum area required for persistence was higher for parasitoids than for herbivores. In concordance with these effects, the percent parasitism of herbivores significantly decreased with area loss and increasing isolation between plots (Kruess and Tscharntke 2000).

These results suggest that higher trophic levels may be more susceptible to habitat fragmentation. Consequently, habitat fragmentation may release herbivorous insects from parasitoid or predator control, leading to increased infestation of host plant populations.

4.5 Conclusions

In this chapter we have presented four specific hypotheses concerning the relationship between habitat fragmentation and host-pathogen or host-herbivore interactions. Some of these hypotheses are backed up by empirical data, others lack empirical support mainly because they have hardly or not been investigated. Given the evidence for the strong effects that herbivores and pathogens can have on host population dynamics and community structure, and for the fact that decreased host population size and connectivity can affect the opportunities for invasion and regulation of host populations by natural enemies, there is no doubt that habitat fragmentation can affect host population viability through effects on interactions with natural enemies. Single-species approaches to PVAs, elegant by their relative simplicity, may therefore have limited predictive power in cases of significant interaction strength between target organisms and their herbivores, pathogens and higher trophic interactions. Unfortunately, generalizations of the strength and direction of the effects of fragmentation through interactions with other organisms are as yet difficult to make. How such interactions should be incorporated is therefore a major challenge for the years to come and will require a wealth of studies in order to come to generalizations for particular groups of organisms and habitats.

Nevertheless, some conclusions can be drawn. First, it is clear that small populations in a fragmented landscape are not necessarily more susceptible to diseases and herbivores as a result of increased inbreeding. The nature of the interaction between hosts, pathogens, herbivores, and other trophic levels is likely to be very complex. The effects of inbreeding will depend on the history of the system, on the way and the scale at which host and pathogen interact, on the traits involved in avoidance, resistance and tolerance, and on the susceptibility to inbreeding of each of these traits. We have shown an example in which inbreeding increases or decreases resistance, depending on host genotype, and in which components of resistance are differentially affected even within a single host genotype. Inbreeding thus does not necessarily result in increased host susceptibility.

Second, in addition to these genetic considerations, habitat fragmentation may result in protection of local host populations from natural enemies through effects on population dynamics. Decreasing host population size and connectivity can reduce the probability of transmission, colonization, and persistence of pathogens and herbivores, e.g. by reaching and modifying thresholds for invasion and persistence. Increasing connectivity in a fragmented landscape therefore has the potential drawback of increasing disease transmission.

Third, it is clear that species interactions, in terms of plant-plant, plant-animal and animal-animal interactions, at multitrophic levels are important aspects of the viability of populations. In some cases, fragmentation appears to have a stronger impact on predators of herbivores than on the herbivores that they exploit, suggesting that local hosts may experience higher herbivore pressure.

In summary, there is an urgent need for studies that will enable us to generalize and make predictions of the possible consequences of habitat fragmentation on population viability as affected by interactions with diseases and herbivory. In our view, including these interactions in PVAs is a complex but essential step towards reliable predictions of plant population viability.

References

Ajala SO (1992) Inheritance of resistance in maize to the spotted stem-borer, *Chilo partellus* (swinhoe). Maydica 37:363–369

Anderson RM, May RM (1979a) Population biology of infectious diseases I. Nature 280:361–367

Anderson RM, May RM (1979b) Population biology of infectious diseases II. Nature 280:455–461

Alexander HM (1990) Dynamics of plant-pathogen interactions in natural plant communities. In: Burdon JJ, Leather SR (eds) Pests, pathogens and plant communities. Blackwell, Oxford, pp 31–45

Antonovics J, Thrall PH, Jarosz AM, Stratton D (1994) Ecological genetics of metapopulations: the *Silene-Ustilago* plant-pathogen system. In: Real LA (ed) Ecological genetics. Princeton University Press, Princeton, pp 146–170

Antonovics J, Thrall PH, Jarosz AM (1997) Genetics and spatial ecology of species interactions: the *Silene-Ustilago* system. In: Tilman D, Kareiva P (eds) Spatial ecology: the role of space in population dynamics and interspecific interactions. Princeton University Press, Princeton, pp 158–184

Augspurger CK (1988) Impact of pathogens on natural plant populations. In: Hutchings MJ, Watson AR (eds) Plant population ecology. Blackwell, Oxford, pp 413–433

Biere A, Antonovics J (1996) Sex-specific costs of resistance to the fungal pathogen Ustilago violacea (*Microbotryum violaceum*) in *Silene alba*. Evolution 50:1098–1110

Biere A, Honders SC (1996) Impact of flowering phenology of *Silene alba* and *S. dioica* on susceptibility to fungal infection and seed predation. Oikos 77:467–480

Booy G, Hendriks RJJ, Smulders MJM, van Groenendael JM, Vosman B (2000) Genetic diversity and the survival of populations. Plant Biol 2:379–395

Boyce M (1992) Population viability analysis. Annu Rev Ecol Syst 23:481–506

Brakefield PM (1989) The variance in genetic diversity among subpopulations is more sensitive to founder effects and bottlenecks than is the mean: a case study. In: Font-

devilla A (ed) Evolutionary biology of transient unstable populations. Springer, Berlin Heidelberg New York, pp 145–161
Burdon JJ (1987) Diseases and plant population biology. Cambridge University Press, Cambridge
Brantjes NBM (1976) Riddles around the pollination of *Melandrium album* (Miller) Garcke (Caryophyllaceae) during the oviposition by *Hadena bicruris* Hufn. (Noctuidae, Lepidoptera). I. Proc Kon Ned Akad Wet Ser C 79:1–12
Burdon JJ, Thompson JN (1995) Changed patterns of resistance in a population of *Linum marginale* attacked by the rust pathogen *Melampsora lini*. J Ecol 83:199–206
Burdon JJ, Ericson L, Müller WJ (1995) Temporal and spatial changes in a metapopulation of the rust pathogen *Triphragmium ulmariae* and its host, *Filipendula ulmaria*. J Ecol 83:979–989
Carlsson U, Elmqvist T (1992) Epidemiology of anther-smut disease (*Microbotryum violaceum*) and numeric regulation of populations of *Silene dioica*. Oecologia 90:509–517
Caro TM, Laurenson MK (1994) Ecological and genetic factors in conservation: a cautionary tale. Science 263:485–486
Clay K (1990) The impact of parasitic and mutualistic fungi on competitive interactions among plants. In: Grace JB, Tilman D (eds) Perspectives on plant competition. Academic Press, San Diego, pp 391–412
Dobson A, Crawley W (1994) Pathogens and the structure of plant-communities. Trends Ecol Evol 9:393–398
Dobson AP, May RM (1986) Disease and conservation. In: Soulé ME (ed) Conservation biology: the science of scarcity and diversity. Sinauer Associates, Sunderland, MA., pp 345–365
Ehrlén J (1995) Demography of the perennial herb *Lathyrus vernus*. II. Herbivory and population dynamics. J Ecol 83:297–308
Ellstrand NC, Elam DR (1993) Population genetic consequences of small population size: implications for plant conservation. Annu Rev Ecol Syst 24:217–242
Ericson L, Burdon JJ, Müller WJ (1999) Spatial and temporal dynamics of epidemics of the rust fungus *Uromyces valerianae* on populations of its host, *Valeriana salina*. J Ecol 87:649–658
Fischer M, Matthies D (1997) Mating structure and inbreeding and outbreeding depression in the rare plant *Gentianella germanica* (Gentianaceae). Am J Bot 84:1685–1692
Fischer M, Matthies D (1998) Effects of population size on performance in the rare plant *Gentianella germanica*. J Ecol 86:195–203
Flor HH (1956) The complementary genic systems in flax and flax rust. Adv Genet 8:29–54
Frankel OH, Soulé ME (1981) Conservation and evolution. Cambridge University Press, Cambridge
Fritz RS, Simms EL (1992) Plant resistance to herbivores and pathogens. University of Chicago Press, Chicago, Illinois, USA
Getz WM, Pickering J (1983) Epidemic models: thresholds and population regulation. Am Nat 121:892–898
Gubbins S, Gilligan CA, Kleczkowski A (2000) Population dynamics of plant-parasite interactions: thresholds for invasion. Theor Pop Biol 57:219–233
Hamilton WD (1980) Sex versus non-sex versus parasite. Oikos 35:282–290
Hamilton WD, Axelrod R, Tanese R (1990) Sexual reproduction as an adaptation to resist parasites (a review). Proc Natl Acad Sci USA 87:3566–3573
Harrison S, Fahrig L (1995) Landscape pattern and population conservation. In: Hansson L, Fahrig L, Merriam G (eds) Mosaic landscapes and ecological processes. Chapman and Hall, London, pp 293–308

Hedrick PW, Miller PS (1994) Rare alleles, MHC and captive breeding. In: Loeschcke V, Tomiuk J, Jain SK (eds) Conservation genetics. Birkhaüser Verlag, Basel, pp 187–204
Hess G (1994) Conservation corridors and contagious disease: a cautionary note. Conserv Biol 8:256–262
Hess G (1996) Disease in metapopulation models: implications for conservation. Ecology 77:1617–1632
Hill AVS, Jepson A, Plebanski M, Gilbert S (1997) Genetic analysis of host-parasite coevolution in human malaria. Philos Trans R Soc Lond B 352:1317–1325
Hobbs RJ (1992) The role of corridors in conservation: solution or bandwagon? Trends Ecol Evol 7:389–392
Holmes EE (1997) Basic epidemiological concepts in a spatial context. In: Tilman D, Kareiva P (eds) Spatial ecology: the role of space in population dynamics and interspecific interactions. Princeton University Press, Princeton, NJ, pp 111–136
Hughes AL (1991) MHC polymorphism and the design of captive breeding programs. Conserv Biol 5:249–251
Ingvarsson PK, Lundberg S (1993) The effect of a vector-borne disease on the dynamics of natural plant populations: a model for *Ustilago violacea* infection of *Lychnis viscaria*. J Ecol 81:263–270
Jarosz AM, Burdon JJ (1991) Host-pathogen interactions in natural populations of *Linum marginale* and *Melampsora lini*: II. Local and regional patterns of resistance and racial structure. Evolution 45:1618–1627
Jarosz AM, Davelos AL (1995) Effects of disease in wild plant populations and the evolution of pathogen aggressiveness. New Phytol 129:371–387
Jennersten O, Nilsson SG, Wastljung U (1983) Local plant populations as ecological islands: the infection of *Viscaria vulgaris* by the fungus *Ustilago violacea*. Oikos 41:391–395
Jürgens A, Witt T, Gottsberger G (1996) Reproduction and pollination in Central European populations of *Silene* and *Saponaria* species. Bot Acta 109:316–324
Kruess A, Tscharntke T (1994) Habitat fragmentation, species loss, and biological control. Science 264:1581–1584
Kruess A, Tscharntke T (2000) Species richness and parasitism in a fragmented landscape: experiments and field studies with insects on *Vicia sepium*. Oecologia 122:129–137
Matheson AC, White TL, Powell GR (1995) Effects of inbreeding on growth, stem form and rust resistance in *Pinus alliottii*. Silvae Genet 44:37–46
McCauley DE, Raveill J, Antonovics J (1995) Local founding events as determinants of genetic structure in a plant metapopulation. Heredity 75:630–636
Menges E (2000) Population viability analyses in plants: challenges and opportunities. Trends Ecol Evol 15:51–56
Nunezfaran J, Cabralesvargas RA, Dirzo R (1996) Mating system consequences on resistance to herbivory and life history traits in *Datura stramonium*. Am J Bot 83:1041–1049
Nunney L, Campbell KA (1993) Assessing minimum viable population size: demography meets population genetics. Trends Ecol Evol 8:234–239
O'Brien SJ, Evermann JF (1988) Interactive influence of infectious disease and genetic diversity in natural populations. Trends Ecol Evol 3:254–259
O'Brien SJ, Roelke ME, Marker A, Newman CA, Winkler D, Meltzer L, Colly JF, Bush M, Wildt DE (1985) Genetic basis for species vulnerability in the cheetah. Science 227:1428–1434
O'Brien SJ, Wildt DE, Bush M, Caro TM, Fitzgibbon C, Aggundey I, Leakey RE (1987) East African cheetahs: evidence for two population bottlenecks? Proc Natl Acad Sci USA 84:509–511

Olff H, Brown VK, Drent RH (1999) Herbivores: between plants and predators. Blackwell Science, Oxford
Oostermeijer JGB (1996) Population viability of the rare *Gentinana pneumonanthe*: the relative importance of demography, genetics and reproductive biology. PhD thesis, University of Amsterdam, Amsterdam
Ouborg NJ (1993a) On the relative contribution of genetic erosion to the chance of population extinction. PhD thesis, University of Utrecht, Utrecht
Ouborg NJ (1993b) Isolation, population size and extinction: the classical and metapopulation approaches applied to vascular plants along the Dutch Rhine system. Oikos 66:298–308
Ouborg NJ, van Treuren R, van Damme JMM (1991) The significance of genetic erosion in the process of extinction II. Morphological variation and fitness components in populations varying in size of *Salvia pratensis* L. and *Scabiosa columbaria* L. Oecologia 86:359–367
Ouborg NJ, Biere A, Mudde CL (2000) Inbreeding effects on resistance and transmission-related traits in the *Silene-Microbotryum* pathosystem. Ecology 81:520–531
Packer A, Clay K (2000) Soil pathogens and spatial patterns of seedling mortality in a temperate forest. Nature 404:278–281
Paul ND, Ayres PG (1986) The impact of a pathogen (*Puccinia lagenophorae*) on populations of groundsel (*Senecio vulgaris*) overwintering in the field. II. Reproduction. J Ecol 74:1085–1094
Peters JC, Shaw MW (1996) Effect of artificial exclusion and augmentation of fungal plant pathogens on a regenerating grassland. New Phytol 134:295–307
Shykoff JA, Bucheli E (1995) Pollinator visitation patterns, floral rewards and the probability of transmission of *Microbotryum violaceum*: a venereal disease in plants. J Ecol 83:189–198
Shykoff JA, Bucheli E, Kaltz O (1997) Anther smut disease in *Dianthus silvester* (Caryophyllaceae): natural selection of floral traits. Evolution 51:383–392
Simberloff DS (1988) The contribution of population and community biology to conservation science. Annu Rev Ecol Syst 19:473–511
Soulé ME, Wilcox BA (1980) Conservation biology: an evolutionary-ecological perspective. Sinauer Associates, Sunderland
Strauss S, Karban R (1994) The significance of outcrossing in an intimate plant-herbivore relationship. I. Does outcrossing provide an escape from herbivores adapted to the parent plant? Evolution 48:454–464
Thompson JN, Burdon JJ (1992) Gene-for-gene coevolution between plants and parasites. Nature 360:121–125
Thrall PH, Burdon JJ (1997) Host-pathogen dynamics in a metapopulation context: the ecological and evolutionary consequences of being spatial. J Ecol 85:743–753
Thrall PH, Antonovics J, Hall DW (1993) Host and pathogen coexistence in sexually transmitted and vector-borne diseases characterized by frequency-dependent disease transmission. Am Nat 142:543–552
Thrall PH, Burdon JJ, Murray BR (2000) The metapopulation paradigm: a fragmented view of conservation biology. In: Young AG, Clarke GM (eds) Genetics, demography and viability of fragmented populations. Cambridge University Press, Cambridge, pp 75–95
Van der Putten WH, Van Dijk C, Peters BAM (1993) Plant-specific soil-borne diseases contribute to succession in foredune vegetation. Nature 362:53–55
Van Treuren R (1993). The significance of genetic erosion for the extinction of locally endangered plant populations. PhD Thesis, State University of Groningen, Groningen
Van Treuren R, Bijlsma R, Ouborg NJ, van Delden W (1991). The significance of genetic erosion in the process of extinction. I. Genetic differentiation in *Salvia pratensis* and *Scabiosa columbaria* in relation to population size. Heredity 66:181–189

Wennström A (1994) Systemic diseases on hosts with different growth patterns. Oikos 69:535–538

Wennström A, Ericson L (1990) The interaction between the clonal herb *Trientalis europaea* and the host specific smut fungus *Urocystis trientalis*. Oecologia 85:238–240

Wright S (1969) Evolutions and the genetics of populations, vol II. The theory of gene frequencies. University of Chicago Press, Chicago

Young AG, Boyle T, Brown AHD (1996) The population genetic consequences of habitat fragmentation for plants. Trends Ecol Evol 11:413–418

Young AG, Clarke GM (2000) Genetics, demography and viability of fragmented populations. Cambridge University Press, Cambridge

Zabel J, Tscharntke T (1998) Does fragmentation of *Urtica* habitats affect phytophagous and predatory insects differentially? Oecologia 116:419–425

5 The Origin and Extinction of Species Through Hybridization

C.A. Buerkle, D.E. Wolf, and L.H. Rieseberg

5.1 Introduction: Consequences of Hybridization

The role of hybridization in the origin, maintenance, and loss of biodiversity has been the subject of speculation and debate for more than two centuries (Linnaeus 1760; Kölreuter 1893; Arnold 1997). Some authors have emphasized the creative role of hybridization in fostering species or community diversity (Linnaeus 1760; Kerner von Marilaun 1894–1895; Lotsy 1916; Stebbins 1942; Anderson 1949; Whitham and Maschinski 1996; Arnold 1997), whereas others have focused on its role as a destructive evolutionary force, contributing to the extinction of rare populations or species (Cade 1983; Rieseberg 1991; Ellstrand 1992; Levin et al. 1996; Rhymer and Simberloff 1996; Carney et al. 2000). Although the emphasis of these authors may vary, most appear to recognize the diversity of possible evolutionary outcomes of hybridization.

Outcomes that may lead to increased biodiversity include the development of extreme or novel adaptations (Rieseberg et al. 1999), and the origin of introgressive races (Abbott 1992) or hybrid species (Gallez and Gottlieb 1982; Rieseberg 1997; Ramsey and Schemske 1998). In addition, it has been suggested that plant hybrid zones may serve as focal points of biodiversity and represent centers of abundance for normally rare species (Martinsen and Whitham 1994; Whitham et al. 1994; Whitham and Maschinski 1996). Alternatively, hybridization may reduce biodiversity when species are assimilated by congeners (Martin et al. 1985; Rieseberg et al. 1989; Hubbard et al. 1992; Carney et al. 2000), or weakened by outbreeding depression (Price and Waser 1979; Templeton 1986), in which mean fitness and population size decline due to gamete wastage in the formation of unfit hybrid individuals. Consequences of hybridization with less direct effects on species diversity include the reinforcement of preexisting reproductive barriers (Butlin and Tregenza 1997; Sætre et al. 1997), the introgression of adaptive or neutral traits across semipermeable hybrid zones (Parsons et al. 1993), and the maintenance of stable hybrid zones (Barton and Hewitt 1985).

Ecological Studies, Vol. 165
C.A. Brigham, M.W. Schwartz (Eds.)
Population Viability in Plants

Several of the reviews listed above make suggestions regarding the ecological and genetic circumstances under which different outcomes of hybridization will be most likely. Unfortunately, these suggestions have rarely been studied quantitatively, making it difficult to evaluate their validity. Exceptions include theoretical studies of polyploid establishment (Roderiguez 1996a,b), the maintenance of stable hybrid zones (Barton and Hewitt 1985), extinction of rare populations through hybridization and assimilation (Huxel 1999; Wolf et al. 2001), and diploid hybrid speciation (McCarthy et al. 1995; Buerkle et al. 2000). However, these studies generally focus on the conditions that favor one particular outcome, rather than the likelihood of the many different outcomes of hybridization.

In order to predict the fate of two hybridizing groups, it is necessary to consider all of the potential outcomes of hybridization. In this chapter, we focus on the major diploid outcomes of hybridization: maintenance of stable hybrid zones, extinction of native species, hybrid speciation, and adaptive trait introgression. A review of conclusions from theoretical studies of each phenomenon is followed by a description of the results of a simulation study designed to examine the likelihood of these outcomes under a variety of parameter values.

5.1.1 Maintenance of Stable Hybrid Zones

Early zoological workers envisioned two possible outcomes of hybridization: (1) the merger of the hybridizing taxa due to extensive introgression, or (2) the cessation of hybridization due to the reinforcement of reproductive barriers (Dobzhansky 1951; Mayr 1963). By contrast, botanists tended to view hybridization as a creative process often leading to the formation of stable hybrid lineages (Anderson 1949; Stebbins 1950). Although the consequences of hybridization envisioned by botanists and zoologists were vastly different, both groups did agree, at least implicitly, that hybrid zones were ephemeral. This assumption appears to have been widely held until the mid-1970s, when a compilation of evidence from empirical hybrid zone studies was used to argue that hybrid zones could be stable for long periods (Moore 1977). Moore suggested that the observed stability was due to a fitness advantage of hybrids in intermediate habitats, often along ecotones. Termed the "bounded hybrid superiority model", this theory predicts strong correlations between habitats and genotypes. The model also assumes that hybrid zones occur along ecotones, leading to the prediction of clinal variation for species-specific characters (Endler 1977; Moore 1977).

Shortly after the appearance of Moore's hybrid superiority model (Moore 1977), Barton (1979) developed the ideas of Bazykin (1969), and provided mathematical evidence that stable hybrid zones could be maintained by a balance between dispersal of parental individuals into the zone and selection against the hybrids that were produced. In this "dynamic equilibrium" or "tension zone" model, hybrids exhibit reduced viability or fertility due to disruptions of co-adapted gene complexes. Thus, selection is independent of the environment, and correlations between habitats and genotypes are not predicted.

Both the bounded hybrid superiority and the dynamic equilibrium models predict a clinal structure for hybrid zones. However, in the early 1980s it was recognized that zones may sometimes be mosaic-like in structure due to adaptation of the parental taxa to patchily distributed habitats (Harrison 1986; Howard 1986). This observation led to the development of the "mosaic hybrid zone model", which predicts strong correlations between genotypes and habitats. Unlike the bounded hybrid superiority model, hybrids are not assumed to be more fit than the parental species in intermediate habitats. Rather, the hybrid zones are thought to be maintained by a balance between dispersal and selection in the same way as envisioned for the dynamic equilibrium model. Thus, clinal variation for diagnostic markers can be predicted on a very local spatial scale and possibly on a very broad geographical scale, but mosaic patterns of variation are predicted at intermediate geographic levels.

Recently, a new verbal model has been proposed which attempts to incorporate aspects of the three preceding models (Arnold 1997). This "evolutionary novelty" model is most similar to the mosaic model in that it suggests that both endogenous (environment-independent) and exogenous (environment-dependent) selection are critical for maintaining hybrid zones. However, it differs from the mosaic model in that not all hybrid individuals are predicted to be less fit than the parental taxa. Rather, hybrid fitness is predicted to be variable and to depend on both hybrid genotype and on habitat. Hybrids are predicted to do particularly well in intermediate or novel habitats, but some hybrid genotypes may even outperform parents in parental habitats. Like the bounded hybrid superiority and mosaic models, strong correlations between habitats and genotypes are predicted in the evolutionary novelty model. However, clinal variation is not necessarily predicted, although clines may be observed in ecotonal habitats or on broad geographical scales.

Convincing empirical evidence has been compiled in support of each of these models, which apply to both plants and animals (Barton and Hewitt 1985; Rand and Harrison 1989; Emms and Arnold 1997; Wang et al. 1997). However, the proportion of natural hybrid zones that meet the expectations of the different models remains the subject of debate (Barton and Hewitt 1989; Arnold 1997; Rieseberg 1998).

The studies described above attempt to provide an explanation for the existence of stable hybrid zones and to demonstrate how hybridization can increase diversity among individuals. However, they tell us much less about the likelihood of different evolutionary outcomes of hybridization or the conditions that favor different outcomes. In contrast, theoretical treatments of extinction through hybridization, hybrid speciation, and adaptive trait introgression discussed in the following two sections are more directly relevant to these issues.

5.1.2 Extinction Through Hybridization with Congeners

When hybrid zones are not stable, the genetically recognizable form of one or both species is likely to be extirpated, at least locally. Over the past decade, several authors have discussed some of the ecological factors that might contribute to extinction of rare plant species through hybridization (Cade 1983; Rieseberg 1991; Ellstrand and Elam 1993; Levin et al. 1996). Factors cited include the strength of reproductive barriers, the vigor and fertility of hybrids, population size, demographic stochasticity, habitat requirements, the diversity of self-incompatibility alleles, and increases in herbivory and disease that might occur if hybrids provide a "bridge" from one species to another. Population genetic models of migration among differentiated populations demonstrate the importance of migration into a population versus the strength of selection against migrants and hybrids (Crow et al. 1990; Barton 1992; Tufto 2000), and that even a small amount of introgression can readily destroy adaptations that depend on many genes (Barton 1992). This is because even a low rate of hybridization leads to a very high proportion of individuals with hybrid ancestry. Another notable finding of population genetic models is that when immigrants have higher viability than the resident genotype, but hybrids have lower viability than either parental type, the immigrant genotype is more likely to take over if selection acts after immigration but prior to mating (i.e., immigration of seeds is more likely to cause extinction of the native genotype than immigration of pollen, Crow et al. 1990).

Only one simulation study has explicitly addressed the ecological conditions influencing local extirpation through hybridization in plants (Wolf et al. 2001). This simulation was designed not only to study the phenomenon of extirpation, but also to be used as a predictive tool much like population viability analysis (PVA) modeling techniques. Rather than modeling the persistence of a genetically homogeneous population, the simulation dealt with the persistence of a unique native genotype. Toward this goal, the model includes a large number of specific, measurable parameters (Table 5.1) and explicitly models the life cycle of annual plants.* The genetic makeup of each generation was determined by the mating and competitive success of each class in the previous generation. The first generation of the simulation represented the first generation of hybridization after the invading species was introduced into sympatry with the native species. Three genotypic classes were recognized:

* We refer to this as the life-cycle based model.

native, invader, and hybrid. The class of each plant was determined by parentage, such that all inter-class crosses produced hybrids, and all offspring of hybrid parents produced hybrid offspring. This mimics a system in which a very large number of loci determines hybrid fitness, and gives a reproductive advantage to the hybrid class, since some offspring of parental individuals will be hybrids, but hybrids will produce no parental offspring. Prezygotic isolation was provided by pollen-tube competition (conspecific pollen grew faster, on average, than foreign pollen) and postzygotic isolation was introduced both through reduced hybrid pollen and seed production and through reduced hybrid competitive ability. The population was assumed to be small enough that seeds and pollen were randomly dispersed.

To identify and prioritize the parameters (Table 5.1) that should be estimated accurately in hybridizing plant populations whose status is of concern, a series of sensitivity analyses was carried out (Wolf et al. 2001). Overall, competitive ability, initial frequency, selfing rate, and ovule production of the native species had the strongest influence on its rate and probability of extinction. Parameter values describing the native species were always ranked higher than the same parameters for the hybrids, because the native species could be extirpated by either hybrids or invaders. Because of this, and because fertility was divided among multiple parameters (pollen and ovule production and pollen-tube growth rate), hybrid fertility was not among the factors

Table 5.1. Parameters employed by the life-cycle-based simulations (Wolf et al. 2001), and values based on data from crop × wild hybridization in sunflowers (*Helianthus annuus*; Wolf et al. 2001), which were used in simulations for Fig. 5.1

Parameter	Native	Hybrid	Invader
Carrying capacity	–100–	–100–	$n^{a,b}$
Initial number of plants	100	0	$n^{a,b}$
Selfing rate	0	0	0
Amount of pollen produced per plant	4143	p^{a}	1223
Number of pollen grains per stigma	10	10	10
Pollen-tube growth on conspecific stigmas	1	1	1
Pollen-tube growth on foreign stigmas	1	1	1
Number of ovules per plant	21	o^{a}	0^{b}
Number of seeds per patch	10	10	10
Seedling competitive ability	1	0.8	–[b]

[a] Variable in Fig. 5.1

[b] In the simulations (Fig. 5.1) using the crop × wild model (Wolf et al. 2001), it was assumed that a parapatric foreign species donated pollen (but not seeds) each generation to the native population, and that the foreign species maintained a stable population size. This was so that competition between the native and invading species would not influence the outcome. The crop × wild model differs from the majority of the life-cycle-based simulations (Wolf et al. 2001), which assume that the hybridizing species are sympatric and contribute both pollen and seeds to the hybridizing population

that best predicted extinction. Finally, it is important to note that analyses based on different parameter sets did not rank the parameters in the same order, and the sensitivity of each parameter often differed dramatically among parameter sets. For instance, when there was no reproductive isolation through pollen-tube growth, native selfing rate was highly ranked (1.5). Whereas, the same parameter was ranked eighth out of ten when reproductive isolation was strong. This demonstrates how all parameters interact to influence the probability and speed of extinction under at least some conditions, and how it is necessary to perform sensitivity analyses on each data set to determine what environmental changes will produce the largest increase in population viability.

When only a single habitat containing two species was modeled, a stable hybrid zone was never achieved – either one of the parental species or the hybrids overtook the habitat, regardless of hybridization. A stable hybrid zone persisted only when two distinct habitats were included in the simulation, and the competitive ability of each species was at least three times higher in its own habitat than in the foreign habitat. Thus, hybridization is less likely to result in extinction if the interacting species have divergent habitat requirements, and hybrids do not compete well in either parental habitat. In fact, many stable hybrid zones exist along habitat boundaries (e.g., Freeman et al. 1999; Kentner and Mesler 2000). According to Barton and Hewitt (Barton 1979; Barton and Hewitt 1985, 1989), hybrid zones can be maintained in the absence of habitat differences, but moving hybrid zones will stall along habitat boundaries. Any factor that can stall the movement of a hybrid zone at the perimeter of a species' range will decrease the likelihood of extinction. Thus, regardless of the underlying reasons for differences in habitats occupied by hybridizing species, divergent habitats are likely to reduce the likelihood of extinction through hybridization.

In one case in which parameter values were obtained from data on two naturally hybridizing species, both of which were native, hybridization actually reversed the outcome of interaction between two species due to an asymmetry in reproductive isolation. When hybridization between the wild sunflowers *Helianthus annuus* and *H. petiolaris* was modeled, *H. annuus* generally went extinct because it was more often the seed parent of F1 hybrids, and thus produced fewer conspecific offspring than *H. petiolaris* did. However, because *H. annuus* produced more seeds in total, it took over the population in the absence of hybridization. This result demonstrates how studies that ignore the possibility of hybridization, or make casual analyses of hybridization, could lead to erroneous conclusions concerning extinction risk.

Probably the most important finding from the simulation study (Wolf et al. 2001) was that extinction of the pure native genotype can occur quickly – in less than five generations even when the invading species and hybrids lack a competitive advantage. The native species can go extinct in less than ten generations even when only pollen from a nearby congener invades the popula-

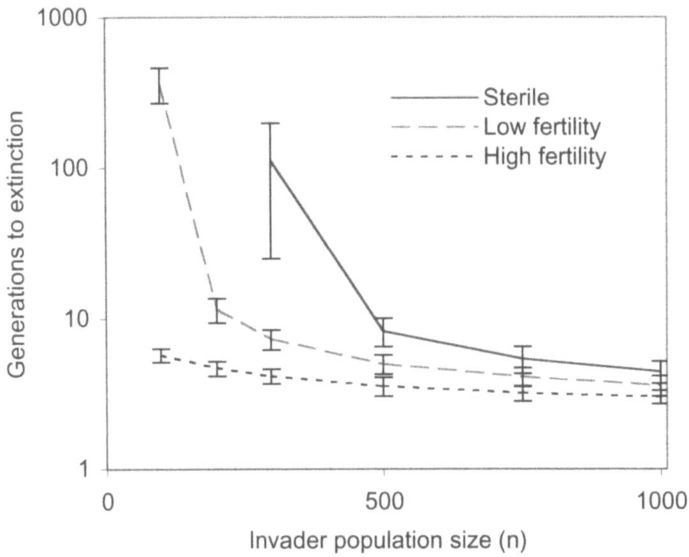

Fig. 5.1. The mean (±SD) number of generations to extinction of the native species through hybridization in the life-cycle-based simulations. The native species never went extinct in the absence of hybridization because only pollen migrated into the population (in simulations allowing seed migration, extinction was even more rapid). The simulations used parameter values obtained from hybridizing cultivated and wild sunflowers (Table 5.1), but the results are generalizable. Sterile hybrids: hybrid pollen (p) and ovule (o) production=0. Low hybrid fertility: p=1050 and o=4. High hybrid fertility: p=4143 and o=21

tion, and hybrids are entirely sterile (Fig. 5.1). Thus, hybridization can be viewed as perhaps the most rapidly acting genetic threat to endangered species. In contrast, extinction through mutational meltdown is likely to occur on a timescale of 100 generations or more (Lynch et al. 1995). Thus, when a sensitive taxon comes into contact with a more common congener, or when any species comes into contact with an invasive congener, risk assessment and prescriptive action must be rapid.

5.1.3 Hybrid Speciation and Adaptive Trait Introgression

Hybrid speciation occurs when novel forms generated by hybridization attain isolation from the parental species and become genetically stabilized. Reproductive isolation is straightforward for hybrids with increased ploidy because hybrids between polyploids and their diploid parents are generally sterile, and many well-described polyploid species are the result of hybridization (reviewed in Grant 1981; Soltis and Soltis 1993). In contrast, homoploid hybrid speciation presents a special problem to evolutionary biology.

Although both forms of hybrid speciation must be initiated while the parental species are in sympatry or parapatry, the process by which homoploid hybrids become reproductively isolated from their parents is less obvious. Nonetheless, the existence of a number of homoploid hybrid species has been confirmed with molecular markers (Rieseberg 1997; Wolfe et al. 1998; Wang et al. 2001).

Two mechanisms have been proposed by which a hybrid form may become stabilized (Grant 1958). The first mechanism is referred to as recombinational speciation and requires that the parental species differ by two or more chromosomal rearrangements (Stebbins 1957; Grant 1958). The chromosomal differences reduce fertility in F1 hybrids and in later generation hybrids that are heterozygous for the rearrangements. However, backcrossing or interbreeding among early-generation hybrids may give rise to some novel, chromosomally balanced (i.e., homozygous) genotypes with restored fertility (underdominant loci could function similarly). Furthermore, hybrids between this novel cytotype and parental species will have reduced fertility. Thus, if the new cytotype can persist as a lineage, it may constitute a novel hybrid species. The second mechanism that may lead to the stabilization of hybrid lineages is "the segregation of a new type isolated by external barriers" (p. 243, Grant 1981). Hybrid species often occupy habitats that are very different from those occupied by their parental species (Kerner von Marilaun 1894–1895; Anderson 1948; Grant 1981; Templeton 1981; Rieseberg 1997; Wang et al. 2001). This is probably because a novel hybrid lineage is more likely to survive if it experiences little competition from its parents, and if divergence in habitat preferences results in some spatial isolation and a reduction of hybridization with the parents. It seems likely that both mechanisms have contributed to the origin of most of the documented homoploid hybrid species (Rieseberg 1997).

Two simulation studies have examined the process by which hybrid species can arise (McCarthy et al. 1995; Buerkle et al. 2000). McCarthy et al. (1995) developed a spatially explicit, individual-based simulation in which two species with different chromosome arrangements were allowed to hybridize. Two outcomes were possible: either a stable hybrid zone was maintained, or a novel hybrid genotype spread throughout the simulated space, supplanting both parental species. The latter outcome was considered hybrid speciation, although it would be recognized as an extinction event in the life-cycle-based model (Wolf et al. 2001). This is known to have occurred in California *Raphanus*, when cultivated radish, *R. sativus*, and jointed charlock, *R. raphanistrum* merged genetically (N. C. Ellstrand, pers. comm.). However, in the documented cases of homoploid hybrid speciation, parental species have continued to coexist with the derived hybrid species (reviewed in Rieseberg 1997). The conditions favoring hybrid speciation and the maintenance of parental species are likely to differ from those that lead to novel lineages through the merger of species' genomes. Thus, Buerkle et al. (2000) developed a computer simulation model to investigate the conditions necessary for the formation and independent evolution of a novel hybrid species and the concurrent maintenance of parental species. This simulation included the chromosomal factors of the McCarthy et al. (1995) simulation, a novel habitat in which hybrids had higher viability than members of the parental species, as well as the possibility of spatial isolation between the hybrid habitat and that of each parental species (Fig. 5.2).

The most frequent outcome of hybridization in this model* was adaptive trait introgression (Fig. 5.3), in which one of the parental species was able to move into the open niche by acquiring advantageous habitat-preference alleles from the other parent. Hybrid speciation was less frequent overall (Fig. 5.3), but did occur in a sizable proportion of the simulations, especially when the sterility barrier isolating the parental species was weak, and when habitat selection strongly favored hybrids in the open niche (i.e., hybrid habitat). By contrast, simulation replicates in which no introgression occurred were rare under the conditions simulated. This is likely due to the presence of an open habitat, in which recombinant genotypes at habitat-preference loci had higher fitness.

* Referred to as the three-habitat multi-locus simulation.

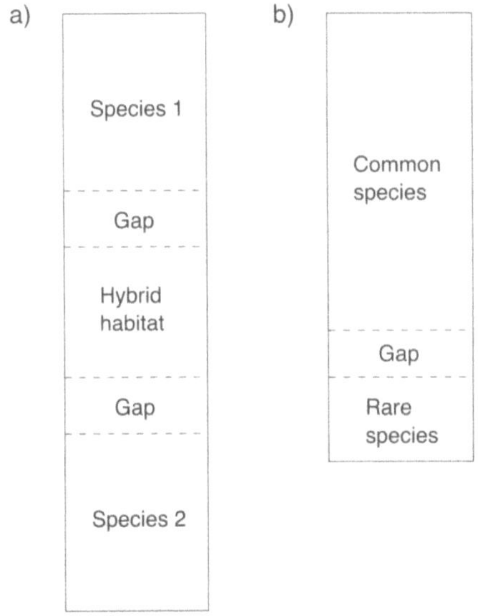

Fig. 5.2a, b. Spatial grids used for the multilocus simulations. The distance of pollen and seed dispersal decline according to geometric distributions with parameters 0.4 and 0.6, respectively (details in Buerkle et al. 2000). To minimize artifacts with respect to dispersal (e.g., edge effects), the grid was assumed to be an elongated cylinder, such that the long edges of the grid were connected and ends of the grid were distant from the hybrid zone. As a result, the grid simulated a transect through a more extensive hybrid zone. Each cell on the grid had a carrying capacity of five individuals. **a** In the three-habitat multilocus simulations of hybrid speciation (Buerkle et al. 2000), the grid consisted of 20 columns by 80 rows. Parental habitats and the distance between them were fixed in size, while the size of the hybrid habitat could be varied to accommodate different sized gaps separating the habitats. **b** In the two-habitat multilocus simulations presented in this chapter, the grid was 20 columns by 60 rows, and gap size ranged from zero to two rows

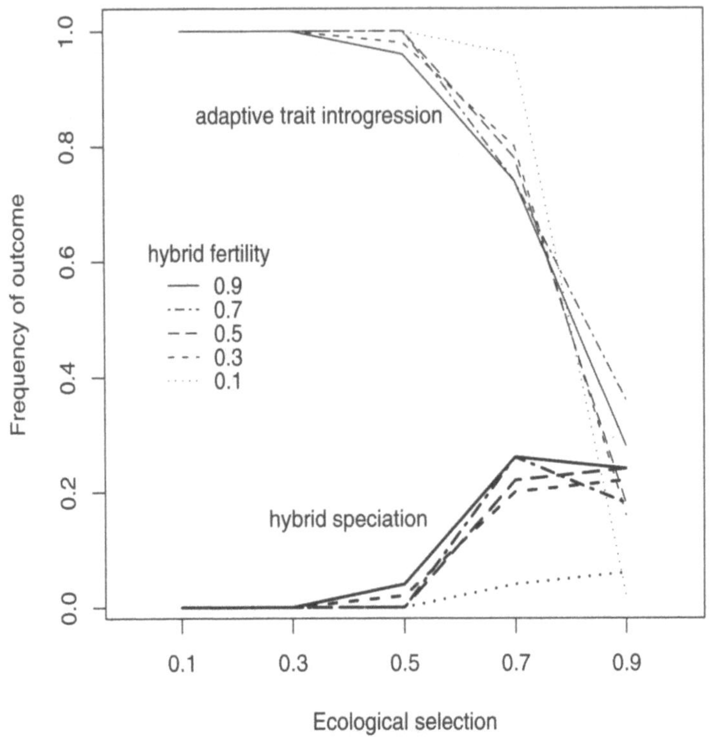

Fig. 5.3. Frequency of three-habitat multi-locus simulations ending in adaptive trait introgression (*light lines*) or hybrid speciation (*heavy lines*). Fifty replicates were run at each combination of hybrid fertility (r) and habitat selection (s) parameters. Geographic isolation between habitats was held constant at eight grid rows on each side of the hybrid habitat. At s=0.9 some replicates ended after 4,000 generations without adaptive trait introgression or hybrid speciation. (Buerkle et al. 2000)

The conditions that favor the origin of new homoploid species are very different from those that facilitate its independent evolution. In particular, observations of gene flow in the simulations showed that high F1 fertility favors the origin of new hybrid species, but also translates into a very weak barrier between the newly derived hybrid lineage and its parents. As a result, the genetic integrity of the new hybrid lineage cannot be maintained in the presence of gene flow with its parents. Thus, for hybrid lineages to evolve independently in sympatry with parental populations, the initial barrier between the parental species must be strong, a requirement that greatly reduces the likelihood of this mode of speciation. Alternatively, fairly strong isolation between the hybrid species and its parents can be achieved via strong habitat selection and significant spatial separation. This simulation result is consistent with earlier predictions that successful hybrid speciation will be most likely following hybrid founder events, in which several hybrids give rise to a new population that is ecologically and spatially isolated from the parental species (Charlesworth 1995; Rieseberg 1997). These predictions also accord well with observations from well-studied hybrid species in nature, all of which appear to be isolated ecologically from their parents (reviewed in Rieseberg 1997). Close to half are spatially isolated as well (Rieseberg 1997).

Finally, it should be emphasized that, although adaptive trait introgression is not equivalent to speciation, it may lead to the establishment of novel forms when open habitats are available for colonization. The simulations show that adaptive trait introgression is a more likely outcome of hybridization than hybrid speciation, and in taxa than are prone to hybridization, it may play a significant role in shaping their evolutionary trajectories (Abbott 1992; Rieseberg and Wendel 1993; Arnold 1997; Dowling and Secor 1997).

5.2 The Relative Frequency Of Extinction, Homoploid Speciation and Other Consequences of Hybridization

Although some theoretical work has investigated the conditions under which several outcomes of hybridization may occur, no study has attempted to estimate the relative likelihood of each outcome. Thus, we have modified the simulation model of Buerkle et al. (2000) to investigate the relative frequency of three outcomes: (1) maintenance of a stable hybrid zone, (2) extinction of the rare species, and (3) homoploid hybrid speciation. The primary difference between this and the earlier model (Buerkle et al. 2000) is that the empty habitat for hybrids was removed and initial populations were separated by less space (Fig. 5.2).* Both models utilized a two-locus model for habitat-specific ecological selection, and a two-locus model (equivalent to two chromosomal rearrangements) for fertility selection. Therefore, we refer to this as the two-habitat multi-locus model.

5.2.1 Methods

We considered hybridization between species whose local populations differed in size, and consequently occupied different proportions of the spatial grid. In our case, the local populations began with 4,800 individuals of the common species (or 4,600 when a small spatial gap was modeled) and 1,200 individuals of the rare species.

The simulation continued for 1,000 generations or until one of the following conditions occurred. It was again possible for a homoploid hybrid species

* The programs for the simulation models described in this chapter were written in the C computer language. Source code is available from the authors upon request.

to arise and we monitored for 500 generations the fate of a population of individuals with a novel, homozygous fertility type, once there were at least 600 plants (half the size of the "rare" species). Alternatively, the common species could expand and lead to the extinction of all individuals with the fertility type of the rare species. In these cases we ended the simulations once the fertility type of the common species represented 95 % of the individuals in the habitat formerly occupied by the rare species. By this time, the remaining 5 % of individuals all had a hybrid genotype at fertility loci, and individuals with the pure fertility type of the rare species were extinct.

Thirty replicates were performed for each combination of: (a) strength of selection on habitat differences between the two species (*s*, 0.1, 0.3, 0.5, 0.7, 0.9); (b) fertility of F1 hybrids (*r*, 0.1, 0.3 0.5 0.7 0.9); and (c) the amount of spatial isolation between the two species (0 or 2 rows on the spatial grid).

5.2.2 Results

The development of a stable hybrid zone was by far the most common outcome of these two-habitat simulations (84 %; Fig. 5.4). Only when the fertility barrier was relatively weak was the rare population replaced by the common one, almost always through adaptive trait introgression. Stronger habitat selection reduced the likelihood of extinction, but its effect was quite small. In fact, even for the strongest habitat selection coefficient (s=0.9), extinction occurred in the majority of replicates when hybrid fertility was high (r=0.9) and there was no spatial isolation. Likewise, spatial isolation reduced the likelihood of extinction. However, substantial spatial isolation was not sufficient to prevent frequent extinction of the rare population when F1 fertility was high (r=0.9) and habitat selection weak (s=0.1). Not surprisingly, if the common species performed even moderately better than the rare species in the latter's native habitat, extinction of the rare population invariably resulted.

Of the 1,500 two-habitat simulations, only 31 (2.1 %) resulted in hybrid speciation (Fig. 5.4). Hybrid speciation occurred under two different sets of conditions. In the absence of spatial isolation, hybrid speciation occurred when F1 fertility (r=0.9) was high and habitat selection strong (s=0.5–0.9). With substantial spatial isolation, hybrid speciation was favored by high F1 fertility (r=0.9), but weak habitat selection (s=0.1–0.3). It is noteworthy that the origin of new hybrid species was always accompanied by the extinction of the rare parental species. In addition, the population size of the derived hybrid species was highly variable and in most cases (24/31) the hybrid species failed to maintain its population size minimally at half that of the size of the initial population of the rare species. If the fate of these populations had been tracked for more time, it is likely that many of them would have eventually been lost and replaced by the fertility type of the common parental species. In other words, replicates that ended in hybrid speciation

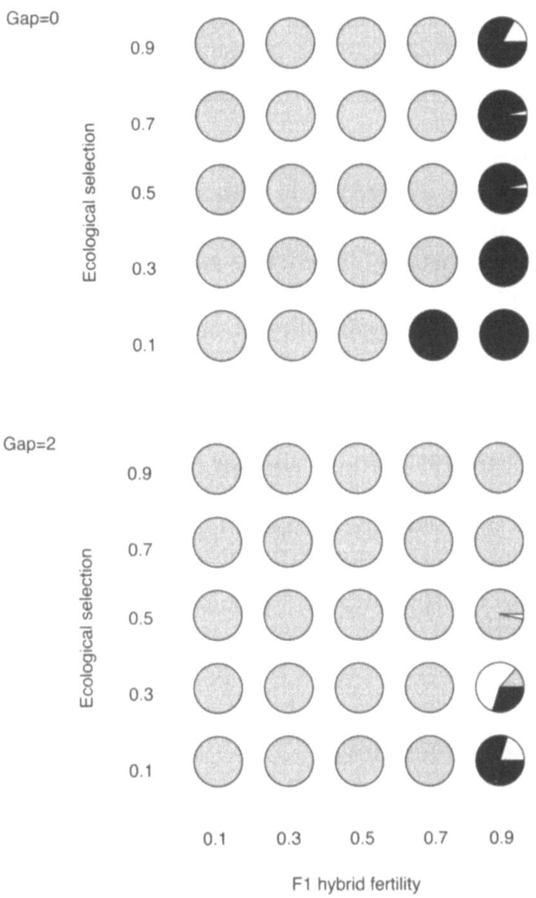

Fig. 5.4. The area in each graph represents the frequency of two-habitat multi-locus simulations resulting in: stable hybrid zone (*gray*), extinction of the rare species and replacement by the common species (*black*), or hybrid speciation (*white*). Results are shown for 30 replicates at each combination of hybrid fertility and habitat selection parameters. Replicates in which the rare species was replaced by the common species were ones in which the fertility type of the common species was able to spread into the habitat of the rare species, and always involved adaptive trait introgression. Instances in which hybrid species arose always led to the extinction of the rare species. In addition, hybrid species lineages were unstable and were unlikely to persist

probably occupy a temporary and intermediate state that leads to the predominance of the common parental species (as do the replicates labeled extinction).

5.2.3 Discussion

The primary conclusion from these simulation experiments is that even a fairly weak sterility barrier greatly reduces the risk of extinction (Fig. 5.4). For example, a 30 % reduction in hybrid fertility essentially eliminates the possibility of extinction due to hybridization, except when habitat selection is weak (s=0.1) and there is no spatial isolation. Habitat selection appears to play a more minor role in reducing extinction risk under the conditions considered in this model. The reason for this involves adaptive trait introgression. That is, under conditions that lead to extinction, the common species could move into the rare environment simply by acquiring, through introgressive hybridization, the habitat preference alleles of the rare species. Spatial isolation can greatly reduce the likelihood of extinction, in part because the frequency of hybridization itself is reduced.

Another important conclusion from these two-habitat simulations concerns the rarity of hybrid speciation as an outcome in these simulation experiments when compared to the three-habitat simulations described earlier (Figs. 5.3, 5.4). This result suggests that the presence of an open niche and greater spatial isolation are critical to successful hybrid speciation. This is not a novel conclusion. The notion that the availability of open habitat is an important prerequisite for hybrid speciation was first suggested by Anton Kerner more than a century ago (Kerner von Marilaun 1894–1895) and has been stressed by modern students of hybridization as well (Grant 1981; Templeton 1981; Arnold 1997; Rieseberg 1997). In the absence of a niche for the hybrids, a new hybrid species simply replaces the rare taxon, and the total number of taxa does not change. Moreover, the new hybrid taxon may be even more susceptible to gene flow and genetic swamping than its rare parent. Thus, in these circumstances, hybrid speciation is likely to be a temporary way-station on the road to extinction.

5.3 Frequency of Outcomes: Overview

In a comprehensive review of introgression in plants, Rieseberg and Wendel (1993) asked whether introgressive hybridization had led to an overall increase or decrease in global biodiversity. The question could not be confidently answered then or now. Nonetheless, with the steady accumulation of case studies in the population and systematic biology literature, combined with recent theoretical work summarized here, many of the possible outcomes of hybridization have been identified and we can begin to rank their importance. In addition, considerable progress has been made in terms of evaluating the relative influence of various ecological and evolutionary parameters on individual outcomes.

Empirical (Abbott 1992; Rieseberg et al. 1995; Rieseberg et al. 1996; Arnold 1997) and theoretical (McCarthy et al. 1995; Buerkle et al. 2000) studies of hybrid speciation indicate that the ecological and genetic conditions favoring hybrid speciation are quite stringent, suggesting that it is a rare speciation mode (Figs. 5.3, 5.4). In particular, it appears that the presence of an open niche in which hybrids are favored is essential to this mode of speciation. The frequency of unoccupied, hybrid-friendly niches in nature is unclear and should be the focus of future study. However, recent studies have shown that the expression of extreme or transgressive phenotypes in segregating hybrids is the rule rather than the exception (Rieseberg et al. 1999), providing a simple mechanism for adaptation of hybrids to niches that differ from those preferred by their parents.

Our simulations suggest that a much more frequent outcome of hybridization is adaptive trait introgression (Figs. 5.3, 5.4). In our model, adaptive trait introgression occurs when alleles underlying habitat preference move from one fertility group to another, allowing a species (a fertility group) to adapt to a different habitat (loci contributing to fertility and ecological performance were unlinked). When there were three habitats (Buerkle et al. 2000), adaptive trait introgression allowed colonization of the novel, empty habitat. However, when there were only two habitats, adaptive trait introgression allowed the common species to move into the habitat of the rare species and led to the extirpation of the fertility type of the rare species. In the two-habitat simulations, this outcome was predominant when hybrid fertility was high, in other words, when intrinsic barriers to introgression were weak. Consequently, in species pairs with weak fertility isolation, adaptive trait introgression is likely to be a pervasive and highly effective means by which invasive species can enter new habitats and displace native taxa.

Until recently, adaptive trait introgression was considered to be unlikely, at least by zoologists (p. 300 in Dobzhansky 1951; p. 133 in Mayr 1963). Part of the reluctance to accept introgressive hybridization as an important evolutionary mechanism stems from the biological species concept, which views species as reproductively isolated entities. Thus, a strict interpretation of the biological species concept would deny the existence of interspecific gene flow, as any taxa that exchange genes would necessarily be defined as semispecies or some lower rank. Often entwined with the biological species concept has been the view that species' genomes are highly coadapted. Thus, alien alleles were considered likely to disrupt internally balanced combinations of genes, leading to reduced fitness (Mayr 1963).

More recently, however, it has become apparent that species' genomes are porous with respect to gene flow, and that hybrid zones represent semi-permeable membranes that restrict the flow of alleles that are negatively selected in hybrids (i.e., those that contribute to reproductive isolation), but facilitate the interspecific flow of neutral or advantageous alleles (Barton and Hewitt 1985; Harrison 1990). Under this paradigm, which has long been held by botanists (Anderson 1949; Stebbins 1957; Heiser 1973; Grant 1981), the genetic architecture of species' barriers is the critical predictor of adaptive trait introgression. This is because alleles that contribute to an adaptation must integrate into a new genetic background before the chromosomal block they are associated with is eliminated by selection (Barton and Hewitt 1985). As a result, if many genes contribute to reduced hybrid fitness, then much of the genome may be resistant to introgression due to linkage. Another population genetic concern relates to the movement of polygenic adaptations across hybrid zones, because polygenes contributing to the trait will become disassociated by recombination. Thus, only alleles that are individually favorable are likely to cross the hybrid zone. Simple predictions that follow include: (1) adaptive trait introgression is most likely among recently evolved species pairs with weak reproductive isolation, (2) simply inherited traits are more likely to successfully transgress species barriers, and (3) chromosome number should be positively correlated with the likelihood of successful introgression.

The final outcome of hybridization that was considered here, a stable hybrid zone, is the most common outcome of the two-habitat comparative simulation (Fig. 5.4) and is likely to be the most common outcome in nature as well. The genetic simulations demonstrate that even with fairly modest levels of hybrid sterility, habitat selection and/or spatial isolation, two species can coexist in the presence of hybridization and gene flow. These predictions are largely confirmed by observations in nature; many hybrid zones are stable and have been in existence for hundreds or thousands of generations (Barton and Hewitt 1985; Harrison 1990; Graham et al. 1995). As discussed earlier, the factors that maintain these stable hybrid zones are less well understood and remain the subject of considerable controversy (Moore 1977; Barton and Hewitt 1985; Harrison 1986; Arnold 1997).

One surprising result of comparing our multi-locus genetic simulations to the life-cycle-based simulations (Wolf et al. 2001), is that they differ in their ranking of the relative importance of habitat selection (i.e., competitive ability of the native species) and hybrid fertility. The genetic model indicates that hybrid fertility is the most critical factor reducing the likelihood of extinction through hybridization, whereas the life-cycle model places a premium on habitat selection. These divergent conclusions are probably the result of differences in the assumptions underlying each of the models.

An important difference between the models is the spatial structure they employed. The life-cycle model assumes that individuals are randomly distributed in space, and therefore pollen distribution between species is also random. In contrast, the spatial grid in the multi-locus model began with each fertility group spatially clustered, and therefore the majority of pollen was distributed to members of the same fertility group. Clustering, combined with fertility barriers, creates positive frequency-dependent selection in each habitat (selection against migrants bearing foreign fertility alleles). However, when mixing is random, habitat differentiation is the only force selecting against migrants from the other habitat. This difference in the spatial configuration is likely to be the primary cause of the differing ranks of habitat selection and hybrid fertility. In actuality, hybrid zones of both types exist, and there is evidence that habitat differentiation is of key importance when the species are closely mixed (Kentner and Mesler 2000), and fertility barriers are important when species are spatially separated (Nichols and Hewitt 1986)

A second factor that is likely to account for differences in the results of the models involves the definition of species. In the multi-locus genetic model, species were defined based on their genotype at fertility loci. That is, a species contains all individuals that have the same homozygous genotype at fertility loci, even if those individuals are of hybrid ancestry and are recombinant at other loci. Fertility selection is a direct consequence of the fertility genotype and, not surprisingly, it is more effective at maintaining a pure population of a given fertility type than is indirect selection from ecological performance. In contrast, the life-cycle model (Wolf et al. 2001) defines species on the basis of racial purity; a single hybrid ancestor disqualifies individuals from membership in either the rare or common taxon.

Both definitions of species are overly simplistic and represent endpoints on the spectrum of possible definitions. Few biologists would argue that species should be defined on the basis of two loci underlying hybrid fertility (as modeled in the multi-locus case), but this model does illustrate the significant variation in the permeability of different portions of the genome. Likewise, it is difficult to argue that any individual with hybrid ancestry is impure and unworthy of conservation. Backcross individuals are often indistinguishable from the parental species, except through detailed molecular marker analysis (Linder et al. 1998).

Unless gene flow into a population is substantial and overwhelms the effects of local selection, components of a foreign species' genome may assimilate into the genetic background of the native species without substantially changing the morphological and ecological attributes of the local population. In terms of conservation, it could be argued that while this is not an optimal outcome, diversity is preserved. In fact, a low level of gene flow may be beneficial to a genetically depauperate population that suffers from fixation of deleterious alleles or from low diversity of self-incompatibility alleles. However, introgression of this type is a slippery slope, because no intrinsic barriers to reproduction between populations exist. The persistence of the rare taxon would depend primarily on continued, divergent selection in the habitats occupied or cessation of gene flow. Otherwise, unique attributes are likely to be lost due to the replacement of native alleles through hybridization. Only those portions of the genome that continue to be protected as the result of divergent selection (direct selection on loci and closely linked regions) will retain unique alleles in the face of hybridization (see discussion of introgression). In this context it is clear that the persistence of a rare taxon will be compromised by hybridization. However, we presently have no criteria for assessing what level of introgression should be tolerated. In part this is because we are just beginning to characterize the genetic architecture of adaptive differences and reproductive isolation between species (Coyne and Orr 1998; Kim and Rieseberg 1999). In addition, it is likely that genetic architectures vary substantially among species pairs.

5.4 Species Conservation Among Hybridizing Taxa

Although much of the information presented in the preceding sections is principally theoretical, it does have very practical implications for the conservation and management of species. Both the simulation and empirical data clearly indicate that hybridization can, in some instances, be a creative evolutionary force. Thus, hybridization should not necessarily be viewed as a negative feature in the conservation of species, particularly when hybridization does not appear to result from human-mediated translocations or disturbance. On the other hand, the size of the parameter space under which hybridization is likely to contribute to species extinction does appear to be much larger than that for the creative outcomes considered herein. For this reason, evidence of human-mediated hybridization involving rare or endangered taxa should be viewed with caution and studied carefully to determine likely outcomes. In the following two sections, we discuss the implications of our simulation models for the conservation of species threatened by hybridization.

5.4.1 Anthropogenic Disturbance and Hybridization

Two important factors to consider in predicting the viability of a hybridizing population is the antiquity of the association between the species involved and the insularity of the habitat. Species that have naturally hybridized and coexisted for hundreds or thousands of years are likely to form a stable hybrid zone. As discussed above, these species often have different habitat requirements and hybridize along habitat boundaries or barriers to dispersal. However, hybrid zones involving non-native species that were introduced by humans, or that have rapidly expanding ranges due to human-induced habitat disturbance, are likely to be a threat to the persistence of native species.

Further, species introduced onto islands and into insular habitats are most likely to encounter species without strong prezygotic barriers to reproduction (Carlquist 1974; Ganders and Nagata 1984) and with restricted ranges (Simberloff 2000). Often the exotic species or hybrids can outperform the native species in its own habitat (Callaway and Josselyn 1992; Carney et al. 2000). In addition, the native species may be threatened even if the invading species and hybrids are competitively inferior due to the absorbing nature of the hybrid state (Wolf et al. 2001). Because extinction of the native form may occur quickly, cases of hybridization between sensitive native species and exotic species may require rapid assessment and action (Abernethy 1994; Wolf et al. 2001).

Many examples of plants whose populations are threatened by hybridization appear to come from island taxa and other isolated biota. The Haleakala greensword (*Argyroxiphium virescens*) was endemic to the upper slopes of the Haleakala volcano on Maui. It is thought that this species became extinct in part due to hybridization with the more common Haleakala silversword (*A. sandwichense*), which now occupies the former range of *A. virescens*. Two hybrid individuals living in the ancestral range contain the only known remnants of the *A. virescens* genome (Carr and Medeiros 1998). The Haleakala ecosystem has been highly damaged by the rooting and grazing of feral pigs, which may have been the primary cause of the decline of *A. virescens* (Stone 1985).

The Catalina Island mountain mahogany (*Cercocarpus traskiae*) is on the verge of genetic assimilation by a more common congener. The vegetation of Santa Catalina Island is highly degraded due to grazing by feral goats, sheep and pigs (Thorne 1967). Hybridization with the more common *C. betuloides* has further decreased the potential for a viable population of *C. traskiae*. Of the adult trees remaining in 1995, there were only seven pure *C. traskiae* and three hybrids. However, the population may be on the way to recovery. After excluding herbivores from the area, there are now over 70 seedlings, most of them pure *C. traskiae* (Rieseberg and Gerber 1995). Additionally, the creation of new *C. traskiae* populations that are remote from *C. betuloides* should promote the long-term survival of this species (Rieseberg and Gerber 1995). Other rare island endemics at risk from hybridization include several Canary Island species in the genus *Argyranthemum* (Brochmann 1984; Bramwell 1990; Levin 2000) and *Arbutus canariensis* (Pascual et al. 1993).

Although rare endemic species are particularly susceptible to genetic swamping from hybridization (Rieseberg and Gerber 1995; Levin 2000), even abundant, mainland plants can be at risk. The invading cordgrass, *Spartina alternifolia*, is still less abundant than the common, native *S. foliosa* in the San Francisco Bay. However, *S. alternifolia* produces 21-fold more viable pollen than the native species (Anttila et al. 1998) and has a higher clonal growth rate than *S. foliosa* (Callaway and Josselyn 1992). Likewise, the hybrids are strong and vigorous. Simulations of hybridization between these species predict that *S. foliosa* will be locally extirpated in 3–20 generations (Wolf et al. 2001).

5.4.2 Conclusion: Implications for Population Viability Analysis and Management

The simulation models illustrate that, in addition to the ecological and genetic factors that are typically considered in PVA, populations may be significantly reduced in number or genetically modified as a result of hybridization. The models also show that for those cases in which geographic contact between species does not result in a stable hybrid zone, a remarkably small number of generations may be sufficient for hybridization to lead to significant population declines or extinction. These results give rise to some suggestions for how the effects of hybridization might be incorporated into PVA and species conservation plans.

Detailed simulation models such as the life-cycle model (Wolf et al. 2001) could be used to model population viability based on a variety of biological parameters. Alternatively, a matrix model of a threatened parental population could incorporate hybridization based on its effects on transition probabilities (Menges 2000; for a review of matrix modeling methods see Chap. 6, this vol.). Hybridization will reduce the production of seeds with genotypes of the threatened parental species. Different rates of reduction could be modeled based on estimates of native and hybrid seed production made in settings with different numbers of native, hybrid, and invading plants as pollen donors. In addition, competition with hybrid or invading plants could reduce recruitment to seedling and reproductive stages.

Some special considerations apply to PVAs involving hybridizing populations. In unstable hybrid zones, the growth rate of the hybrid population is likely to be non-linear and may accelerate as the frequency of hybrids or the invading species increases (Wolf et al. 2001). Thus, in those cases that are of the most concern to conservation (i.e., hybrid zones that have arisen recently), parameter estimates and simple matrix models will not be accurate or useful for long-term predictions. To address this complexity, it may be possible to develop separate matrices for each of the parental species and for a hybrid class, and to incorporate interaction of the matrices into a projection model. Overall, if the growth rate of the hybrid or invading populations is greater than one, especially if the growth rate is greater than that of the native species, hybridization is likely to be a significant threat, even if the native species has a growth rate greater than one during the time of study.

Because of the rapidity with which hybridization may overcome a native population, it may be desirable to remove the invading species (and possibly hybrids) as soon as the threat is detected, while studies of population viability are carried out in less sensitive populations. Certainly it is critical to determine whether hybridizing populations can remain stable over several generations, both with respect to size and space occupied. Appropriate management actions may include destruction of the invading species and hybrids, habitat improvement, or transplantation of the native species to new, more protected locations (Rieseberg and Gerber 1995). If the hybridizing populations coexist in a variety of ecological contexts and yield a range of parameter estimates, matrix models may be useful in determining the most promising management strategy.

The genetic models raise additional complications for the management of populations threatened by hybridization. In stable hybrid zones, the integrity of species may be maintained over many generations. In contrast, in the case of invasions or novel hybridization, significant introgression may occur rapidly, possibly without obvious phenotypic changes. Thus, as discussed above, it is probably advisable to manage populations in a way such as to minimize hybridization and introgression, at least until the stability of hybrid zones and parental populations can be ascertained. Otherwise, in the case of recent or novel contact between species and in the absence of divergent selection, hybridization will likely lead to significant erosion of differences between species and to the loss of unique features of the threatened taxon.

A second implication of the genetic models is that in most cases recognizable hybrids should be treated as lost from the threatened population (as was done in the life-cycle model, Wolf et al. 2001). In extremely small populations, hybrids may be a significant repository of the genome of the rare taxon (e.g., the Haleakala greensword; Carr and Medeiros 1998). But in less extreme situations, hybrids may be a threat to the persistence of the rare taxon, because their mixed, intermediate genetic composition makes them conduits for further genetic introgression between parental species.

Acknowledgments. We are grateful to Tim Bell, Marlin Bowles, Naoki Takebayashi, and an anonymous reviewer for carefully reading earlier versions of the manuscript and for making very helpful suggestions for improvements. L.H.R.'s work on hybridization has been funded by the National Science Foundation and the United States Department of Agriculture.

References

Abbott RJ (1992) Plant invasions, interspecific hybridization and the evolution of new plant taxa. Trends Ecol Evol 7:401–404
Abernethy K (1994) The establishment of a hybrid zone between red and sika-deer (genus *Cervus*). Mol Ecol 3:551–562
Anderson E (1948) Hybridization of the habitat. Evolution 2:1–9
Anderson E (1949) Introgressive hybridization. John Wiley, New York
Anttila CK, Daehler CC, Rank EN, Strong DR (1998) Greater male fitness of a rare invader (*Spartina alterniflora*, Poaceae) threatens a common native (*Spartina foliosa*) with hybridization. Am J Bot 85:1597–1601
Arnold ML (1997) Natural hybridization and evolution. Oxford University Press, New York
Barton NH (1979) The dynamics of hybrid zones. Heredity 43:341–359
Barton NH (1992) On the spread of new gene combinations in the third phase of Wright's shifting-balance. Evolution 46:551–557
Barton NH, Hewitt GM (1985) Analysis of hybrid zones. Annu Rev Ecol Syst 16:113–148
Barton NH, Hewitt GM (1989) Adaptation, speciation and hybrid zones. Nature 341:497–503
Bazykin AD (1969) Hypothetical mechanism of speciation. Evolution 23:685–687
Bramwell D (1990) Conserving biodiversity in the Canary Islands. Ann MO Bot 77:28–37
Brochmann C (1984) Hybridization and distribution of *Argyranthemum coronopifolium* (Asteraceae-Anthemideae) in the Canary Islands. Nord J Bot 4:729–736
Buerkle CA, Morris RJ, Asmussen MA, Rieseberg LH (2000) The likelihood of homoploid hybrid speciation. Heredity 84:441–451

Butlin RK, Tregenza T (1997) Is speciation no accident? Nature 387:551–552

Cade TJ (1983) Hybridization and gene exchange among birds in relation to conservation. In: Schonewald-Cox CM, Chambers SM, MacBryde B, Thomas WL (eds) Genetics and conservation: a reference for managing wild animal and plant populations. Benjamin-Cummings, Menlo-Park, CA, pp 288–310

Callaway JC, Josselyn MN (1992) The introduction and spread of smooth cordgrass (*Spartina alterniflora*) in south San Francisco Bay. Estuaries 15:219–226

Carlquist S (1974) Island biology. Columbia University Press, New York

Carney SE, Wolf DE, Rieseberg LH (2000) Hybridization and forest conservation. In: Boyle T, Young A, Boshier D (eds) Forest conservation genetics: principles and practice. CIFOR and CSIRO Publishers, Collingwood, Australia, pp 167–182

Carr GD, Medeiros AC (1998) A remnant greensword population from Pu`u `Alaea, Maui, with characteristics of *Argyroxiphium virescens* (Asteraceae). Pacific Sci 52:61–68

Charlesworth D (1995) Evolution under the microscope. Curr Biol 5:835–836

Coyne JA, Orr HA (1998) The evolutionary genetics of speciation. Philos Trans R Soc Lond B Biol Sci 353:287–305

Crow JF, Engels WR, Denniston C (1990) Phase three of Wright's shifting-balance theory. Evolution 44:233–247

Dobzhansky T (1951) Genetics and the origin of species, 3rd edn. Columbia University Press, New York

Dowling TE, Secor CL (1997) The role of hybridization and introgression in the diversification of animals. Annu Rev Ecol Syst 28:593–620

Ellstrand NC (1992) Gene flow by pollen: implications for plant conservation genetics. Oikos 63:77–86

Ellstrand NC, Elam DR (1993) Population genetic consequences of small population size: implications for plant conservation. Annu Rev Ecol Syst 24:217–242

Emms SK, Arnold ML (1997) The effect of habitat on parental and hybrid fitness: reciprocal transplant experiments with Louisiana irises. Evolution 51:1112–1119

Endler JA (1977) Geographic variation, speciation, and clines. Princeton University Press, Princeton

Freeman DC, Wang H, Sanderson S, McArthur ED (1999) Characterization of a narrow hybrid zone between two subspecies of big sagebrush (*Artemisia tridentata*, Asteraceae). VII. Community and demographic analyses. Evol Ecol Res 15:487–502

Gallez GP, Gottlieb LD (1982) Genetic evidence for the hybrid origin of the diploid plant *Stephanomeria diegensis*. Evolution 36:1158–1167

Ganders FR, Nagata KM (1984) The role of hybridization in the evolution of Bidens on the Hawaiian Islands. In: Grant WF (ed) Plant biosystematics. Academic Press, Toronto

Graham JH, Freeman DC, McArthur ED (1995) Narrow hybrid zone between two subspecies of big sagebrush (*Artemisia tridentata*: Asteraceae). II. Selection gradients and hybrid fitness. Am J Bot 82:709–716

Grant V (1958) The regulation of recombination in plants. In: Exchange of genetic material: mechanisms and consequences. Cold Spring Harbor Symposium on Quantitative Biology, vol 23. Cold Spring Harbor, New York , pp 337–363

Grant V (1981) Plant speciation. Columbia University Press, New York

Harrison RG (1986) Pattern and process in a narrow hybrid zone. Heredity 56:337–349

Harrison RG (1990) Hybrid zones: windows on evolutionary process. Oxford Surv Evol Biol 7:69–128

Heiser Jr CB (1973) Introgression re-examined. Bot Rev 39:347–366

Howard DJ (1986) A zone of overlap and hybridization between two ground crickets. Evolution 40:34-43

Hubbard AL, McOrist S, Jones TW, Boid R, Scott R, Easterbee N (1992) Is survival of European wildcats *Felis silvestris* in Britain threatened by interbreeding with domestic cats? Biol Conserv 61:203–208
Huxel GR (1999) Rapid displacement of native species by invasive species: effects of hybridization. Biol Conserv 89:143–152
Kentner EK, Mesler MR (2000) Evidence for natural selection in a fern hybrid zone. Am J Bot 87:1168–1174
Kerner von Marilaun A (1894–1895) The natural history of plants, vols 1 and 2. Blackie and Son, London
Kim SC, Rieseberg LH (1999) Genetic architecture of species differences in annual sunflowers: implications for adaptive trait introgression. Genetics 153:965–977
Kölreuter JG (1893) Vorläufige Nachricht von Einigen das Geschlecht der Pflanzen betreffenden Versuchen und Beobachtungen, nebst Fortsetzungen 1, 2 und 3. In: Pfeffer W (ed) Ostwald's Klassiker der exakten Wissenschaften. W Engelmann, Leipzig
Levin DA (2000) The origin, expansion, and demise of plant species. Oxford University Press, New York
Levin DA, Francisco-Ortega JK, Jansen RK (1996) Hybridization and the extinction of rare plant species. Conserv Biol 10:10–16
Linder CR, Taha I, Seiler GJ, Snow AA, Rieseberg LH (1998) Long-term introgression of crop genes into wild sunflower populations. Theor Appl Gen 96:339–347
Linnaeus C (1760) Disquisitio de sexu plantarum, ab Academia Imperiali Scientiarum Petropolitana praemio ornata. Amoenitates Acad 10:100–131
Lotsy JP (1916) Evolution by means of hybridization. M Nijhoff, The Hague
Lynch M, Conery J, Burger R (1995) Mutation accumulation and the extinction of small populations. Am Nat 146:489–518
Martin MA, Shiozawa KD, Loudenslager EJ, Jensen JN (1985) Electrophoretic study of cutthroat trout populations in Utah. Great Basin Nat 45:677–687
Martinsen GD, Whitham TG (1994) More birds nest in hybrid cottonwood trees. Wilson Bull 106:474–481
Mayr E (1963) Animal species and evolution. Harvard University Press, Cambridge, MA
McCarthy EM, Asmussen MA, Anderson WW (1995) A theoretical assessment of recombinational speciation. Heredity 74:502–509
Menges ES (2000) Population viability analyses in plants: challenges and opportunities. Trends Ecol Evol 15:51–56
Moore WS (1977) An evaluation of narrow hybrid zones in vertebrates. Q Rev Biol 52:263–267
Nichols RA, Hewitt GM (1986) Population-structure and the shape of a chromosomal cline between two races of *Podisma pedestris* (Orthoptera, Acrididae). Biol J Linn Soc 29:301–316
Parsons TJ, Olson SL, Braun MJ (1993) Unidirectional spread of secondary sexual plumage traits across an avian hybrid zone. Science 260:1643–1646
Pascual MS, Ginovés JRA, del Acro Aguilar M (1993) *Arbutus* _ androsterilis, a new interspecific hybrid between *A. canariensis* and *A. unedo* from the Canary Islands. Taxon 42:789–792
Price MV, Waser NM (1979) Pollen dispersal and optimal outcrossing in *Delphinium nelsoni*. Nature 277:294–297
Ramsey J, Schemske D (1998) Pathways, mechanisms and rates of polyploid formation in flowering plants. Annu Rev Ecol Syst 29:467–501
Rand DM, Harrison RG (1989) Ecological genetics of a mosaic hybrid zone: mitochondrial, nuclear, and reproductive differentiation of crickets by soil type. Evolution 43:432–449

Rhymer JM, Simberloff D (1996) Extinction by hybridization and introgression. Annu Rev Ecol Syst 27:83–109

Rieseberg LH (1991) Hybridization in rare plants: insights from case studies in *Cercocarpus* and *Helianthus*. In: Falk DA, Holsinger KE (eds) Genetics and conservation of rare plants. Oxford University Press, New York, pp 171–181

Rieseberg LH (1997) Hybrid origins of plant species. Annu Rev Ecol Syst 28:359–389

Rieseberg LH (1998) Molecular ecology of hybridization. In: Carvalho GR (ed) Advances in molecular ecology. NATO Science Series, IOS Press, Amsterdam, pp 459–487

Rieseberg LH, Gerber D (1995) Hybridization in the Catalina Island mountain mahogany, (*Cercocarpus traskiae*). Conserv Biol 9:199–203

Rieseberg LH, Wendel JF (1993) Introgression and its consequences in plants. In: Harrison RG (ed) Hybrid zones and the evolutionary process. Oxford University Press, New York, pp 70–109

Rieseberg LH, Zona S, Aberbom L, Martin T (1989) Hybridization in the island endemic, Catalina mahogany. Conserv Biol 3:52–58

Rieseberg LH, Van Fossen C, Desrochers A (1995) Hybrid speciation accompanied by genomic reorganization in wild sunflowers. Nature 375:313–316

Rieseberg LH, Sinervo B, Linder CR, Ungerer M, Arias DM (1996) Role of gene interactions in hybrid speciation: evidence from ancient and experimental hybrids. Science 272:741–745

Rieseberg LH, Archer MA, Wayne RK (1999) Transgressive segregation, adaptation, and speciation. Heredity 83:363–372

Roderiguez DJ (1996a) A model for the establishment of polyploidy in plants. Am Nat 147:33–46

Roderiguez DJ (1996b) A model for the establishment of polyploidy in plants: viable but infertile hybrids, iteroparity, and demographic stochasticity. J Theor Biol 180:189–196

Sætre GP, Moum T, Bures S, Král M, Adamjan M, Moreno J (1997) A sexually selected character displacement in flycatchers reinforces premating isolation. Nature 387:589–592

Simberloff D (2000) Extinction-proneness of island species-causes and management implications. Raffles Bull Zool 48:1–9

Soltis DE, Soltis PS (1993) Molecular data and the dynamic nature of polyploidy. Crit Rev Plant Sci 12:243–275

Stebbins GL (1942) The genetic approach to problems of rare and endemic species. Madroño 6:241–272

Stebbins GL (1950) Variation and evolution in plants. Columbia University Press, New York

Stebbins GL (1957) The hybrid origin of microspecies in the *Elymus glaucus* complex. Cytol Suppl 36:336–340

Stone SC (1985) Alien animals in Hawaii's native ecosystems: towards controlling the adverse effects of introduced vertebrates. In: Stone SC, Scott JM (eds) Hawaii's terrestrial ecosystems: preservation and management. Cooperative National Park Resources Studies Unit, University of Hawaii, Manoa. p 263

Templeton AR (1981) Mechanisms of speciation-a population genetic approach. Annu Rev Ecol Syst 12:23–48

Templeton AR (1986) Coadaptation and outbreeding depression. In: Soulé ME (ed) Conservation biology: the science of scarcity and diversity. Sinauer, Sunderland, MA, pp 105–116

Thorne RF (1967) A flora of Santa Catalina Island, California. Aliso 6:1–77

Tufto J (2000) Quantitative genetic models for the balance between migration and stabilizing selection. Genet Res 76:285–293

Wang H, McArthur ED, Sanderson SC, Graham JH, Freeman DC (1997) Narrow hybrid zone between two subspecies of big sagebrush (*Artemisia tridentata*: Asteraceae). IV. Reciprocal transplant experiments. Ecology 51:95–102
Wang XR, Szmidt AE, Savolainen O (2001) Genetic composition and diploid hybrid speciation of high mountain pine, *Pinus densata*, native to the Tibetan Plateau. Genetics 159:337–346
Whitham TG, Maschinski J (1996) Current hybrid policy and the importance of hybrid plants in conservation. In: Maschinski J, Hammond D (eds) Southwestern rare and endangered plants: proceedings of the second conference. USDA Forest Service, Rocky Mountain Forest and Ranger Expt Station, Ft Collins, CO, pp 103–112
Whitham TG, Morrow PA, Potts BM (1994) Plant hybrid zones as centers of biodiversity: the herbivore community of two endemic Tasmanian eucalypts. Oecologia 97:481–490
Wolf DE, Takebayashi N, Rieseberg LH (2001) Predicting the risk of extinction through hybridization. Conserv Biol 15:1039–1053
Wolfe AD, Xiang QY, Kephart SR (1998) Diploid hybrid speciation in *Penstemon*. Proc Natl Acad Sci USA 95:5112–5115

II. Modeling Approaches for Population Viability Analysis

6 Approaches to Modeling Population Viability in Plants: An Overview

C.A. Brigham and D.M. Thomson

6.1 Introduction

Population modeling, and especially the concept of population viability analysis (PVA), has played a fundamental role in the development of conservation biology as a discipline (Chap. 1, this Vol.). Early applications of PVA largely focused on the use of demographic models to identify minimum viable population sizes, but the term has since broadened to encompass a much wider range of approaches (Beissinger and Westphal 1998; Chaps. 6, 9, this Vol.). Typically, PVA involves the development of models to address questions about the extinction risks facing one or more populations or to explore the factors influencing population persistence. What role PVA can or should play in conservation and management is a question that continues to stimulate a great deal of debate. Some authors argue that PVA is an effective and important conservation tool (Schemske et al. 1994; Brook et al. 2000), while others caution that PVA methods can be easily misused or may provide unreliable guidance to managers (Beissinger and Westphal 1998). In spite of this debate, there is no question that both the number of PVA approaches available to practitioners and their application to conservation problems is growing rapidly.

In this chapter, we aim to provide a concise tour of the sometimes-bewildering array of models that have been used for PVA in general and for plants in particular, focusing on the kinds of problems and data for which these different model structures are most appropriate. We build on previous reviews (Beissinger and Westphal 1998; Groom and Pascual 1998; Morris et al. 1999; Menges 2000a,b) by emphasizing recent theoretical and empirical work relevant to PVA for plants, including some methods that have received little attention in the plant literature. Finally, we suggest future directions for the application of modeling to the conservation and management of plant populations.

Ecological Studies, Vol. 165
C.A. Brigham, M.W. Schwartz (Eds.)
Population Viability in Plants

PVA approaches have been classified based on a variety of characteristics, including whether they are analytical or simulation-based, deterministic or stochastic, and spatially structured or non-spatial (Beissinger and Westphal 1998; Groom and Pascual 1998). In this chapter, we emphasize the importance of data availability in determining when certain model structures may be appropriate, and describe three main types of models based on their relative complexity and the kinds of data they require (Morris et al. 1999). First, the simplest PVA models are unstructured models that describe changes in the total size of a population (or some fraction of a population, such as all the females). This kind of model does not account for differences in mortality or reproduction among individuals in the population and is often based on data consisting of a time series of population counts. Second, age- or stage-structured models incorporate variability among individuals of different ages, or more commonly for plants, sizes. These models use information on the mortality and reproduction of individuals in different age or size classes to predict population growth or persistence. Both unstructured and stage-structured models are typically applied to single populations. Third, a variety of spatially structured modeling approaches have been used to address viability in multiple populations or habitats connected by movement. Spatial models range from methods that use information about patch occupancy or turnover to make predictions about persistence, to more detailed models that explicitly incorporate information about the movement of individuals between patches. In addition to these three major types of models, we review approaches to incorporating genetic effects on viability into PVA. Finally, we discuss some recent trends in the PVA literature and potential future directions for PVA work.

A general point of agreement among reviews of PVA methodologies over the last decade has been that predictions of extinction probabilities and times to extinction are likely to have very wide confidence bounds even for the most data-rich PVAs (Beissinger and Westphal 1998; Morris et al. 1999). Recent theoretical studies reinforce this message, suggesting that many years of data may be required for PVA models to provide good estimates of extinction risk (Ludwig 1996, 1999; Fieberg and Ellner 2000). Many authors have suggested that PVA should be thought of primarily as a tool for asking questions about relative viability (e.g., rankings of extinction probabilities among multiple populations), or the importance of different limiting factors or management alternatives for population growth, rather than as a way of predicting the fate of a population (Beissinger and Westphal 1998; Groom and Pascual 1998; Fieberg and Ellner 2000; Coulson et al. 2001). For this reason, in comparing different model structures we focus on how each approach can be applied in addressing issues of relative viability and management.

6.2 Unstructured Models

So many classic PVA studies have been based on stage- or age-structured models (Menges 1990) that PVA is sometimes treated as almost synonymous with matrix modeling approaches. The majority of published PVAs for plants in particular have used matrix models (Menges 2000a). Although there is a long tradition of applying unstructured models to management problems for animal populations, few of these methods have been used for rare plants (Crone and Gehring 1998; Groom and Pascual 1998; Morris et al. 1999). One major advantage of unstructured models is that they may be less data-intensive than alternative methods, since they can often be parameterized using simple population counts. This is particularly important given that many existing monitoring programs for rare species generate estimates of population size or density rather than the detailed demographic data that are needed to create a matrix model, such as survivorship and fecundity rates (Morris et al. 1999). Although the plant PVA literature has primarily emphasized matrix methods, developing approaches for using the sparser data sets that are more commonly available to managers is one of the major challenges to making PVA more widely applicable in conservation (Crone and Gehring 1998).

6.2.1 The Diffusion Approximation

One unstructured approach to PVA that has received increasing attention recently is the diffusion approximation method, which is based on a large body of theoretical literature but has been brought to a wider audience mostly by the work of Dennis et al. (1991) and is also discussed in Chap. 7 (this Vol.). This technique uses a time series of population counts to estimate the mean and variance of the stochastic population growth rate. With some assumptions about the underlying process of population growth, this information can then be used to generate a range of predictions about population persistence, including extinction probabilities and times to extinction (Fig. 6.1). This method has minimal data requirements compared to methods for development of a stage-structured model, and requires only a small number of straightforward calculations. Moreover, the method can be used in the face of some commonly encountered data limitations, such as missing census dates.

Like all simple models, the diffusion approximation relies on assumptions that may limit its applicability in some management contexts (Dennis et al. 1991; Morris et al. 1999). First, this method assumes that all the observed variability in population counts results from true environmental variation in growth rates, when in reality observed variability will represent a combination of environmental stochasticity and observation error. This assumption means that when observation error is substantial the diffusion approximation

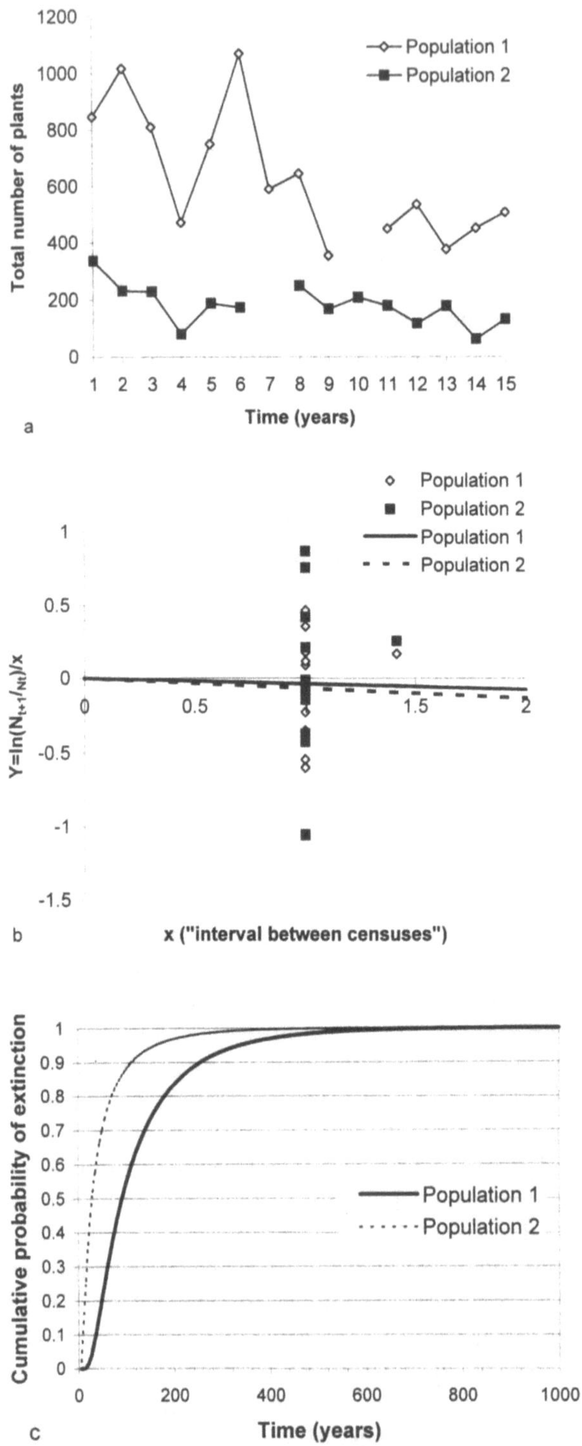

Fig. 6.1a–c. Application of an unstructured population viability analysis (PVA) approach, the diffusion approximation, to population monitoring data for the rare California endemic *Fritillaria pluriflora* (data from Homestake Mining Company 2001). **a** Counts over a 15-year period for two populations of *F. pluriflora*. For each population, 1 year was removed from the data set to illustrate how the diffusion approximation can be used with missing censuses. **b** Regression analysis to calculate the mean and variance of the stochastic growth rate for each population (see Dennis et al. 1991 for details). The fitted regression lines estimate the mean growth rate, while the scatter about the line represents an estimate of the variance. **c** The cumulative distribution functions of extinction times, which show the total probability for each population of falling below an extinction threshold of five plants in a given number of years. A comparison of the two curves allows the relative likelihood of extinction for these two populations to be evaluated, with the smaller and more variable population 2 at higher risk

may overestimate the degree to which environmental stochasticity affects the population. Second, the method assumes that counts represent a constant fraction of the total population. This assumption may be especially problematic for plant species that have difficult-to-observe stages, such as seed banks, bulbs, or dormant individuals. For species with a substantial seed bank, fluctuations in adult numbers result not only from changes in total population size but from year-to-year variation in the fraction of the population in the above-ground stages. If counts of above-ground individuals alone are used as estimates of population size, this will lead to overestimates of the variability in population growth. Finally, the diffusion approximation, in its most basic form, does not account for complex influences on population growth and persistence, such as density dependence and catastrophes. Although methods have been developed to incorporate both density dependence (Dennis and Taper 1994) and catastrophes (Lande 1993; Mangel and Tier 1993) into unstructured models, using these more complicated models will require a longer time series of data because more parameters must be estimated.

A number of recent theoretical papers have explored the use of diffusion approximation under different scenarios, including how the method performs when these simplifying assumptions are violated (Ludwig 1999; Fieberg and Ellner 2000; Meir and Fagan 2000; Chap. 7, this Vol.). Encouragingly, this work suggests that the predictions the diffusion approximation makes about population persistence are resilient to violations of the underlying assumptions. The method appears to be fairly robust even to substantial observation error (Meir and Fagan 2000). Similarly, violations of the assumption that counts represent a constant fraction of the total population appear to have little effect on predictions of the diffusion approximation in at least some situations, although this is not likely to be the case for very short-lived species (Chap. 7, this Vol.). The most serious potential limitation to the application of the diffusion approximation is that a large number of years of census data may be needed to estimate extinction probabilities (Ludwig 1996, 1999; Fieberg and Ellner 2000). One study suggests that the method can accurately project future probabilities of extinction only over a time horizon 10–20 % as long as the data set used to develop the model (Fieberg and Ellner 2000). For example, a 100-year monitoring data set would be needed to make good predictions about the likelihood of population persistence over the next 10–20 years. Other recent work, however, suggests that this is probably a worst-case assessment, and that in many situations the diffusion approximation could be usefully applied with much shorter time series of data (Meir and Fagan 2000; Chap. 7, this Vol.). Still, at least 10 years of census data are probably needed to usefully apply diffusion approximation (Morris et al. 1999).

Applications of the diffusion approximation method to management problems have increased dramatically over the last decade, but it remains almost unused in the plant PVA literature (Chap. 7, this Vol., but see Morris et al. 1999). Only a handful of published PVAs for plants have used unstructured modeling approaches of any kind (Menges 2000a). Not all of these studies were based on population count data; in some cases, more detailed information on germination rates, survivorship, and fecundity were collected, but then summarized in a simpler unstructured model (Watkinson 1990; Lonsdale et al. 1998). This approach is especially common for annual plants. However, at least one of these studies used unstructured models based only on population count data (Crone and Gehring 1998). Using 6 years of population counts for several sites, Crone and Gehring (1998) compared the fit of several unstructured models that differed in their inclusion of complications such as density dependence and spatial variability. This approach is one example of how unstructured models can be used to explore the potential importance of different factors in driving observed population trends, even when data are too limited to be confident in predictions about population persistence.

6.2.2 When to Use Unstructured Models

The major strength of unstructured population models from the perspective of conservation and management is that they require relatively little data and could potentially be applied to monitoring many data sets that consist only of population counts. The lack of attention to unstructured models in the plant PVA literature is somewhat surprising, since one of the most common criticisms of PVA models is that their heavy data requirements make them impractical for most species of conservation concern. These data requirements are especially problematic when PVA is used to compare multiple populations or species. Unstructured models may therefore be the most useful approach to ranking or comparing the relative viability of different populations. The relative simplicity of unstructured approaches also lends itself to the fitting of multiple models that differ in their assumptions (e.g., density-independent vs density-dependent growth) as a way of generating hypotheses about the factors affecting a population (Crone and Gehring 1998). However, this simplicity comes at a price: a lack of the detailed biology used to develop more complicated models, which makes unstructured approaches much less useful for exploring management alternatives (Morris et al. 1999).

Finally, unstructured models are less appropriate for plant species with certain life histories, especially annuals or other species whose population dynamics are strongly influenced by a seed bank. Population counts may also be much less useful to assess viability in long-lived species such as trees. This is because for such species changes in demographic rates that could have important implications for viability may be much slower to manifest as changes in overall population size; that is, population counts tend to be highly correlated between years because they are strongly influenced by the history of the population (Morris et al. 1999). Unstructured approaches are probably best suited to relatively short-lived species for which multiple population turnovers can be captured over a time scale of 10–20 years.

6.3 Stage-Structured Models

The types of model most commonly used in PVA, especially for plants, are stage- or age-structured (matrix) models. In contrast to unstructured models, matrix models account for differences in rates of reproduction and mortality among individuals of different ages or sizes. As a result, they can include much more biological reality than unstructured models, for example the presence of reproductively immature individuals in the population, or lower mortality and higher reproduction among larger individuals. Including this additional biological reality also requires more intensive data collection: regularly spaced censuses of marked individuals tracking their mortality and reproduction. An additional problem for plant populations is estimating germination rates, and seed-sowing experiments are often needed for this purpose. With these data, the population is then divided into age, stage or size classes that differ in their vital rates (survivorship and fecundity). The vital rates are used to calculate transition probabilities that can be summarized in the form of a matrix describing how the number of individuals in each class changes from one year to the next. The tools of matrix algebra allow us to then generate a range of predictions about population growth and structure (Caswell 2001; Fig. 6.2).

Matrix models are especially useful not only because they incorporate biological details that often have important implications for viability, but because they generate a broader range of predictions than unstructured models. The simplest matrix models assume density-independent and deterministic population dynamics leading to either exponential growth or decline. These models predict the long-term deterministic growth rate, lambda (λ), but also the stable stage distribution of the population and the reproductive values of individuals in different stage classes. Matrix models generate short-term, transient population dynamics in addition to equilibrium population growth rates, although these predictions of transient dynamics have rarely been used in conservation planning (Fox and Gurevitch 2000). More complex models incorporating stochastic variability in vital rates can be used to generate extinction probabilities.

Perhaps most importantly, matrix models allow us to evaluate how changes in vital rates influence population growth through sensitivity analysis. For this reason, they can be powerful tools for exploring the effects of different management strategies on population viability. One early PVA used sensitivity analysis to show that changes in juvenile survivorship influence population growth of sea turtles much more strongly than do changes in egg or

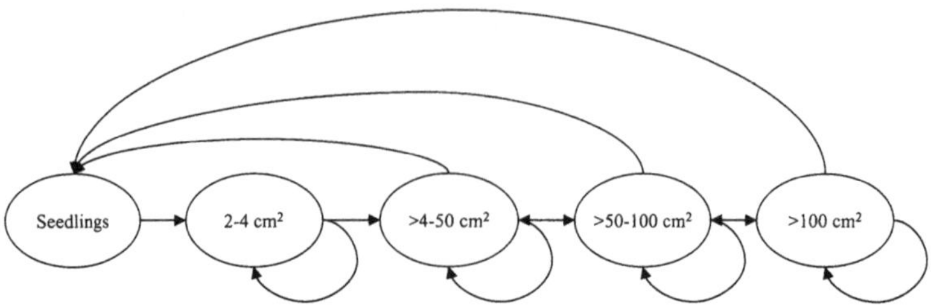

FROM

TO	Seedlings	2-4 cm^2	>4-50 cm^2	>50-100 cm^2	>100 cm^2
Seedlings	0	0	0.06	0.05	0.04
2-4 cm^2	0.14	0.25	0	0	0
>4-50 cm^2	0.85	0.35	0.68	0.15	0.01
>50-100 cm^2	0	0	0.21	0.55	0.13
>100 cm^2	0	0	0.05	0.12	0.81

= M

$$[N_{t+1}] = [M]\,[N_t] = \lambda[N_t]$$

Fig. 6.2. A stage-structured matrix model for the rare Colorado endemic *Aletes humilis* (Morris et al. 1999). The life cycle diagram at the *top* shows the division of the population into five stage classes, including a seedling category, and four adult classes based on plant diameter. *Arrows* indicate the directions of possible transitions between classes, including stasis and shrinkage in addition to growth and reproduction. Note that for simplicity not all possible transitions (e.g. growth across more than one class) are shown. The probabilities of these different transitions are summarized in the matrix (*M*). The population trajectory can be predicted by multiplying the vector of population sizes with the matrix *M* at each time step. Once the population reaches its stable stage distribution, this is equivalent to multiplying by the dominant eigenvalue of the matrix (λ), which represents the geometric growth rate

hatchling mortality (Crouse et al. 1987). This study was pivotal in redirecting conservation efforts for sea turtles from a focus on beach protection to a greater emphasis on lowering fisheries by catch of juvenile and adult turtles. For plants, matrix models have been used to explore the effects on population viability for a variety of factors such as burning (Gross et al. 1998; Menges and Dolan 1998; Kaye et al. 2001), grazing (Lesica 1995), harvesting (Olmsted and Alvarez-Bullya 1995; Nantel et al. 1996) and trampling (Maschinski et al. 1997; Gross et al. 1998).

6.3.1 Developing a Matrix Model

The increasing availability of package programs, such as RAMAS, VORTEX, and ALEX, has played an important role in the increasing popularity of matrix modeling approaches in conservation, and has made these methods much more widely accessible. Many of these programs can be used to develop quite complicated models, including factors such as stochasticity, density dependence, catastrophes and even spatial structure. For this reason, careful consideration of both species life history and the available data is critical to the appropriate use of these tools. Several recent studies emphasize that the many choices about model structure made in developing a matrix-based PVA can substantially alter model predictions. For example, the results of incorporating density dependence depend on the underlying functional form, with different PVA packages potentially generating different predictions using the same data set (Mills et al. 1996). A full discussion of the many options and approaches for developing and analyzing more complicated matrix models is beyond the scope of this review, but we briefly summarize some of the major issues and key references below.

One of the most basic decisions in constructing a matrix model is how to divide the population into different classes. Models for plants have traditionally been size- rather than age-based, because many plant species are difficult to age, and size is often a better predictor of individual fates. However, in some cases, the use of alternative or multiple predictor variables, such as age and size, may be more appropriate (Werner and Caswell 1977). Methods for estimating age-based vital rates from stage-structured matrices are now available (Cochrane and Ellner 1992), creating a greater range of options in model construction. For stage-based models, decisions about how many classes to divide the population into can potentially affect model predictions. Using a small number of classes may lead to the lumping of individuals that vary substantially in vital rates, and can also allow biologically unrealistic transitions (e.g., a three-class seedling, juvenile, adult model for trees would allow a small probability of seedling to large adult transitions in only 3 years). On the other hand, increasing the number of classes reduces sample sizes for estimating vital rates and their variances. Several formal approaches to setting class boundaries have been proposed (Vandermeer 1978; Moloney 1986), but most studies use breakpoints that seem biologically reasonable based on regressions of survivorship and fecundity against size. Some recently developed methods avoid this issue by incorporating continuous variation in vital rates rather than dividing the population into classes (Easterling et al. 2000).

Although the important role stochastic variation plays in population viability has long been recognized (e.g., Shaffer 1983), until recently relatively few matrix-based plant PVAs incorporated stochasticity (Menges 2000a, but see Bierzychudek 1982). The use of stochastic matrix models for plants is growing (Gross et al. 1998; Menges and Dolan 1998; Kaye et al. 2001). However, including stochasticity raises some additional problems and alternatives in model construction. First, several different approaches have been used, including random draws of matrices from different years (Bierzychudek 1982; Menges and Dolan 1998), drawing individual matrix elements from estimated distributions of vital rates (Gross et al. 1998), and analytical methods based on the means and variances of vital rates (Benton et al. 1995, for an animal example). A recent study by Fieberg and Ellner (2001) provides guidance in selecting from these methods. Their work suggests that assembling matrices by drawing from distributions for vital rates (what they term the "parametric matrix approach") may be a more robust approach than draws of whole matrices, even though this method generally requires some assumptions about the shape of the underlying distributions. This is largely because the random matrix draws method, although the most commonly used in plant PVA, makes more rigid assumptions about the correlation structure among matrix elements (see below).

A second issue is that stochastic models require relatively long-term demographic data sets. In theory it is possible to construct a deterministic matrix model based on 2 years of data (one transition), and a stochastic one with three or more years of data. In practice, basing a stochastic matrix model on less than 5–10 years of data is rarely justified and even these relatively long-term data sets are likely to underestimate the true range of variation in vital rates (Morris et al. 1999). For example, the persistence of some plant species may depend on episodic years of strong recruitment tied to environmental factors such as El Niño events, so that a 5- to 10-year study might capture few or even none of these critical years. Unfortunately, few demographic data sets for plants encompass even 5 years (Menges 2000a).

A further concern is that correlation among vital rates in between-year variability can strongly influence estimates of viability, with positive correlations leading to more variable predictions of population growth (Doak et al. 1994). It is reasonable to expect that vital rates will often respond similarly to changes in environmental conditions; for example, drought years may cause reductions in survivorship for more than one stage class, or both increased mortality and reduced fecundity. Accounting for these positive correlations will lead to more variable predictions of population growth than obtained with a model that does not include correlation, by reinforcing the differences between growth rates in good and bad years. The problem is that obtaining realistic estimates of correlation structure for most matrix applications would require many more years of data than are typically available (e.g., more transitions than the number of parameters in the matrix) (Fieberg and Ellner 2001). In the absence of such data, the best approach is probably to explore a range of assumptions about correlation structure in order to provide more realistic confidence bounds on model predictions. Correlation likely has stronger impacts on estimates of extinction probabilities than on other model predictions, so that exploring alternative assumptions is especially important in reporting extinction risks (Fieberg and Ellner 2001).

Density dependence is rarely included in PVA for plants (Menges 2000a), although tools are available for incorporating density dependence into matrix models (DeKroon et al. 2000; Oostermeijer 2000; Caswell 2001). This is partly because density dependence is generally assumed to be relatively unimportant for the rare species that are the focus of most PVAs, and partly because parameterizing and analyzing density-dependent models are more difficult. However, in some cases accounting for density dependence may be important in addressing viability, especially in evaluating the effects of management. Different methods have been used for incorporating density dependence, for example, the use of a population ceiling, and the assumptions about how density dependence operates can have important effects on model predictions (Mills et al. 1996). Demographic stochasticity is another factor typically ignored in plant PVA (Menges 2000a). One straightforward way to account for demographic stochasticity is to use bootstrapping techniques in estimating vital rates (Kalisz and McPeek 1992), although the effects on model predictions are not likely to be substantial unless population sizes are fairly small.

6.3.2 Sensitivity and Elasticity Analysis

In addition to the potential complexities and alternatives in model construction, a range of approaches is available for the analysis of matrix models. In particular, alternative methods for sensitivity analysis have been a major focus of the recent theoretical literature (Mills et al. 1999; Caswell 2000; DeKroon et al. 2000; Grant and Benton 2000; Wisdom et al. 2000). Sensitivity analysis explores the influence of different vital rates on population growth and is often the model prediction most directly applied to management. Early stage-structured PVAs typically used analytical methods for calculating deterministic sensitivities and elasticities, which, respectively, represent the absolute or relative change in λ with changes in a particular vital rate (Caswell 2001). However, this approach is based on several important assumptions that may lead to an unrealistic picture of how management could influence population growth (Mills et al. 1999). First, the calculation of analytical sensitivities relies on the assumption that changes in the vital rates are small and uncorrelated. A related issue is that sensitivities and elasticities are specific to a given set of matrix parameters; that is, changing one or more of the vital rates will also change the sensitivities of other elements. Finally, vital rates with high sensitivities do not necessarily correspond to the life history stages that are currently limiting population growth or that are the most productive targets for management. For example, in long-lived species such as trees, population growth is typically most sensitive to changes in adult survivorship. Adult survivorship in such species also tends to be higher and less variable than other vital rates, so that managing for increased survivorship nevertheless may be very difficult or even biologically impossible. Further, if seedling recruitment is low or nonexistent, management aimed at increasing recruitment may be far more important in promoting population persistence even in long-lived species (Chap. 9, this Vol.).

A number of recent developments in the PVA literature are aimed at addressing some of these limitations. Methods have been developed to extend analytical sensitivities to more complicated models, incorporating stochasticity or density dependence (Caswell 1996; Grant and Benton 2000). In addition, simulation-based approaches are increasingly being used to investigate how larger or correlated changes in vital rates affect population growth. These methods typically involve generating a large number of parameter sets by drawing from the range of observed variation in vital rates, then using multiple regression techniques to explore the relationship between different vital rates and λ(Wisdom and Mills 1997; Wisdom et al. 2000) or the probability of extinction (McCarthy et al. 1995; McCarthy 1996). An added advantage is that these approaches account for the fact that not all vital rates are equally variable in evaluating the importance of different effects on λ. However, this kind of sensitivity analysis depends on our estimates of past variability in vital rates to predict how future changes may influence the population, and these estimates may be uncertain (Cross and Beissinger 2001). A similar but alternative approach to partitioning effects of vital rates on $\lambda$is the Life Table Response Experiment (LTRE) method (Horvitz et al. 1997; Caswell 2001), which uses data from multiple populations or experimental treatments.

Along similar lines, a promising and increasingly common approach that addresses some limitations of sensitivity analysis is the use of experiments in combination with modeling to explore the effects of different management strategies (Groom and Pascual 1998; Menges and Dolan 1998; Oostermeijer 2000; Kaye 2001). The comparison of matrices for treated and control populations (e.g. burned and unburned) is probably a more realistic way to rank management options than sensitivity analysis. One advantage of using models in combination with experimental work is that modeling allows for an integrated evaluation of how effects on different life history stages combine to influence population growth and persistence. Stage-structured models can be especially useful for investigating the roles of periodic disturbances such as fire in population persistence, for example by using parameters from burned and unburned plots to simulate effects of changing fire frequency.

It should be noted here that matrix models are not the only stage-structured PVA models that can be used to explore management alternatives. Several recent PVA studies on responses to fire management have developed other methods, such as the application of behavioral modeling approaches like stochastic dynamic programming (SDP) (McCarthy et al. 2001b). Some of these alternative model structures can more flexibly incorporate the effects of different factors, such as time since last fire or population size, which may play an important role in determining the consequences of management decisions. These approaches are still similar to matrix models in requiring demographic data for individuals in both treated and control plots, and in the calculation of growth rates or extinction probabilities to rank management strategies.

6.3.3 When to Use Stage-Structured Models

In some ways, plants are more amenable to the application of stage-structured models than animals, because marking and relocating individuals is much easier. This greatly facilitates the measurement of survivorship rates, which for animals is typically complicated by the need to estimate both survival and recapture probabilities. However, plant life histories also present some challenges to the construction of matrix models (Menges 2000a, b). Many of these, such as dormancy, clonal reproduction or the presence of long-lived stages, can be problematic regardless of the model structure used. For example, as discussed in the previous section, population counts may provide poor estimates of viability for long-lived plants. Yet building matrix models can also be difficult for such species, because estimating mortality for the biggest individuals that rarely die requires large sample sizes or many years of data collection. This is especially a problem because these long-lived stages often have the highest reproductive values and the greatest influence on population growth. Given these limitations, an advantage of stage-structured models is that the influence of poor vital rate estimates for particular life history stages can be explicitly evaluated (e.g., Chap. 7, this Vol.). However, in some cases so much important information may be missing that the construction of a model is simply not warranted.

In summary, the popularity of stage-structured models in PVA reflects their many strengths: matrix models are relatively simple to use, yet flexible, and they provide a variety of useful predictions about population growth and viability. Perhaps the most important advantage of matrix models is the ease with which they can be used to evaluate different management approaches, although methods such as SDP provide useful alternatives. Plant species with certain life history complications, such as seed dormancy or long-lived stages, provide challenges to the application of matrix models. Nevertheless, stage-structured approaches such as matrix models may still provide the most robust approach to PVA for these species, because the effects of different assumptions about vital rates that are difficult to measure can at least be explicitly evaluated.

The major drawback to stage-structured models from a conservation and management perspective is the amount of data required. Almost as many years of data collection may be needed to parameterize a stochastic matrix model as for unstructured models like the diffusion approximation, and demographic data is often much more labor-intensive to acquire than simple population counts (Morris et al. 1999). It is also not clear how good the predictive power of these more complicated PVA models really is. Recent work testing the performance of stage-structured PVA models provides some evidence for the reliability of matrix models in predicting population trajectories (Bierzychudek 1999; Brook et al. 2000), but the strength of this evidence has been questioned (Coulson et al. 2001; McCarthy et al. 2001a). The predictions matrix models make about management are likely to be more robust than predictions of extinction risk, but this assumption remains essentially untested. It may be even more difficult to validate or test the predictions PVA models make about management than their predictions of growth rates (Coulson et al. 2001).

6.4 Spatially Structured Models

An appreciation for the importance of space in ecological interactions has been accumulating over the past two decades (Ricklefs and Schluter 1993; Tilman and Pacala 1997). Spatial structure can impact persistence in numerous ways, such as spatial mediation of competition, metapopulation dynamics, spatial mediation of herbivory or predation, and spatial effects on genetics, just to name a few. How to incorporate spatial effects into population viability modeling and how important such effects are have been topics of some debate. Modeling of metapopulation dynamics has received a lot of attention, yet it is unclear how common they are in nature. Some authors have suggested that metapopulation dynamics may be common for naturally patchy plants (Erikkson 1996; Husband and Barrett 1996). Although spatially structured PVAs are still uncommon, there are enough of them in the literature to warrant a review of different modeling approaches.

A variety of spatially structured modeling methods have been applied to problems in conservation and management, but these approaches can be broadly grouped based on the way space is treated in the models and the kinds of data required to parameterize them (for a review of spatially structured plant examples and disturbance see Chap. 11, this Vol.). Some, such as metapopulation models, predict the occupancy of discrete habitat patches through time. These patch-based approaches vary in the extent to which dispersal is specified in the model. In some cases, patch colonization and extinction probabilities are estimated indirectly (logistic regression and incidence function), while other methods directly incorporate information on dispersal. Additionally, for PVA a common approach is to use megamatrix models, in which matrices are constructed for different populations and then linked by a dispersal function (see review of this method in Chap. 11, this Vol.). More complicated, spatially explicit models track the locations and densities of individuals across a landscape, but require detailed data on dispersal or movement behavior. Spatially structured PVA of any kind has rarely been used for plants (but see examples in Harrison and Ray 2002) and the few that exist are typically spatially implicit. Below, we focus on relatively simple patch models because they probably represent a more useful way of thinking about spatial structure in rare plant populations. Moreover, the kinds of detailed dispersal data required for more complicated models are available for very few species, let alone rare plants.

6.4.1 Metapopulation Models

The term metapopulation has been used to mean several different things. Generally, metapopulations are a collection of populations linked by immigration, emigration, and extinction (Hanski and Gilpin 1997). Metapopulation models make predictions about the persistence of a system of habitat patches, and usually involve measuring patch area, isolation, and occupancy to estimate rates of colonization and extinction (Fig. 6.3). Some models require measured rates of dispersal while others calculate this rate from observed occupancy. All metapopulation models require identification of appropriate habitat and an ability to assess whether the habitat is occupied or not. Both of these requirements can be difficult for plants. Biologists often identify plant habitat based on whether the plant is present, and assume that unoccupied habitat is inappropriate. If the exact habitat requirements of a species are not known, determination of unoccupied habitat can be quite difficult. Additionally, plants with seed banks or underground dormant stages such as bulbs or corms may occupy a site but be difficult to detect. Further, some metapopulation modeling approaches (see the logistic regression model below) require identification of extinctions and recolonizations, and dormant stages may inflate estimations of these phenomena.

Levins (1969) formulated the classic metapopulation approach, which assumes that all patches are identical in habitat quality, are equally connected, and that the metapopulation is at equilibrium. In nature, patches typically differ in quality and connectivity and metapopulations are often in flux, thus these assumptions are too restrictive and this original formulation has not been widely used. A recent deterministic version of this model that relaxes some of these assumptions allows for spatial variation in patch area and connectivity and is thus more realistic (Hanski and Ovaskainen 2000).

A second, widely used metapopulation modeling approach is the incidence function model (IFM, see Hanski 1994, 1999; Sjogren-Gulve and Hanski 2000). This approach uses presence/absence data along with patch area and distance between patches to calculate the probability that a given patch will be occupied during a given time step in the model. Patch-specific rates of extinction and colonization are calculated based on observed occupancy, patch area, and

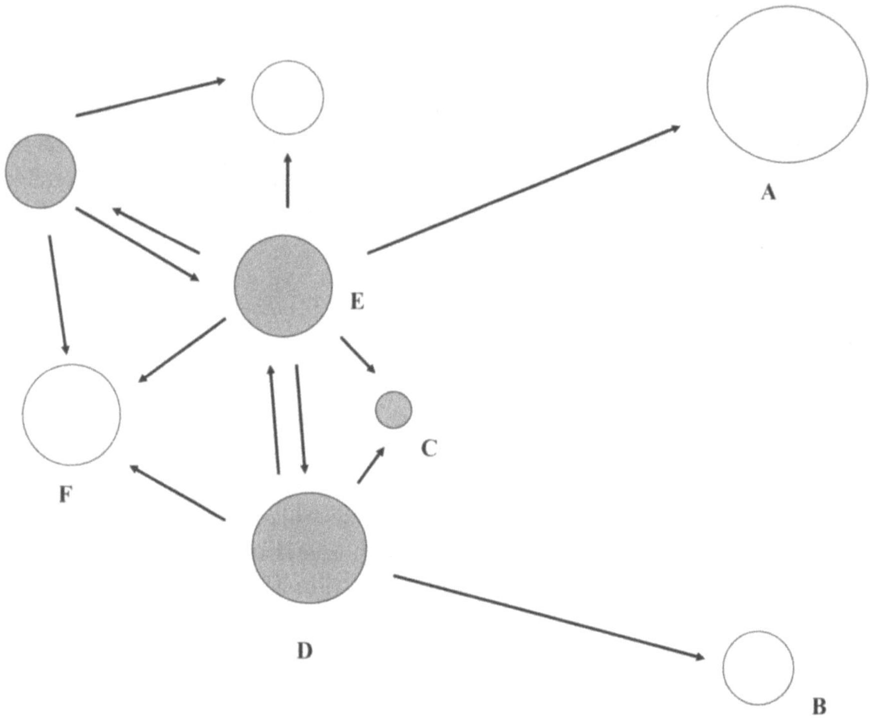

Fig. 6.3. *Circles* represent suitable habitat patches for a species. *Shaded circles* are occupied patches ; *open circles* are unoccupied. *Arrows* represent migration. All metapopulation models specify the size and isolation of patches while some (such as logistic regression) may also include other habitat factors that determine suitability. In this diagram both *A* and *B* are unoccupied and equally distant from occupied patches, however *A* is more likely to be colonized than *B* because it is larger. Note that unoccupied patches generate no dispersers (have no *arrows* originating from them). This figure also includes a rescue effect. Patch *C* is occupied due to migration from *D* and *E* but has no dispersers originating from it because it is a population sink (the habitat is really too small to support a self-sustaining population). At the current time-step, *F* is unoccupied but is likely to be colonized because of the proximity of many occupied patches

isolation. The extinction rate is assumed to be dependent on patch area alone, not isolation (unless a rescue effect is included). The exact relationship between patch area and extinction is parameterized from survey data. The model also requires specification of a patch area below which extinction probability equals one. Colonization can be a function of distance from a mainland population or a sum of the probability of colonists from all surrounding patches based on their areas and distances from the target patch. For this model, the rates of immigration or dispersal distances do not have to be measured, as these are inferred from observed patch occupancy.

Although an IFM can be parameterized from a single survey of populations, relying on a single snapshot of patch occupancy typically gives poor results (Kindvall 2000). If several years of survey data are available, the predictive abilities of IFM models appear to be quite similar to those of the logistic regression approach (see below). Incidence functions may also be as precise as more data-intensive metapopulation models. For instance, an IFM was as accurate as a data-intensive RAMAS-GIS-based model according to a case study using crickets (Kindvall 2000). IFM has been applied to the dynamics of Florida scrub plants by Quintana-Ascencio and Menges (1996) but not in a strict PVA approach.

The logistic regression model (LRM- (Sjogren-Gulve 1991; Sjogren-Gulve and Ray 1996) is another way to model metapopulation dynamics. This approach, like IFM, uses presence/absence data, patch area, and isolation, but can also incorporate other environmental or demographic factors that may affect colonization or extinction. LRM is based on a state transition model in which two separate logistic regression models are run, one to predict extinction and one to predict colonization. In each model, the researcher can include any factors that might influence colonization or extinction with either categorical or continuous predictor variables. Once specified, the logistic regressions are used to project patch occupancy into the future. The future status of patches is calculated using multiple logistic regression, based on the environmental predictor variables associated with each patch and the sum of its colonization and extinction probabilities. These probabilities can be summed for all patches to predict the future state of the metapopulation.

Similar to IFM, LRM is based solely on occupancy and does not keep track of the population sizes within a patch, although population size could be incorporated into the logistic regression as a predictor variable. The model assumes a steady-state metapopulation and can include spatial or temporal correlation among patches. [For further details on this modeling approach see the overview in Sjögren-Gulve and Hanski (2000) or specific models described in Sjögren-Gulve and Ray (1996).] This method has been used to create metapopulation-based PVAs for five serpentine seep plants in California (Harrison and Ray 2001). Harrison and Ray (2001) also review when metapopulation approaches should be considered with plants and the special concerns of plant metapopulations (e.g., pollinator-mediated effects).

These metapopulation models can be used to calculate a minimum viable metapopulation (MVM) or a minimum amount of suitable habitat (MASH). Further details on these approaches can be found in Sjögren-Gulve and Hankski (2000).

6.4.2 Spatially Structured Matrix Models

A quite different approach to modeling spatial effects than those described above is to incorporate spatial variability into a matrix modeling approach. Valverde and Silvertown (1997) created a metapopulation model using this approach for *Primula vulgaris*. This model is similar to IFM and LRM in that it examines occupancy of habitat patches and includes colonization, migration, and extinction; however, it differs from IFM and LRM by including population dynamics within patches. With *Primula vulgaris*, Valverde and Silvertown estimated new colonizations based on seed production and dispersal curves. Matrix modeling can also be used to incorporate spatial dependence of vital rates without assuming metapopulation dynamics. Menges and Dolan (1998) used this approach in their PVA of *Silene regia*. This model incorporates the effects of spatial differences in habitat quality by creating different transition matrices for each population modeled. The authors used stage-based matrix models to create viability estimates and extinction times for different populations with differing demographic rates. Such stage-based matrix models differ from Valverde and Silvertown's approach and IFM and LRM by modeling only occupied patches and ignoring colonization of new sites. Furthermore, these matrix modeling approaches are very data-intensive (requiring demographic data for each of the populations modeled).

6.4.3 When to Use Spatially Structured Models

In recent years, great progress has been made in developing spatially structured models. The variety and sophistication of these modeling approaches has now far outstripped the available data, especially for rare plants. Many spatial models require estimates of dispersal probabilities and distances that are very difficult to obtain; worse, for these models dispersal estimates are generally the single most important factor determining model predictions (Ruckelhaus et al. 1997; Beissinger and Westphal 1998).

Models that do not require estimates of dispersal distances, for instance, many metapopulation approaches, have other restrictions. In general, metapopulation models should be used with species that have numerous small, discrete populations in order to obtain a large enough sample of patches. Many formulations of metapopulation models (e.g., Levins' model, logistic regression, and incidence function) ignore local dynamics and instead focus on turnover rates throughout the metapopulation. Species modeled with this approach should have dynamics dominated by extinction and colonization events, not local dynamics.

Before choosing a spatially structured approach, it is relevant to ask whether it is appropriate to invest the time and resources required to obtain the information necessary required. In a recent review, Harrison and Ray (2002) argued that although there is evidence for plant metapopulation dynamics in nature, plants may be especially robust to spatial effects due to their buffering capacities (e.g., seed banks, clonal reproduction) and short-distance seed dispersal. They concluded, with caution, that single population approaches instead of metapopulation approaches may be appropriate for most plant species. Morris et al. (1999) suggested that for many species immigration may be either high enough to model subpopulations as a single population or low enough to model each population as independent. However, spatial differences in vital rates may need to be included in many PVAs to obtain realistic outcomes. As with all modeling issues, a good knowledge of the natural history of the species being modeled is necessary in order to determine whether a spatially explicit modeling approach is warranted. Finally, even in cases in which we suspect that spatial dynamics may play an important role in viability, the data necessary to develop a spatially structured PVA will often not be available. For this reason, spatial models may be most valuable in PVA for plants as a heuristic tool.

6.5 Genetics in Population Viability Analysis

Although the potential genetic effects of small population size are a focus of substantial concern and interest in the conservation literature (e.g., Barrett and Kohn 1991; Ellstrand and Elam 1993), few PVAs include genetics (Menges 2000a). The number of studies of genetic diversity in rare populations of plants has skyrocketed over the past decade; however, it is still difficult to assess the relationship between genetics, population declines, and fitness (for recent reviews see Ellstrand and Elam 1993; Chaps. 2, 3, this Vol.).

There are several different approaches for incorporating genetics into a PVA. Burgman and Lamont (Burgman and Lamont 1992) used a theoretical relationship between declines in population size and inbreeding effects to examine potential impacts of inbreeding depression on viability of *Banksia cuneata*. In this model, inbreeding depression based on the population size of the previous year modifies the number of seedlings in each model iteration. The authors assumed a negative linear relationship between the inbreeding coefficient and the number of seedlings. The model contains a number of unrealistic assumptions (the effective population size is equal to the number of adults in the population, inbreeding effects are linear) but it is a start at incorporating genetics and assessing the relative strength of genetic vs other effects on population persistence.

If data are available on the relationship between population size, genetic diversity, and fitness, these data can be used to generate a more explicit model of genetic effects. Unfortunately, such data seldom exist. An exception is the grassland herb *Gentiana pneumonanthe*, which has been intensively studied by J.G. Oostermeijer and colleagues. Oostermeijer (2000) has used this wealth of data to examine genetic vs demographic effects on viability. Four models are constructed that include a range of population effects, from no effects of small population size to inclusion of fecundity effects, inbreeding effects, and the combination of the two. Comparing the results of the four different model types, Oostermeijer found that reproductive effects alone did not have a significant effect on model outcomes, but inbreeding depression effects alone and inbreeding depression plus reproductive effects significantly reduced the mean time to extinction. Inbreeding effects were strongest at small starting population sizes (below 250 plants). Models including inbreeding depression showed smaller peak population sizes, more rapid extinction, and lower λ than the simple model formulation. The combined inbreeding depression and reproductive effect model decreased the mean time to extinction by 6 years in comparison to the model with population size effects.

A different method of assessing the impacts of genetics on population viability is to correlate the genetic diversity of different populations with their viability. Menges and Dolan (1998) used this approach to examine the impact of genetics on *Silene regia*. In this case, demographic rates were collected from populations differing in size, geographic location, genetic diversity, and management regime (time since burning, mown/not mown, etc.). The authors assumed that any effects of these different factors would be reflected in measured demographic rates. They used a matrix model with different transition matrices for each population and then related each factor to model outcomes (e.g., λ, mean time to extinction) through a MANOVA. This approach is data-intensive but founded solidly in reality. One drawback to this approach is that it is static and does not incorporate the effects of potential further declines in genetic diversity as population size declines.

In general, there are now a variety of ways to examine the effects of genetics on population viability. These range from the extremely data-intensive and particular (e.g., Menges and Dolan 1998; Oostermeijer 2000), to the theoretical (e.g., Burgman and Lamont 1992). Thus, for species that may be experiencing inbreeding depression or other negative genetic consequences of small population size it is possible to include these effects in PVA; however, such inclusion may not be necessary for all species (see discussion in Chap. 3, this Vol.).

6.6 Future Directions

As many other authors have pointed out, the biggest limitation of traditional PVA approaches is that detailed demographic data and long-term population trends are likely to be unavailable for many, or even most, species of conservation concern (Beissinger and Westphal 1998; Morris et al. 1999; Burgman et al. 2001). It is often argued for this reason that PVA is useful mostly for developing detailed case studies on a few species that could contribute to improved management and decision-making for those that are less well-studied. Some classic PVA studies have played an important heuristic role in changing approaches to conservation and management. For example, Shaffer's (1983) models for grizzly bears helped focus attention on the potential for stochastic variation to create substantial extinction risk even in growing populations (Beissinger and Westphal 1998). Similarly, Crouse et al.'s (1987) PVA on sea turtles clarified the importance of adult survivorship in the population viability of long-lived species. Even simple models often generate results that would not have been obvious otherwise, and that help to improve our biological intuition (Morris et al. 1999).

Beyond this heuristic value, the development of methods to address conservation problems using limited data deserves greater attention in the PVA literature. Several avenues of recent research suggest some directions for future work. First, developments over the last decade in techniques for the analysis of count data and population trends suggest that these approaches could provide a useful alternative to more data-intensive methods, if applied with an appreciation for their limitations.

A second promising avenue is to use data sets or models for well-studied species to develop rules of thumb based on life history or ecological traits that could be applied in the absence of population or demographic data. Along these lines, Fagan et al. (2001) used long-term (15 years or more) population time series to explore the relative roles of growth rate, variability, and carrying capacity in the population dynamics of different taxa. Their results suggested a potential for using traits such as age of first reproduction and body size in predicting broad patterns of population dynamics and extinction risk. Similarly, several studies have used stage-structured models to explore the relationship between life history strategies and matrix elasticities for both animals (Heppell et al. 2000) and plants (Silvertown and Franco 1993; Silvertown et al. 1996). Although the use of elasticity patterns to draw broad conclusions about life history tradeoffs has been criticized (Shea et al. 1994), these approaches do hold the promise that some results of matrix-based PVA could be predicted based on biological traits without carrying out a PVA at all. The need to develop decision tools and planning processes in conservation that are alternatives to PVA is increasingly recognized (Burgman et al. 2001), and this type of analysis provides one avenue by which PVA could contribute to such planning frameworks.

A related strategy is to use information from well-studied species to improve model development for more limited data sets. For example, a lack of long-term data to estimate vital rate variability and correlation structure is a major obstacle to the application of stochastic matrix models in PVA. Life history theory suggests that tradeoffs should constrain matrix parameters in ways that may be predictable enough to at least target the range of assumptions that need be explored in model development. In this vein, Pfister (1998) found evidence from published matrix models supporting the theoretical prediction that stages with the strongest influence on population growth should be more buffered against variability. Fieberg and Ellner (2001) have suggested that another way to improve stochastic matrix models is to search for relationships between vital rate variability and environmental drivers such as weather. This could provide an approach for improving the incorporation of stochasticity into stage-structured PVA with demographic data sets of short duration.

None of these methods are clear solutions to the challenge of carrying out conservation planning and management with few data, and in many cases will provide only a broad caricature of the information that a more detailed analysis could generate. Given the problems inherent even in data-rich PVAs, it is unrealistic to expect that limited data will provide more than limited answers. Nevertheless, when data are available, it is hard to justify ignoring methods that could help maximize our use of relevant information (Morris et al. 1999).

6.7 Conclusions

In spite of the many limitations of PVA approaches and ongoing debate over their value, PVA will almost certainly continue to play a major role in conservation and management. In this chapter we have discussed some of the many modeling methods applied to PVA, and their strengths and shortcomings. The most fundamental step in developing a PVA is choosing a modeling approach appropriate to species life history and, even more importantly, the quantity and quality of data available (Morris et al. 1999). This step will only become more critical as the number and complexity of available modeling tools continue to grow. It is important to keep in mind that the most complicated model is often not the best model. Although it is tempting to include in PVA as many as possible of the factors that we suspect are important to population persistence, such as stochasticity, density dependence, or spatial structure, the data are often not available to justify the construction of such complicated models. An important concern is that the development of these complicated models can imply or create a false sense of security in predictions that are supported by only sparse data (Beissinger and Westphal 1998). Further, adding greater complexity to a model can make it more difficult to disentangle how different factors and assumptions affect model predictions. Simple models can often provide guidance in addressing management questions, even when they leave out some factors that we think may be important (Beissinger and Westphal 1998).

Plant life histories present many complications in developing a PVA, such as seedbanks, long-lived stages, and substantial spatial and temporal variance in demographic rates, although tools are available to address some of these problems. For plants with difficult life histories (e.g., long-lived dormant stages), it is probably best to choose a model structure that allows inclusion of this complexity, even when data are limited. With such an approach it is at least possible to explore different hypotheses and to see how model outcomes vary with assumptions about the missing stage (e.g., Chap. 7, this Vol.). Finally, in spite of their complex life histories, plants also provide some unique opportunities for the application of PVA. Many critics of PVA have advocated a greater emphasis on experimental approaches that directly explore the factors limiting a population, as an alternative to a reliance on modeling (Beissinger and Westphal 1998; Groom and Pascual 1998). Rare plant populations are much more amenable to small-scale experimental studies than many rare animals, as evidenced by the recent growth in the plant PVA literature of studies using experimental data in conjunction with modeling. Creating a stronger link between experimental and modeling approaches represents both an important opportunity and challenge for future plant PVA.

Acknowledgements. We would like to thank Mark Schwartz, Dan Doak, Bret Elderd, Felipe Dominguez-Lozano, Jason Hoeksema and an anonymous reviewer for comments and discussions that improved this manuscript.

References

Barrett SCH, Kohn JR (1991) Genetic and evolutionary consequences of small population size in plants: implications for conservation. In: Falk DA, Holsinger KE (eds) Genetics and conservation of rare plants. Oxford University Press, Oxford, pp 3–30

Beissinger SR, Westphal MI (1998) On the use of demographic models of population viability in endangered species management. J Wildl Manage 62:821–841

Benton TG, Grant A, Clutton-Brock TH (1995) Does environmental stochasticity matter: analysis of red deer life-histories on Rum. Evol Ecol 9:559–574

Bierzychudek P (1982) The demography of jack-in-the-pulpit, a forest perennial that changes sex. Ecol Monogr 52: 335–351

Bierzychudek P (1999) Looking backwards: assessing projections of a transition matrix model. Ecol Appl 9:1278–1287

Brook BW, O'Grady JJ, Chapman AP, Burgman MA, Akcakaya HR, Frankham R (2000) Predictive accuracy of population viability analysis in conservation biology. Nature 404:385–387

Brook BW, Lim L, Harden R, Frankham R (1997) Does population viability analysis software predict the behaviour of real populations? A retrospective study on the Lord Howe Island Woodhen (*Tricholimnas sylvestris* (Sclater). Biol Conserv 82:119–128

Burgman MA, Lamont BB (1992) A stochastic model for the viability of *Banksia cuneata* populations: environmental, demographic, and genetic effects. J Appl Ecol 29:719–727

Burgman MA, Possingham HP, Lynch AJJ, Keith DA, McCarthy MA, Hopper SD, Drury WL, Passioura JA, Devries RJ (2001) A method for setting the size of plant conservation target areas. Conserv Biol 15:603–616

Caswell H (1996) Second derivatives of population growth rate: calculations and applications. Ecology 77:870–879

Caswell H (2000) Prospective and retrospective perturbation analyses: their roles in conservation biology. Ecology 81:619–627

Caswell H (2001) Matrix population models: construction, analysis and interpretation. Sinauer Associates, Sunderland, MA, 722 pp

Cochrane ME, Ellner S (1992) Simple methods for calculating age-based life history parameters for stage-structured models. Ecol Monogr 63:345–363
Coulson T, Mace GM, Hudson E, Possingham H (2001) The use and abuse of population viability analysis. Trends Ecol Evol 16:219–221
Crone EE, Gehring JL (1998) Population viability of *Rorippa columbiae*: multiple models and spatial trend data. Conserv Biol 12:1054–1065
Cross PC, Beissinger SR (2001) Using logistic regression to analyze the sensitivity of PVA models: a comparison of methods based on African wild dog models. Conserv Biol 15:1335–1346
Crouse DT, Crowder LB, Caswell H (1987) A stage-based population model for Loggerhead sea turtles and implications for conservation. Ecology 68:1412–1423
DeKroon H, Plaiser A, van Groenendael J (1987). Density-dependent simulation of the population dynamics of a perennial grassland species, *Hypochaeris radicata*. Oikos 50:3–12
DeKroon H, van Groenendael J, Ehrlen J (2000) Elasticities: a review of methods and model limitations. Ecology 81:607–618
Dennis B, Taper ML (1994) Density dependence in time series observations of natural populations: estimation and testing. Ecol Monogr 64:205–224
Dennis B, Munholland PL, Scott JM (1991) Estimation of growth and extinction parameters for endangered species. Ecol Monogr 61:115–144
Doak D, Kareiva P, Klepetka B (1994) Modeling population viability for the desert tortoise in the Western Mojave Desert. Ecol Appl 4:446–460
Doak DF, Thomson DM, Jules EK (2002)PVA for plants: understanding the demographic consequences of seed banks for population health. In: Beissinger S, McCullough D (eds) Population viability analysis. University of Chicago Press, Chicago
Easterling MR, Ellner SP, Dixon PM (2000) Size-specific sensitivity: applying a new structured population model. Ecology 81:694–708
Ellstrand NC, Elam DR (1993) Population genetic consequences of small population size: implications for plant conservation. Annu Rev Ecol Syst 24:217–242.
Erikkson O (1996) Regional dynamics of plants: a review of evidence for remnant, source-sink, and metapopulations. Oikos 77:248–258
Fagan WF, Meir E, Predergast J, Folarin A, Kareiva P (2001) Characterizing population vulnerability for 758 species. Ecol Lett 4:132–138
Fieberg J, Ellner SP (2000) When is it meaningful to estimate an extinction probability? Ecology 81:2040–2047
Fieberg J, Ellner SP (2001) Stochastic matrix models for conservation and management: a comparative review of methods. Ecol Lett 4:244–266
Fox GA, Gurevitch J (2000) Population numbers count: tools for near-term demographic analysis. Am Nat 156:242–256
Gerber LR, DeMaster DP (1999) A quantitative approach to Endangered Species Act classification of long-lived vertebrates: application to the North Pacific humpback whale. Conserv Biol13:1203–1214
Grant A, Benton TG (2000) Elasticity analysis for density-dependent populations in stochastic environments. Ecology 81:680–693
Groom MJ, Pascual MA (1998) The analysis of population persistence: an outlook on the practice of viability analysis. In: Fiedler PL, Kareiva PF (eds) Conservation biology, 2nd edn. Chapman and Hall, New York, pp 4–27
Gross KG, Lockwood JR, Frost C, Morris WF (1998) Modeling controlled burning and trampling reduction for conservation of *Hudsonia Montana*. Conserv Biol 12:1291–1302
Hanski I (1994) A practical model of metapopulation dynamics. J Anim Ecol 63:151–162
Hanski I (1999) Metapopulation ecology. Oxford University Press, Oxford, 313 p

Hanski I, Gilpin ME (1997) Metapopulation biology; ecology, genetics and evolution. Academic Press, San Diego, 512 p

Hanski I, Ovaskainen O (2000) The metapopulation capacity of a fragmented landscape. Nature 404:755–758

Harrison S, Ray C (2002) Plant population viability and metapopulation-level processes. In: Beissinger S, McCullough D (eds) Population viability analysis. University of Chicago Press, Chicago, p 577

Heppell S, Caswell H, Crowder LB (2000). Life histories and elasticity patterns: perturbation analysis for species with minimal demographic data. Ecology 81:654–665

Homestake Mining Company (2001) MacLaughlin Mine: annual monitoring report. 1 July 2000–30 June 2001. Homestake Mining Company, Napa County, California

Horvitz C, Schemske DW, Caswell H (1997) The relative "importance" of life-history stages to population growth: prospective and retrospective analyses. In: Tuljapurkar S, Caswell H (eds) Structured population models in marine, terrestrial and freshwater systems. Chapman and Hall, New York, pp 247–271

Husband BC, Barrett SCH (1996) A metapopulation perspective in plant population biology. J Ecol 84:461–469

Kalisz S, McPeek M (1992) Demography of an age-structured annual: resampled projection matrices, elasticity analyses, and seed bank effects. Ecology 73:1082–1093

Kaye TN (2001) Population viability analysis of endangered plant species: an evaluation of stochastic methods and an application to a rare prairie plants. PhD Thesis, Oregon State University, Corvallis, OR

Kaye TN, Pendergrass KL, Finley K, Kauffman JB (2001) The effect of fire on the population viability of an endangered prairie plant. Ecol Appl 11:1366–1380

Kindvall O (2000) Comparative precision of three spatially realistic simulation models of metapopulation dynamics. Ecol Bull 48:101–110

Lande R (1993) Risks of population extinction from demographic and environmental stochasticity and random catastrophes. Am Nat 142:911–927

Lesica P (1995) Demography of *Astragalus scaphoides* and effects of herbivory on population growth. Great Basin Nat 55:142–150

Levins R (1969) Some demographic and genetic consequences of environmental heterogeneity for biological control. Bull Entomol Soc Am 15:237–240

Lindenmayer DB, Ball I, Possingham HP, McCarthy MA, Pope ML (2001) A landscape-scale test of the predictive ability of a spatially explicit model for population viability analysis. J Appl Ecol 38:36–48

Lonsdale WM, Braithwaite RW, Lane AM, Farmer J (1998) Modelling the recovery of an annual savanna grass following a fire-induced crash. Aust J Ecol 23:509–513

Ludwig D (1996) Uncertainty and the assessment of extinction probabilities. Ecol Appl 6:1067–1076

Ludwig D (1999) Is it meaningful to estimate a probability of extinction? Ecology 80:298–310

Mangel M, Tier C (1993) A simple direct method for finding persistence times of populations and application to conservation problems. Proc Natl Acad Sci USA 90:1083–1086

Maschinski J, Frye R, Rutman S (1997) Demography and population viability of an endangered plant species before and after protection from trampling. Conserv Biol 11:990–999

McCarthy MA (1996) Extinction dynamics of the helmeted honeyeater: effects of demography, stochasticity, inbreeding and spatial structure. Ecol Model 85:51–163

McCarthy MA, Burgman MA, Ferson S (1995) Sensitivity analysis for models of population viability. Biol Conserv 73:93–100

McCarthy MA, Possingham HP, Day JR, Tyre AJ (2001a) Testing the accuracy of population viability analysis. Conserv Biol 15:1030–1038

McCarthy MA, Possingham HP, Gill AM (2001b) Using stochastic dynamic programming to determine optimal fire management for *Banksia ornata*. J Appl Ecol 38:585–592

Meir E, Fagan WJ (2000) Will observation error and biases ruin the use of simple extinction models? Conserv Biol 14:148–154

Menges ES (1990) Population viability analysis for an endangered plant. Conserv Biol 4:52–62

Menges ES (2000a) Applications of population viability analyses in plant conservation. Ecol Bull 48:73–84

Menges ES (2000b) Population viability analyses in plants: challenges and opportunities. Trends Ecol Evol 15:51–56

Menges ES, Dolan RW (1998) Demographic viability of populations of *Silene regia* in Midwestern prairies: relationships with fire management, genetic variation, geographic location, population size, and isolation. J Ecol 86:63–78

Mills LS, Hayes SG, Baldwin C, Wisdom MJ, Citta J, Mattson DJ, Murphy K (1996) Factors leading to different viability predictions for a grizzly bear data set. Conserv Biol 10:863–873

Mills LS, Doak DF, Wisdom MJ (1999) Reliability of conservation actions based on elasticity analysis of matrix models. Conserv Biol 13:815–829

Moloney KA (1986) A generalized algorithm for determining canopy size. Oecologia 69:176–180

Morris W, Doak D, Groom M, Kareiva P, Fieberg J, Gerber L, Murphy P, Thomson D (1999) A practical handbook for population viability analysis. The Nature Conservancy Press, New York

Nantel P, Gagnon D, Nault A (1996) Population viability analysis of American ginseng and wild leek harvested in stochastic environments. Conserv Biol 10:608–621

Olmsted I, Alvarez-Bullya ER (1995) Sustainable harvest of tropical trees: demography and matrix models of two palm species in Mexico. Ecol Appl 5:484–500

Oostermeijer JGB (2000) Population viability analysis of the rare *Gentiana pneumonanthe*: the importance of genetics, demography and reproductive biology. In: Young AG, Clarke GM (eds) Genetics, demography and viability of fragmented populations. Cambridge University Press, Cambridge, pp 313–333

Pfister CA (1998) Patterns of variance in stage-structured populations: evolutionary predictions and ecological implications. Proc Natl Acad Sci USA 95:213–218

Possingham HP, Davies I (1995) ALEX: a model for the viability analysis of spatially structured populations. Biol Conserv 73:143–150

Quintana-Ascencio PF, Menges E (1996) Inferring metapopulation dynamics from patch-level incidence of Florida scrub plants. Conserv Biol 10:1210–1219

Ricklefs RE, Schluter D (1993) Species diversity in ecological communities: historical and geographical perspectives. The University of Chicago Press, Chicago, 416 p

Ruckelhaus M, Hartway C, Kareiva P (1997) Assessing the data requirements of spatially explicit dispersal models. Conserv Biol 11:1298–1306

Schemske DW, Husband BC, Ruckelshaus MH, Goodwillie C, Parker IM, Bishop JG (1994) Evaluating approaches to the conservation of rare and endangered plants. Ecology 75:584–606

Shaffer ML (1983) Determining minimum viable population sizes for the grizzly bear. Int Conf Bear Res Manage 5:133–139

Shea K, Rees M, Wood SN (1994) Trade-offs, elasticities, and the comparative method. J Ecol 82:951–957

Silvertown J, Franco M (1993) Comparative plant demography: relationship of life-cycle components to the finite rate of increase in woody and herbaceous perennials. J Ecol 81:465-476

Silvertown J, Franco M, Menges E (1996) Interpretation of elasticity matrices as an aid to the management of plant populations for conservation. Conserv Biol 10:591–597

Sjögren-Gulve P (1991) Extinction and isolation gradients in metapopulations: the case of the pool frog (*Rana lessonae*). Biol J Linn Soc 42:135–147

Sjögren-Gulve P, Hanski I (2000) Metapopulation viability analysis using occupancy models. Ecol Bull 48:53–72

Sjögren-Gulve P, Ray C (1996) Using logistic regression to model metapopulation dynamics: large-scale forestry extirpates the pool frog. In: McCullough DR (ed) Metapopulations and wildlife conservation. Island Press, Washington, DC, pp 111–137

Tilman D, Pacala S (1997) Spatial ecology. Princeton University Press, Princeton, 368 pp

Valverde T, Silvertown J (1997) A metapopulation model for *Primula vulgaris*, a temperate forest understory herb. J Ecol 85:193–210

Vandermeer J (1978) Choosing category size in a stage projection matrix. Oecologia 32:79–84

Watkinson AR (1990) The population dynamics of *Vulpia fasciculata*: a nine-year study. J Ecol 78:196–209

Werner P, Caswell H (1977) Population growth rates and age versus stage-structured models for teasel (*Dipsacus sylvestris* Huds.). Ecology 58:1103–1111

Wisdom MJ, Mills LS, Doak DF (2000) Life stage simulation analysis: estimating vital-rate effects on population growth for conservation. Ecology 81: 628–641

Wisdom M, Mills L (1997) Sensitivity analysis to guide population recovery: prairie-chicken as an example. J Wildl Manage 61:302–312

7 The Problems and Potential of Count-Based Population Viability Analyses

B.D. ELDERD, P. SHAHANI, and D.F. DOAK

7.1 Introduction

The field of conservation biology is focused on protecting species, and thus populations, from declines leading to extinction. This goal is a clear one; however, knowing when a population is in danger of extinction is the tricky first step to any process of species protection. To evaluate the conservation status of a population, biologists increasingly use a suite of methods collectively known as population viability analysis (PVA: Beissinger and Westphal 1998; Soulé 1987). With these mathematical models, we can use data from a population to predict whether it is on average declining, recovering, or persisting at a fairly constant size. By quantifying variability in population growth and decline, these models can also yield predictions of various measures of extinction risk, including mean time to extinction and the ultimate probability of extinction.

For the most part, traditional methods of PVA involve the use of age- or stage-structured matrix models to project populations forward through time and to estimate population growth rates (e.g., Crouse et al.1987; Menges 1990; Chap. 6, this Vol.). These models are particularly useful for PVAs, because they are built upon detailed biological knowledge of a target population and thus support sensitivity analyses that can help inform decisions about the life-stages and vital rates on which to focus our time and resources. However, with these advantages come heavy data requirements that make proper parameterization of matrix models difficult, or at times impossible (Beissinger and Westphal 1998; Chap. 6, this Vol.). Estimating growth rates for each age or life stage of the population is laborious, requiring careful and detailed work to enumerate the fates of individual plants – tagging and measuring plants of a range of stage classes, and repeatedly revisiting them to estimate transition rates for several years. Worse, especially for long-lived species, mortality rates are often difficult to estimate accurately with sample sizes that are adequate for the study of growth and reproduction, requiring even larger numbers of

Ecological Studies, Vol. 165
C.A. Brigham, M.W. Schwartz (Eds.)
Population Viability in Plants

plants to be followed. Finally, estimating recruitment rates generally requires setting up further experimental plots (i.e., seed addition experiments). Assembled into a population matrix, these demographic data are extremely useful in guiding effective management. However, this utility hinges on pre-existing data being available, or on the resources and time to conduct the fine-scale studies needed to acquire diverse demographic data. This latter situation is most common for plant studies, since few rare plants have been studied thoroughly enough to allow a full demographic characterization without further work.

Thus, in many circumstances, data limitations severely hamper our ability to use the usual demographic approach to PVA. As an alternative to this approach, population count data can be used to predict future population trajectories and to evaluate extinction risks through the use of diffusion approximation PVA models (Dennis et al. 1991). These models are highly useful in that the type of data that they require - counts of relative population sizes through time - are more commonly available and are easier (though requiring no fewer years) to gather when not already in hand. With census data on changes in the number of individuals in the population or some subset of the population, we can estimate the mean and variance in stochastic growth rate. By approximating population growth as a diffusion process, we can then make predictions about extinction risk using metrics such as the mean, median, and modal times to extinction, the median population size at some time in the future, and the ultimate probability of extinction (Dennis et al. 1991; Gerber et al. 1999; Morris et al. 1999; Chap. 6, this Vol.).

This general density-independent diffusion-approximation (DA) method of PVA, elucidated by Dennis et al. (1991), holds considerable promise as a way to rigorously analyze population viability when demographic approaches are not feasible. However, its usefulness as a conservation planning tool relies upon numerous poorly explored factors, including the length of the time series available and the reliability of census estimates. Use of this method (Nicholls et al. 1996; Lima et al. 1998; Gerber et al. 1999), as well as attacks on its utility (Ludwig 1996, 1999; Fieberg and Ellner 2000), have become increasingly common but there are still few accessible reviews of the approach or syntheses of the method's problems and potential. In this chapter, we will first describe the basic DA approach, and then assess the quality of results it provides, evaluating and addressing some of the criticisms that it has faced in recent years. We will also ask how this approach performs when faced with a common problem for many plant species, the presence of an important but unseen life stage such as seeds in a seed bank.

7.1.1 A Genealogy of Count-Based PVA

The construction and use of population dynamic models has a long history, with many mathematicians and ecologists contributing advances and techniques that have enabled the development of diffusion-based extinction-time models. Here, we present a brief history of some of the work that contributed most directly to the formulation of the density-independent diffusion-based PVA models on which we focus. (We encourage readers who are totally unfamiliar with these methods and models to skip straight to the next section, which describes how to implement the approach.)

The first stochastic population model that seems to have contributed substantially to the formulation of current diffusion-based models was developed by Feller in 1939 (translation appears in Oliveira-Pinto and Conolly 1982). This model drew attention to the importance of considering temporal variation in birth and death rates for making population projections. Later models were used to predict extinction probabilities for populations with this kind of demographic stochasticity in birth and death rates (Kendall 1949).

Goel and Richter-Dyn (1974) demonstrated that, for unstructured populations, such discrete birth-death processes could be approximated by a continuous diffusion process (a random walk approximation with random deviations around a central tendency to grow or shrink), allowing more powerful mathematical analyses to bear on the prediction of extinction risk and timing. Several diffusion-based models were then developed that incorporated either environmental and/or demographic stochasticity into the population growth process (Gillespie 1972; May 1973; Capocelli and Ricciardi 1974; Karlin and Levikson 1974; Turelli 1978; Leigh 1981; Tier and Hanson 1981). Three of these models also included methods to derive extinction time estimates as a function of variance in either birth and death rates (Leigh 1981) or population size (Capocelli and Ricciardi 1974; Leigh 1981; Tier and Hanson 1981) and have contributed substantially to the development of diffusion process models used for PVAs today.

Parallel to this development of unstructured models (models that do not include age- or stage-based differences between individuals) to predict extinction times, a rich literature developed on how stochastic variation in age-specific demographic rates would influence the behavior of matrix models (Cohen 1977, 1979; Tuljapurkar and Orzack 1980; Tuljapurkar 1982). Of particular importance was work by Tuljapurkar (1982) that provided approximations for the mean and variance of stochastic population growth for matrix models. This development allowed the behavior of complex populations, with their stage-specific demographic rates (including means, variances, and correlations between the variability in these rates), to be summarized in the same measures of population change used by simple, unstructured models for extinction times.

Lande and Orzack (1988) used these more biologically realistic age-structured models to test how well unstructured diffusion process models represent the dynamics of populations with complex life histories. In particular, Lande and Orzack (1988) showed that the mean and variance in growth rate of an age-structured population, approximated by Tuljapurkar's method, are similar to the estimates derived if the population is "simplified" and modeled by a diffusion process such as that developed by Capocelli and Ricciardi (1974). The use of the diffusion process allows for the calculation of all the extinction risk measures previously computable for only very simple models of population growth. In other words, Lande and Orzack (1988) showed that an estimate of overall population dynamics, which can be derived from simple census data (repeated counts of all or part of a population across several years), could be used to give good approximations of the growth rate and extinction risk of a complex, stage-structured population.

7.1.2 The Basics of Count-Based PVA

Although the theoretical developments just described created the potential to use DAs of population growth to assess extinction risks from count data, a clear set of methods to do so was still lacking. Dennis et al. (1991) (and similar work by Braumann 1983) made these mathematical advances useful in conservation biology by showing how to arrive at unbiased maximum likelihood estimates of mean instantaneous stochastic growth rate (μ, the stochastic equivalent to r in a deterministic exponential growth model) and variance in stochastic growth (σ^2) using simple linear regression methods*. While the method of Dennis et al. (1991) ignores the effects of demographic stochasticity, it incorporates the generally more important effects of environmental stochasticity. Of particular importance is that the method can also handle complications arising from unequally spaced population counts, as well as other potential problems associated with real data sets (Dennis 1989). These developments have yielded a method of conducting PVA that is straightforward and uses the type of data that are most often collected on species of conservation concern (Morris et al. 1999).

Using a series of three steps, the DA method of Dennis et al. (1991) translates population count data into predictions of growth rate and extinction risk. As an example, consider a hypothetical population of Hooker's fairybell (*Disporum hookeri*), a common lily (Table 7.1). Values for μ and σ^2 are esti-

* In a technical sense, Dennis et al. (1991) uses a diffusion process based on simple Brownian motion to describe the population trajectory of a density-independent population for which numbers are driven by both an underlying growth rate and by environmental stochasticity. The method uses simple census counts and regression analysis to estimate the infinitesimal mean and variance of this diffusion process, defined as μ and σ_2.

mated by first transforming census data (dates and population counts) so that they can be described by a linear model of the rate of population change across a time interval versus the length of the time interval. In particular, raw census data should be transformed using the following equations:

$$x = \sqrt{t_j - t_i}$$

$$y = \frac{\ln\left(N_j - N_i\right)}{x}$$

where N_i and N_j are adjacent census counts from years t_i and t_j. Thus, x is a measure of the time elapsed between two censuses and y is a measure of population growth over this interval (Table 7.1). These transformations are necessary to equalize the variances in population growth over intervals of different lengths, allowing a regression of y on x to fit the assumptions of a simple linear model. Indeed, the next step is to perform a linear regression of y on x (setting the y-intercept at zero; no growth can occur with no change in time), which yields estimates of μ and σ^2. The slope of the best-fit line gives an estimate of μ, and the variance of the individual data points about this line gives an estimate of σ^2 (Fig. 7.1; Dennis et al. 1991). From a standard regression table, the slope, or μ, is thus given by the regression coefficient (or x-coefficient), and the variance, or σ^2, is given by the mean square residual from the regression analysis. For the Hooker's fairybell data, μ=0.085 and σ^2=3.22.

Table 7.1. Census data for a hypothetical population of *Disporum hookeri*. Here, x is a transformation of the time intervals between censuses $\left(x = \sqrt{t_j - t_i}\right)$, and y is a log-transformation of population growth between two censuses ($y=\ln(N_j-N_i)/x$). Note the skipped years, which lead to larger values of x

Year	Count	x	y
1982	129		
1983	341	1	0.97207
1985	597	1.414214	0.396004
1986	172	1	-1.24442
1989	356	1.732051	0.419986
1990	142	1	-0.9191
1991	476	1	1.209591
1992	9	1	-3.96819
1994	477	1.414214	2.80742
1995	934	1	0.67196
1996	198	1	-1.55121
1997	465	1	0.85377

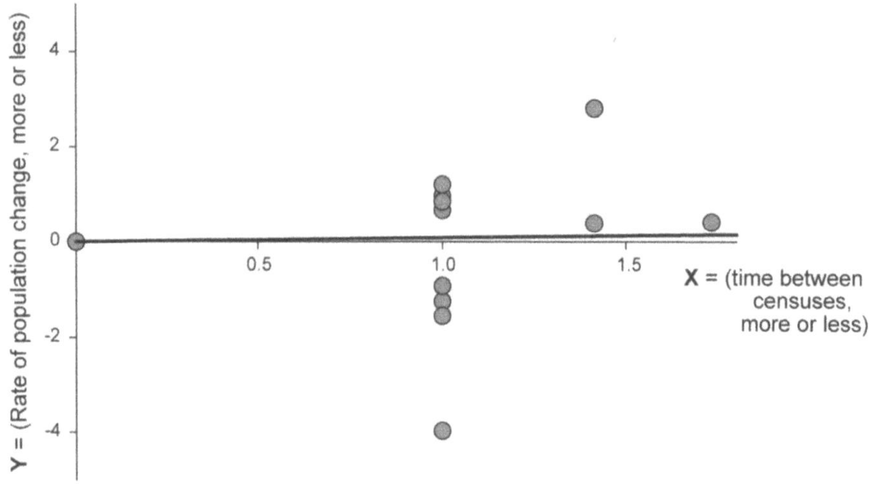

Fig, 7.1. Linear regression of y on x for Hooker's fairybell data, where y is the log-transformation of population growth between two censuses, and x is a transformation of time between censuses (equations given in text). The slope of the regression line gives an estimate of μ for the population, and the scatter of points about this line gives σ^2; these values are then used for further calculations in the diffusion approximation (DA) method

These values, together with a starting population size and pseudo-extinction threshold (the population size at which the population is considered critically endangered or essentially extinct), can be used to calculate various extinction risk measures. Incorporating information on the number of censuses and the length of the time series of data (in years) also allows calculation of confidence intervals about the estimates (Dennis et al. 1991). For Hooker's fairybell, we chose a starting population size of 465, which was the size of the population in the last census, and a pseudo-extinction threshold of four individuals. Conducting the various calculations described by Dennis et al. (1991) using these values yields an ultimate probability of extinction of 0.76, a mean time to extinction of 56 years, and a median time to extinction of 13 years for those trajectories in which the population does go extinct. Thus, this population has a fairly high probability of extinction, and extinction is likely to occur fairly soon; if this was a rare species, we might decide to invest effort into managing this population for recovery. The full results of the approach are best summarized by the cumulative distribution function (CDF) of extinction times (Fig. 7.2). Such a CDF can be used to infer values for multiple extinction measures, such as median time to extinction and the probability of extinction by a given time (100 years, for example). For more information on how to perform the calculations to yield these extinction risk estimates, see Dennis et al. (1991) and Morris et al. (1999). As noted above, the method of Dennis et al. (1991) yields not just these best-fit predictions, but also confidence intervals around all estimates, allowing careful assessment of

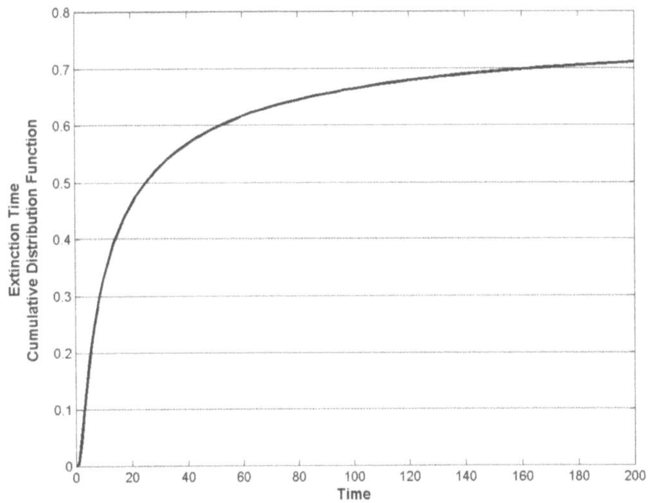

Fig. 7.2. Cumulative distribution function (CDF) for the probability of extinction for Hooker's fairybell. The *line* shows the probability of extinction on or before a certain time in the future. Here, we present the uncorrected CDF, which asymptotes at the ultimate extinction probability for the population (=0.76 in this case). Note that many authors present the conditional extinction time CDF, which shows the probability of extinction on or before a certain time *conditional* on extinction ever occurring. These conditional CDFs must asymptote to 1.0

our confidence in the extinction predictions generated by the DA (e.g., Gerber et al. 1999). Recent work by Holmes (2001) extended this method by providing new estimation methods for DA modeling when observation errors are especially high.

7.1.3 Problems and Criticisms of the DA Method of PVA

The minimum amount of data required and ease of parameterization of the DA model are major advantages that enable its use in the wide variety of situations in which only census data are available. The DA method is thus becoming increasingly popular, with rapid growth in the number of studies implementing it to estimate extinction risk. It has now been used to assess the viability of populations of over 60 different species in published studies (see Table 7.2), and many more in unpublished work. Interestingly, almost all of these applications have been restricted to mammalian and avian taxa; the only exceptions to this that we are aware of are its application to predict viability for a single plant population (Knowlton's cactus, Morris et al. 1999).

Despite this growth in popularity, there are a number of limitations to the DA method of PVA. The model assumes that annual population growth rates are log-normally distributed (or equivalently, that instantaneous per capita

Table 7.2. Examples of uses of the diffusion approximation (DA) method of population viability analysis (PVA) in recent studies

Species	Years of data	Source
Mammals		
Alabama beach mouse (2 populations)	7–11	Oli et al. (2001)
Blue wildebeest	10	Nicholls et al. (1996)
Cricetidae rodent (*Akodon olivaceus*)	5	Lima et al. (1998)
Cricetidae rodent (*Phyllostis darwini*)	5	Lima et al. (1998)
Didelphidae marsupial (*Thylamys elegans*)	5	Lima et al. (1998)
Eland	10	Nicholls et al. (1996)
Giraffe	10	Nicholls et al. (1996)
Grizzly bear	29	Dennis et al. (1991)
Impala	10	Nicholls et al. (1996)
Kudu	10	Nicholls et al. (1996)
North Pacific gray whale	19	Gerber et al. (1999)
Perdido Key beach mouse (2 populations)	7	Oli et al. (2001)
Roan antelope	10	Nicholls et al. (1996)
Sable antelope	10	Nicholls et al. (1996)
Tsessebe	10	Nicholls et al. (1996)
Warthog	10	Nicholls et al. (1996)
Waterbuck	10	Nicholls et al. (1996)
White rhinoceros	10	Nicholls et al. (1996)
Zebra	10	Nicholls et al. (1996)
Birds		
Breeding birds (35 spp. in the UK)	11–19	Gaston and Nicholls (1995)
California condor	16	Dennis et al. (1991)
Central Florida Red-cockaded woodpecker	12	Morris et al. (1999)
Kirtland's warbler	39	Dennis et al. (1991)
Laysan finch	20	Dennis et al. (1991)
Lesser prairie chicken	12	Morris et al. (1999)
North Carolina Red-cockaded woodpecker	11	Morris et al. (1999)
Palila	19	Dennis et al. (1991)
Puerto Rican Parrot	21	Dennis et al. (1991)
White stork	16	Engen and Sæther (2000)
Whooping crane	51	Dennis et al. (1991)
Plants		
Knowlton's cactus	11	Morris et al. (1999)

growth rates are normally distributed) and that the "noise" of environmental stochasticity is small, without catastrophes or other large changes in population growth rates from year to year. However, the distribution that the growth rate takes will vary depending upon the biology of the species being studied and the range of environmental variation that it faces. In particular, assuming that population growth rates should conform to a normal distribution omits the possibility of infrequent years that are catastrophes or bonanzas, which can have extreme effects on population dynamics, despite their infrequency (Mangel and Tier 1994; Ludwig 1996, 1999; but see Lande 1993 and Mangel and Tier 1993 for models that include catastrophes). Second, the model assumes that population growth in one time interval is not correlated with subsequent growth. However, in nature it is very common that environmental conditions correlate through time – for example, in many parts of the world droughts occur in prolonged sequences. Even more basic, the age or stage structure of a population provides a record of recent events that will always create some autocorrelation in population dynamics (Lande and Orzack 1988). Third, the model is density-independent, while either positive or negative density dependence clearly operates for many populations. Elaborations of the basic DA approach have been developed that deal with some of these problems. These include models that incorporate the effects of density dependence (Turchin and Taylor 1992; Turchin 1993; Dennis and Taper 1994; Foley 1994), catastrophes (Lande 1993; Mangel and Tier 1993), and spatial structure (Possingham and Davies 1995). Even though these models guard against some of the most obvious problems in the simple DA approach, they do so with the cost of heavier data demands and incomplete predictions of extinction-time distribution. Although some of these potential pitfalls have been addressed, particularly the use of bootstrapped simulations to provide parameter estimates for density-dependent populations (Dennis and Taper 1994), the limitations and complexity of these methods make them much more difficult to use or interpret with the sparse data usually available in conservation settings especially with respect to extinction time estimates. Due to these limitations, and given that many threatened species are likely to experience relatively density-independent growth, throughout the rest of this chapter we will focus solely on the use and analysis of the basic density-independent DA approach, evaluating its ability to accurately predict population growth rates and extinction times for plant populations.

A particular problem in using the DA method for plants is the invisibility of most seed banks. The DA method does not require absolute population counts, but does presume that a constant fraction of the total population is counted each year, so that measures of changes in the observed population accurately estimate rates of total population change. However, if a radically different fraction of the entire population is uncounted seeds versus counted adult plants in each year (as will be true for many annuals, in particular), the resulting estimates of population growth rates can be highly inaccurate, potentially skewing estimates of extinction risk. The challenges that seed banks may pose to PVA have been raised by previous authors (Higgins et al. 2000; Efford 2001; Doak et al. 2002; Chap. 6, this Vol.) but thus far, explorations of this issue have left more questions than answers.

While all of these issues are potentially important, it is not clear whether any of them doom the utility of the DA model. Like any other model, DA approaches simplify the real complexity of population dynamics, but this does not necessarily make them less useful. However, more fundamental aspects of the DA model have recently received criticism, calling into question the general usefulness of this method for predicting extinction risk. As Ludwig (1996, 1999) pointed out, it is difficult to know how much variation in population growth rate from year to year is due to the effects of environmental stochasticity versus observation error. Ludwig (1999) condemned the DA method primarily for this reason, claiming that this problem creates extremely wide confidence intervals around extinction risk estimates, rendering them largely unusable. He also criticized the lack of density dependence in the DA model, claiming that this both biases estimates of the population growth rate and artificially decreases the breadth of the confidence intervals around these estimates (Ludwig 1999).

Fieberg and Ellner (2000) examined the DA model to determine how much data would be needed to derive useful estimates of extinction probability. They found that the model's predictions of extinction probability over 100 years are highly sensitive to changes in the mean population growth rate (r where r is stochastic population growth on the log scale, with mean μ and standard deviation σ). Because of this need for accuracy in estimates of r, they conclude that (even with perfect data) predictions of extinction probability can only be reliably made for very short time horizons – 10–20 % of the number of years that censuses were conducted – making the DA model of extremely limited value in predicting extinction risk.

However, the work of other authors counters some of these criticisms. Meir and Fagan (2000) explored the impact of observation error in population counts in degrading the accuracy of extinction predictions. They partitioned this error into two types: bias in measurements (leading to systematic over- or underestimates) and random error in measurements. They found that overestimation bias has little effect on predictions of extinction dynamics, with 400 % overestimation required to create significant error in predictions of extinction probability. The patterns for underestimation bias and for random observation error are also encouraging; it is only when both the population growth rate and the variance due to stochasticity are low that extinction probability predictions may deteriorate in quality. So, for scenarios in which there is little risk of extinction (high r), or a high risk of extinction (low r and high σ), predictions of extinction risk do not suffer when data includes random observation error. However, in more ambiguous cases (low r and low σ) we may need to be more cautious in our interpretation of model results, as they are much more sensitive to observation error. This difference in the results obtained by Fieberg and Ellner (2000) and Meir and Fagan (2000) is likely due to the specific scenarios that they examined and the questions they asked. Fieberg and Ellner (2000) focused on the more ambiguous cases, in which r is zero or close to zero, and σ is also low. Meir and Fagan (2000) examined a wider range of scenarios, and while they did find results similar to Fieberg and Ellner's for populations with low r and low σ values, they found more positive results for other scenarios. In addition, Meir and Fagan only explored the effects of observation error on relative predictive power, and did not examine the absolute accuracy of extinction predictions (with or without observation errors).

In a second defense of PVA models, Brook et al. (2000) responded to critiques of Ludwig and others by analyzing the performance of demographic stage-structured PVA models, comparing model predictions to the actual trajectories of populations after the end of the data collection period used to parameterize the models. They found that, for the most part, true population trajectories correlate well with PVA predictions. This result is encouraging and suggests that, at least when comparing a range of populations with differing dynamics, PVA models may do a good job of ranking relative risk or general population trends. These demographic PVA models, while more complicated than DA models, share many of the same simplifying assumptions, and these results indicate that these features do not doom the resulting predictions to be hopelessly biased or imprecise.

Taken as a whole, the past literature provides some support for the utility of the DA approach, but still suggests that without a long time-series of data the predictions of DA models may either be uselessly broad or quite inaccurate. However, the analyses that have come to these conclusions have generally asked about the exact precision of the model predictions, which is at best a poor approximation to the way PVA results are most often, or at least most reasonably, used. More often than not, DA models have been used to make more qualitative judgements of risk, to assess relative risk of different populations, or to evaluate the effectiveness of conservation efforts on a whole host of species in a protected or managed area (Table 7.2). Additionally, efforts to ascertain the reliability of the method have only analyzed the forecasting of extinction risk based on estimation of μ and σ^2. In doing so, these approaches have overlooked the fact that much of the variability in extinction risk comes from differences in initial population size, the third piece of information that comes from a set of count data. Finally, past work has generally not asked about specific life history features that may hinder or aid the utility of the DA method.

Next, we describe the modeling approach that we used to address these questions. Unlike past work that evaluates the absolute accuracy of DA estimates, we emphasize the comparative use of DA results (which of a suite of populations has the highest, and which the lowest, growth rate or extinction risk?). We also try to better simulate the use of data sets by real practitioners; in particular, we include initial population size (the final size of the census data available) as a piece of information to be gleaned from census data and used to project future risk. Finally, we look at the problem of unseen life history stages such as seed banks, that may alter the utility of DA-based PVAs for plants.

7.2 Methods

To examine whether the DA approach can provide useful information when based upon a reasonable amount of data, we constructed a simulation model to compare DA predictions with a known population process. This modeled or "true" population is stage-structured and is governed by a density-independent stochastic transition matrix. All simulations were initiated with 500 individuals arranged in the stable stage class vector for the mean matrix of that simulation. Both survival and fecundity rates were allowed to vary between years according to assigned means and variances. Matrix elements involving growth and survival were drawn from a beta distribution (i.e., a probability distribution bounded by 0 and 1), and fecundity rates from a log-normal distribution. In all simulations we bounded total survival in each year of each class by 1.0, proportionately rescaling the stochastically chosen matrix elements for a stage if their sum exceeded one. The correlations between the vital rates of the population were also varied. We report results for simulations using a correlation coefficient of either 0.08 or 0.80 between all variables.

Each simulation consisted of an initial 50-year "past" period, over all or part of which census data were collected to estimate future viability, followed by a "future" period in which we continued to simulate the population to observe its fate. The future period was set at 50 years, or until the population hit a pseudo-extinction threshold of four individuals, for all simulations. We chose a 50-year time horizon to predict population performance as this seems a reasonable period over which to make management decisions and over which useful predictions of population health might be possible.

All simulations reported here are based on survival and fecundity estimates for a perennial monocot, *Calochortus obispoensis* (Fiedler 1987), whose estimated vital rates yield a deterministic rate of increase (λ) of 1.02 (Table 7.3). To evaluate how well the DA method can predict extinction risk under a range of different circumstances, we tested its performance using "true" populations that spanned a range of growth rates but were all based upon *C. obispoensis* vital rates. To create differing population dynamics, we altered both the mean and variance of a single matrix element ($a_{2,2}$, the survival, without growth, of stage class 2; Tables 7.3, 7.4). The nine resulting matrices differ in both the mean and variability of population growth (Table 7.4). In particular, the range of variances used yielded populations that differed in their annual dynamics, from populations that experience very little change in stage 2 survivorship ($a_{2,2}$) from year to year, to others that experience high variation in stage 2 survivorship between years (with very good years and very bad years being more common than "moderate" years).

Each set of simulations consisted of 5,000 runs using a single combination of the mean and variance in $a_{2,2}$. We varied the number of annual censuses upon which viability predictions were made from 5 years to 50 years. To gen-

Table 7.3. The average matrix(±1 standard deviation) for *Calochortus obispoensis* derived from Fielder (1987). The mean and variance of the $a_{2,2}$ matrix element (in *bold*) were varied away from these estimated values to create simulations with differing dynamics (see Table 7.4)

		From stage:		
		1	*2*	*3*
	1	0	0	1.73 (1.493)
To stage:	*2*	0.50 (0.490)	**0.95 (0.0141)**	0.60 (0.346)
	3	0.03 (0.045)	0.03 (0.0141)	0.38 (0.353)

Table 7.4. Parameters varied factorially (census period, demographic rate, and variance) for comparing diffusion approximation estimates to model growth rates and extinction times. The λ for the average matrix (a) as well as the population μ and σ^2 (b, c) are shown for all combinations of mean $a_{2,2}$ and Var($a_{2,2}$)

a.

Census periods (years)	Mean demographic rate $a_{2,2}$	Variance Var($a_{2,2}$)	λ (average matrix)
5	0.8750	0.0002	0.9605
10	0.9314	0.0054	1.0074
20	0.9596	0.0119	1.0314
50			

b.

		μ	
Mean $a_{2,2}$	*Var($a_{2,2}$)* *0.0002*	*Var($a_{2,2}$)* *0.0054*	*Var($a_{2,2}$)* *0.0119*
0.8750	-0.118	-0.138	-0.156
0.9314	-0.012	-0.045	-0.065
0.9596	0.030	-0.006	-0.022

c.

		σ^2	
Mean $a_{2,2}$	*Var($a_{2,2}$)* *0.0002*	*Var($a_{2,2}$)* *0.0054*	*Var($a_{2,2}$)* *0.0119*
0.8750	0.072	0.125	0.177
0.9314	0.057	0.096	0.148
0.9596	0.048	0.083	0.143

erate the same distribution of population sizes at the junction of the past and future parts of the simulations (i.e., the initial population size for predictions of future viability), regardless of the census interval, we always simulated 50 past years, as noted above. Census data were collected for the appropriate number of years prior to year 50 of the simulation. For instance, a 5-year census would be conducted from year 46 to 50 in the simulation whereas a 10-year census period would include years 41 to 50.

During the census interval, the model stores accurate census data for the population each year. To estimate the population's growth rate, its probability of extinction for 50 years, as well as the population's mean, median, and modal times to extinction we used the techniques outlined by Dennis et al. (1991). These estimates were then compared to the "true" population's behavior for each simulation. Note that in estimating extinction risks we used the census data not only to calculate μ and σ^2, but also to assign an initial population size (the final population size censused).

We also examined whether the DA method can accurately rank the relative viability of a suite of populations that do in fact differ in risk. To compare real and estimated rankings of viability, we simulated 5,000 sets of nine different populations whose growth rates were determined by separate combinations of the demographic and variance parameters in Table 7.4. We then used Spearman rank correlations to compare the ranks of the estimated stochastic population growth rates (μ) with the ranks of the realized growth rates and, similarly, the rankings of the predicted median extinction times and the realized extinction times. For all populations that did not go extinct during the simulation or whose predicted time to extinction was greater than the 50-year forecast period, we set extinction time to 51 years before conducting the Spearman rank analysis, making these values ties.

To explore the effect of unseen seed banks on the accuracy of extinction risk predictions, we ran further simulations treating the smallest size class (stage 1 in Table 7.3) as an invisible stage that could not be censused. In *Calochortus obispoensis* this smallest stage is in fact composed of small, grass-like plants, not seeds, but for our exploration of an unseeable stage, this difference is not important. In exploring the effects of an unseeable class on DA predictions, we first used the original parameter estimates from Fiedler (1987). We then tested four modifications of the basic matrix that included: (1) allowing persistence of stage 1 individuals by setting $a_{1,1}$=0.5 and reducing $a_{2,1}$ to 0.3; and/or, (2) decoupling variations in stage 1 demography from other stages by setting correlations between stage 1 rates ($a_{1,1}$, $a_{1,2}$, and $a_{1,3}$) and all other rates equal to zero.

7.3 Results

7.3.1 Predictions of Population Growth

We first asked whether the DA method would usually provide the correct *qualitative* prediction of population growth or decline. For most simulations, the DA provided a reasonable estimate of the "true" structured population's growth rate and thus its health. Figure 7.3 shows the results from a single set of population runs using the lowest demographic and variance rate. Areas in the upper right and lower left of the graph, delineated by the gray lines, correspond to regions where the model's prediction and results coincided. The upper left and lower right portion of the graph are areas where the model's prediction and results had opposite signs (e.g., the model predicted that the

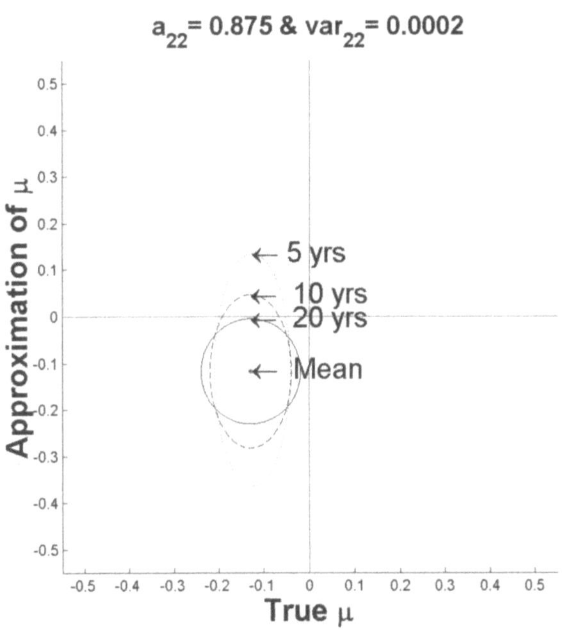

Fig. 7.3. The 95 % confidence interval (CI) ellipses for the approximation of μ as calculated by the diffusion approximation compared to actual or "true" μ realized over 50 simulation years. This plot contains the results for the mean demographic rate $a_{2,2}$=0.875 and its variance ($v_{2,2}$)=0.0002. The *ellipses* show the combined CIs for the real and estimated variation in μ across all simulations. *Ellipses* are plotted for the 5-year (*dotted ellipse*), 10-year (*dashed ellipse*) and 20-year (*solid ellipse*) census periods, as is the mean (*center asterisk*) of the simulation. The *gray horizontal* and *vertical zero lines* are used for reference; areas within the *ellipses* in the *upper right* and *lower left* of the plots represent points where the approximation and "true" μ are equivalent in sign (i.e., either both are positive or both are negative). Thus, these are areas where the model correctly predicted whether the population is growing or declining

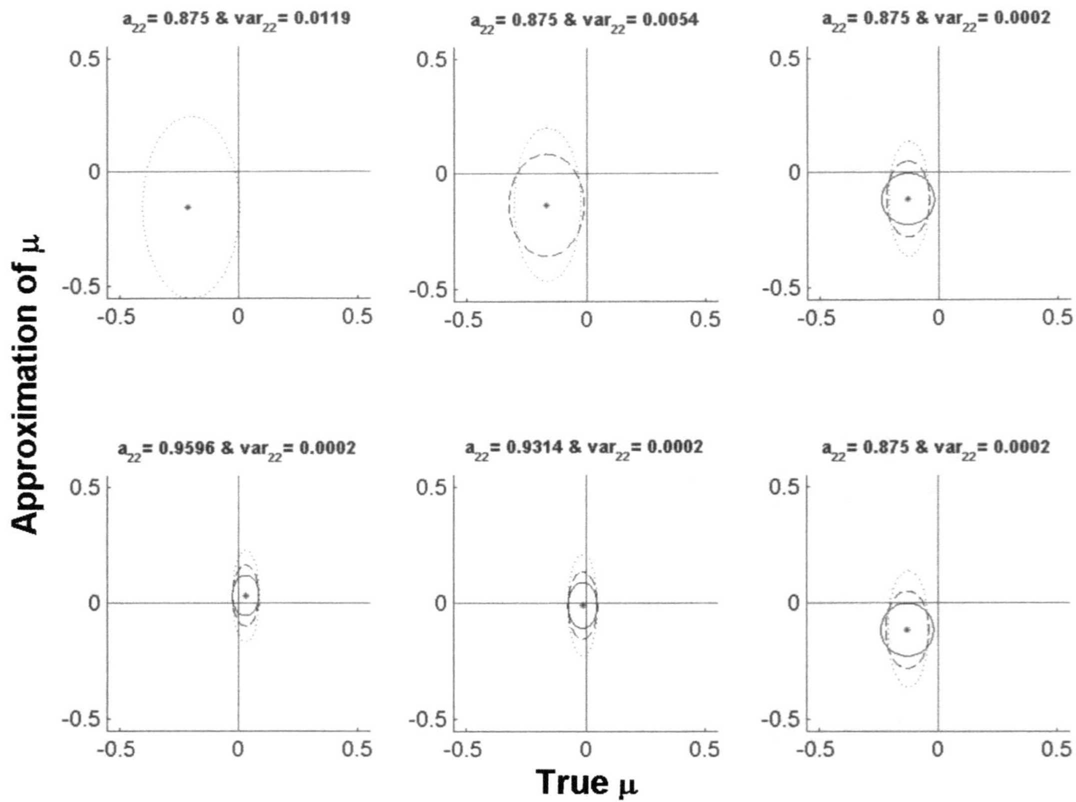

Fig. 7.4. The 95 % confidence interval (CI) ellipses for the approximation of μ as calculated by the diffusion approximation compared to actual or "true" μ realized over 50 simulation years. Within each subplot, the CI for the 5-year (*dotted ellipse*), 10-year (*dashed ellipse*) and 20-year (*solid ellipse*) census periods are shown along with the mean (*center asterisk*) of the simulation. The *horizontal and vertical gray lines* are zero lines used for reference; areas within the *ellipses* in the *upper right* and *lower left* of the plots represent points where the approximation and "true" μ are equivalent in sign (i.e., either both are positive or both are negative). The *upper row* of plots includes the results from simulations with mean $a_{2,2}$=0.875 and three variance rates. The two *upper right* plots do not contain the 20-year census period and the *far upper right* plot does not contain the 10-year census period ellipse due to non-normal distributions caused by high extinction rates. The *lower row* of plots presents the results from simulations using three different mean $a_{2,2}$ values and a single variance rate

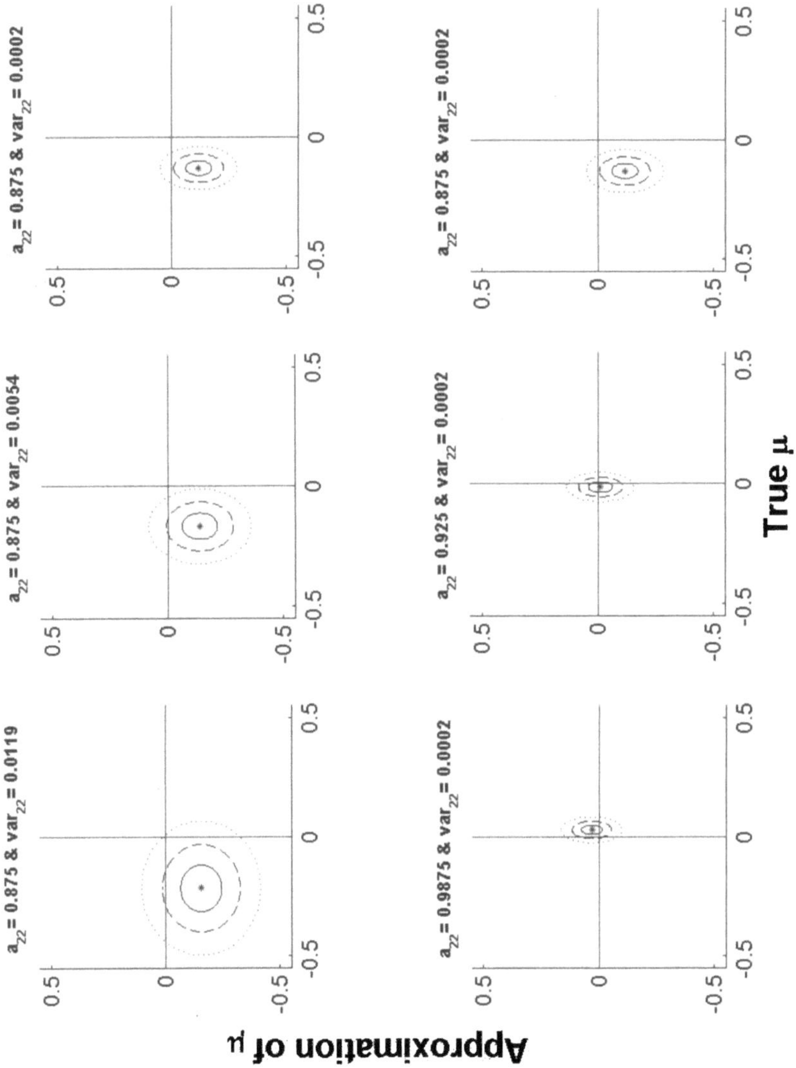

Fig. 7.5. The 95 % (*dotted ellipse*), 80& (*dashed ellipse*), and 50 % (*solid ellipse*) confidence interval (CI) ellipses and the mean (*center asterisk*) for the approximation of μ compared to actual or "true" μ for the simulation modeled. All ellipses assume a 10-year census period and are for the same mean and variance of $a_{2,2}$ used in Fig. 7.4

population should be growing; whereas the population was actually declining). Over the 5,000 replicate runs for all demographic and variance rates used, the mean predicted μ and the mean true (or realized) μ, for the 50 "future" years of the simulations, were almost identical (Fig. 7.4). Furthermore, the degree of uncertainty (difference in upper and lower confidence limits, or the confidence interval, CI) for estimated μ was comparable with the range of true uncertainty in future population trajectories (CIs of true μ) with even 10 years of data. More important is whether the 95 % CIs of predicted values largely contain values of the same sign as that of the "true" μ, indicating good qualitative predictions about long-term population growth or decline. While this predictive power was weak with 5years of census data, as the census period increased, the approximation did a good job of predicting population growth or decline, at least when the mean true μ was of large absolute value (Fig. 7.4). However, if "true" μ is close to zero, the DA predictions were much less reliable (i.e., the signs for the predicted μ and the "true" μ were switched). It is important to note, however, that over a 50-year future period, the *true* dynamics of these 5,000 populations from the different simulation runs range between growth and decline. Thus, predicting the health of a single population with $\mu \approx 0$ will always be difficult, not due so much to estimation problems as to the inherent uncertainty of vacillating dynamics over limited time horizons.

The 95 % CIs encompass all but the most extreme predicted or realized population growth rates. In asking about the basic usefulness of forecasting using the DA method, it is also worthwhile to consider narrower confidence limits. In Fig. 7.5, we plotted 50, 80, and 95 % CIs for 10 years of census data. The 80 % CIs for most of our sets of simulations largely encompass only qualitatively correct values of μ. This was even true for populations that experience a considerable amount of variation; in the simulation with the largest variance, the 80 % CI predictions were essentially all of the correct sign. This result further supports the utility of the DA method in making qualitative assessments of population viability (or the lack thereof).

7.3.2 Predictions of Extinction Risk

To examine how well the DA predicted extinction risk, we first conducted a logistic regression of whether or not the population went extinct over the 50-year "future" (or forecast) period versus the DA prediction of probability of extinction. Before this analysis was done, we examined the histograms generated by the data to verify that the assumptions of the logistic equation were not violated (e.g., most of the data points were located at either end of the distribution (i.e., they consisted mostly of ones and zeros)). Once verified, this analysis was done across all demographic and variance rates and repeated for each census period. The results show that, over many replicates, the DA predicts the probability of extinction reasonably well (Fig. 7.6). However, it tended to underestimate the probability of extinction for populations that had an extremely low chance of extinction and overestimate extinction rates for all other situations. Those populations that had low probabilities of extinction likely went extinct due to a series of extremely bad years; this type of dramatic or catastrophic drop in population size and a population's subsequent extinction were known beforehand not to be well predicted by DA (Dennis et al. 1991). As the probability of extinction rose above 10–20 %, the DA yielded more conservative estimates of extinction probability for all census periods smaller than 50 years. Again, this inaccuracy is likely to be due to the inability of the DA approach to incorporate occasional extreme years (with good years having the largest effect in these cases). Not surprisingly, as the census period increased, estimated extinction probabilities became more accurate.

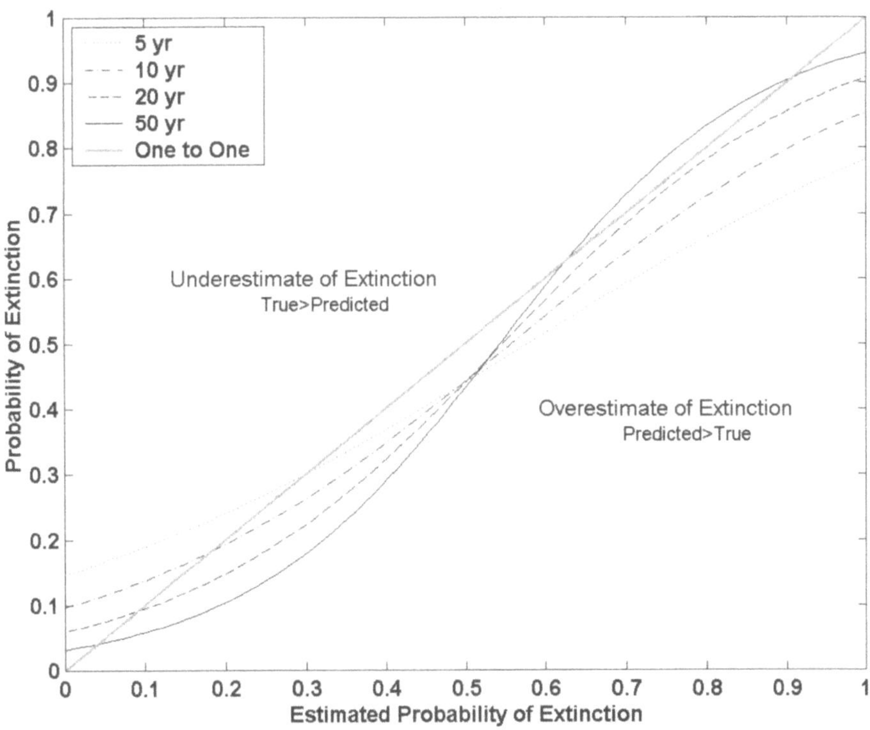

Fig. 7.6. The best fit lines of logistic regressions of the predicted probability of extinction according to the diffusion approximation versus whether or not the population went extinct over 50 years. The *one-to-one line* represents a perfect fit between "true" extinction probability and forecasted extinction probability. Areas above the one-to-one line represent regions where the model underestimated probability of extinction (i.e., liberal estimates) and areas below represent regions where the model overestimated the probability of extinction (i.e., conservative estimates)

Since the DA can, on average, give realistic estimates of extinction probabilities, how well did it predict extinction times? To answer this, we regressed the mean, median, and modal times to extinction for all populations that went extinct during the forecast period of the model against the "true" time to extinction. Although none of these three measures of extinction time relate exactly to the time to extinction, which is conditional on extinction occurring over a short time horizon (50 years for our simulations), they are the three most widely used measures of risk estimated from the DA method. The median and modal extinction time estimates were able to account for a considerable amount of variance in extinction times (Fig. 7.7). The amount of variance explained increases and asymptotes as the census period increased. However, the mean time to extinction, except when estimated with 50 years of census data, accounted for little of the variance in time to extinction. Although the predicted median and modal extinction times do provide useful estimates of extinction risk, it is worth noting that they do *not* provide good precision in estimating the "true" conditional extinction times we observed over our 50-year time horizon (personal observation). In particular, the median overesti-

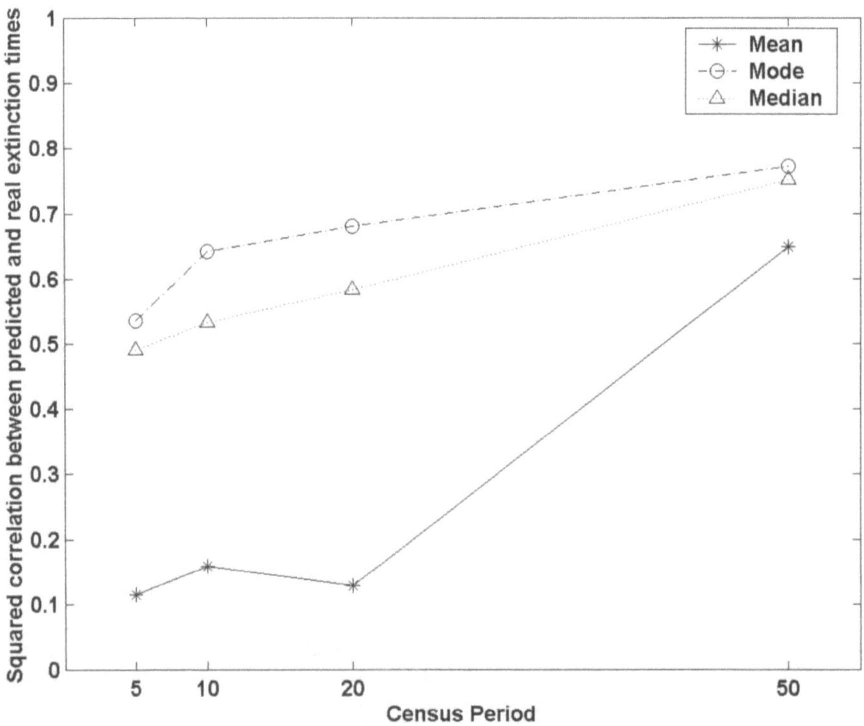

Fig. 7.7. The amount of variance explained (r^2) by a regression of "true" extinction time on either the mean, median, or modal predicted extinction time for populations that went extinct during the simulation. Values are plotted for all census periods

mates, and the mode underestimates, time to extinction. This is not surprising, given the typically skewed distribution of extinction times (Fig. 7.2), but it does show that the simple summary statistics derived from a DA analysis should be interpreted cautiously when assessing the likely timing of extinction over short time horizons.

7.3.3 Ranking Relative Risk

In addition to estimating μ for single populations, DA predictions can be used to rank populations with respect to the amount of extinction risk they face, relative to other populations. To gauge ranking accuracy, we correlated the ranking of "true" versus estimated μs for sets of nine simulated populations, each with different combinations of the mean and variance of matrix element $a_{2,2}$ (Table 7.4). Correlation between true and estimated rankings were positive for over 75% of samples for even a 5-year census period and increased with greater lengths of census data (Fig. 7.8A). Although these correlations are often far from perfect, they do suggest that even moderate amounts of census data can be useful in ranking populations for the potential for future growth. We replicated these simulations using a low correlation in variation of different demographic rates; the resulting decline in population variability substantially improves the power of the DA predictions to rank populations (Fig. 7.8B).

If the same correlation analysis is conducted for predicted and actual time to extinction (calculated from μ, σ^2, and initial population size), the correlation between the DA estimates and "true" outcomes is much stronger (Fig. 7.9). Even with high correlations in matrix element variation and only 5 years of census data, the median rank correlation is over 0.75. With increases in census period or decreases in covariance, the DA method's ability to correctly rank extinction times of populations increased even further. These results emphasize that the realistic use of initial population size estimates, along with μ and σ^2, provides the DA method with considerable predictive power, even over time spans much longer than the period over which census data were collected.

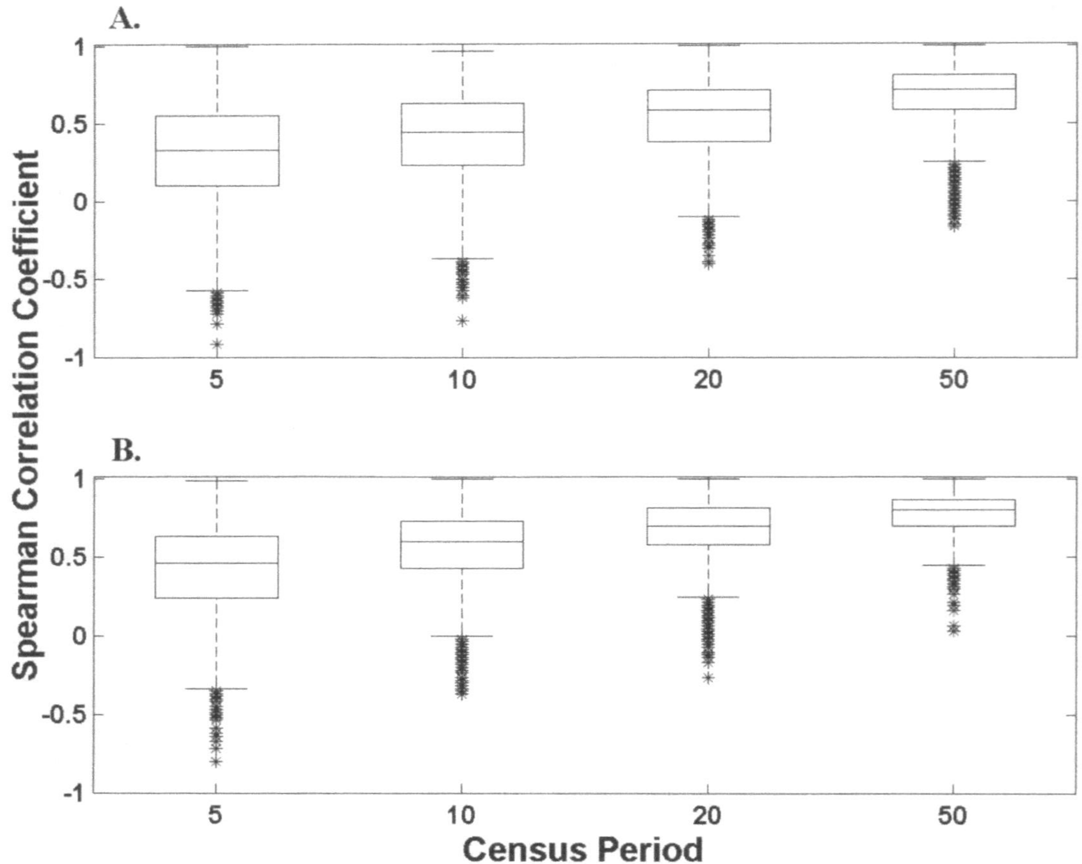

Fig. 7.8A, B. A box plot of Spearman rank correlation coefficients of "true" versus predicted μ for all census periods. Each *box* represents the upper and lower quartiles of data and is divided by a *line* representing the median. The box's whiskers extend to 1.5 interquartile ranges beyond the box's data. *Asterisks* mark all values beyond the whisker. The upper plot (**A**) contains data for simulations run with high correlation (r=0.80) between demographic rates. The lower plot (**B**) contains data for simulations run with low correlation (r=0.08) between demographic rates

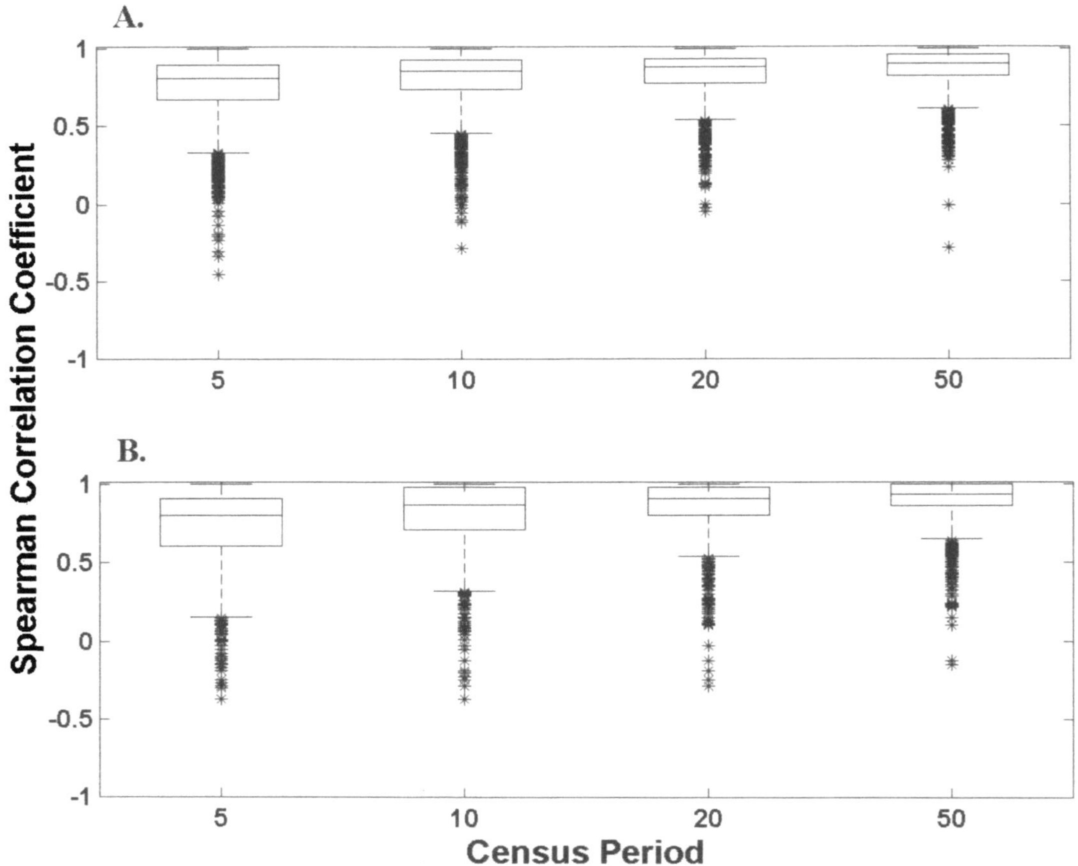

Fig. 7.9A, B. A box plot of Spearman rank correlation coefficients of extinction times (predicted versus actual) for all census periods. Each *box* represents the upper and lower quartiles of data and is divided by a *line* representing the median. The box's whiskers extend to 1.5 interquartile ranges beyond the box's data. *Asterisks* mark all values beyond the whisker. The upper plot (**A**) contains data for simulations run with high correlation (r=0.80) between demographic rates. The lower plot (**B**) contains data for simulations run with low correlation (r=0.08) between demographic rates

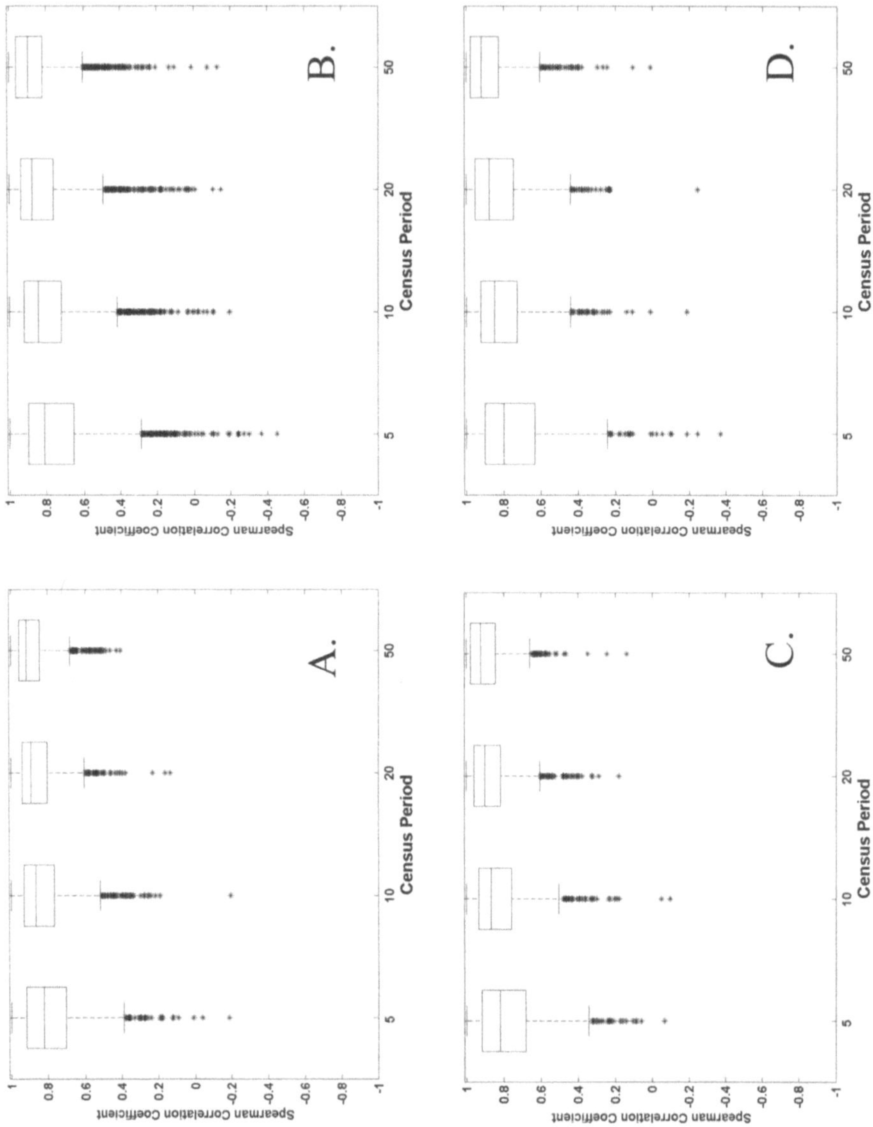

Fig. 7.10A, B. Box plots of Spearman rank correlation coefficients of extinction times for simulations in which seedlings (class 1) were not included in censuses. Plots A and B contain data for simulations run with high correlation (r=0.80) between all demographic rates. **A** Plot data are for simulations run with a perennial class 1; **B** plot data are for simulations run with the original, annual class 1. **C, D** Plots contain data for simulations run with zero correlation (r=0) between class 1 demographic rates ($a_{1,1}$, $a_{1,2}$, $a_{1,3}$) and demographic rates for other stages, but high correlation (r=0.80) between all other rates. **C** Plot data are for simulations run with a perennial class 1; **D** plot data are for simulations run with the original, annual class 1

7.3.4 Effects of an Unseen Stage

For all of the simulations of different life history variations, the predictive power of the DA method was virtually unaffected by the ability to census individuals in the smallest size class. The different variants of the size class vital rates resulted in stable size class distributions that included anywhere from 14 to 18% of the population as part of the "seed bank." Thus, even if up to one sixth of the population could not be censused, there was little difference in estimates of population health between simulations in which we were able to census this class versus those in which we were not. In particular, we found little change in the ability to rank either μ or extinction times when unable to census the smallest class (Fig. 7.10). This result held true for simulations run with the original transition matrix, and also for those in which stage 1 survivorship was increased, regardless of the level of correlation in demographic rates for seed bank stage with those for the other stages. Even making class 1 both perennial and uncorrelated in temporal variation with other stages did not change the lack of effect of seeing this stage (Fig. 7.10C). This rather surprising result seems to arise due to the high annual survivorship of the adult class (stage 3), which makes the observed part of the population the most stable. Clearly, this result needs to be tested for a range of different life histories, but it echoes Meir and Fagan's (2000) analysis, showing that observation errors of population size must be large before they markedly degrade DA predictions.

7.4 Conclusion

Overall, our results indicate that in many circumstances the DA method can provide useful information for PVA practitioners with reasonable amounts of data. In particular, the DA approach does a good job of estimating μ when the true value is not near zero. In contrast, populations with a discrete growth rate near one (i.e., μ is near zero) will be highly affected by the vagaries of environmental stochasticity. In these situations, it would be difficult to have much faith in any prediction of μ or extinction probability, as the fate of the population will largely be determined by difficult-to-estimate variability. If one examines the confidence intervals around the mean (e.g., Fig. 7.4), it becomes evident that the predictive ability of the model should be called into question in these situations. Some critics of the DA method have focused on this weakness by analyzing populations with growth rates at or near equilibrium (Fieberg and Ellner 2000). For these situations, the DA method is indeed a poor predictor, although other methods are likely to be just as bad (Ludwig 1996, 1999; Chap. 6, this Vol.). It is also important to emphasize that part of this poor predictive power arises from the inherent uncertainty of short-term outcomes when environmental variability is large. The key question to ask in a PVA is usually not what the true, long-term population behavior is, but rather what the range of likely outcomes is over a defined time horizon. With μ close to zero, these outcomes can span a wide range of values, just as can the estimated values (Fig. 7.5).

The DA method also does a reasonable job of predicting extinction risk. However, the estimates of extinction provided by the DA method are overly optimistic for populations that have an overall low probability of extinction. Thus, our results confirm that the method contains limitations when trying to forecast rare extinction events (Dennis et al. 1991). We also found that with limited data the method overestimates extinction probabilities for populations that have a high probability of extinction. Thus, the DA method overestimates extinction risk for the most endangered populations; however, it is also with regard to these populations that biologists would most want to be conservative or cautious in developing management strategies.

Not surprisingly, with more census data, predictions of extinction probability become more accurate. However, our results indicate that useful forecasts can be made without enormous amounts of data. Generally, 10 years of census data were sufficient to yield useful predictive power about extinction times (Fig. 7.7). This finding supports the results of Gerber et al. (1999), who studied the quality of DA predictions achieved with various amounts of census data, as compared with the known population dynamics from further censuses, for the North Pacific humpback whale. They found that predictions made with 11 years of census data were substantially more accurate than predictions made with even 8 years of data, but that the quality of predictions did not improve drastically when informed by further years of census data. Although their study focused on a rapidly recovering population with very low temporal variation in population growth, our results suggest that such limited data are useful in many other circumstances as well.

By far, our strongest results were from the ranking of the relative population growth and risk of extinction among a suite of populations with varying growth rates and variances. We found that the rankings of estimated μ values correlated reasonably well with the true rankings. More impressively, rankings of extinction predictions correlated extremely well with true rankings of extinction times. This result is in sharp contrast with more dismal analyses of the DA method (Ludwig 1996, 1999; Fieberg and Ellner 2000). The explanation for this difference is partly due to our emphasis on the ranking of risk rather than predictive precision. In addition, by tailoring our simulations closely to the real use of census data, we included the powerful effects of initial population size in determining extinction risk, which previous analyses have not.

However, in interpreting our ranking results, it is important to bear in mind the wide range of μ and σ^2 values of the populations we were ranking (Table 7.4). Attempts to rank the viability of populations with much more similar dynamics will inevitably be less successful. Still, of the nine populations used for our ranking of relative risk, there were only three different deterministic growth rates, with the rest of the differences generated by changes in variability. Thus, our results show that the DA method can not only differentiate between populations that on average grow or decline, but also between populations that experience different amounts of variation. Critics have charged that the DA method can not reasonably assess population health when confronted with a range of environmental stochasticity. In contrast, we found that the method is robust, at least in assessing the relative health of a group of separate populations.

Although the DA approach is being used with increasing frequency (Table 7.2), it has rarely been applied to plants (but see Morris et al. 1999). The biggest challenge in applying the method to plant populations seems likely to be that population censuses will often not include all stages of the population, due to the difficulty of surveying a plant population's seed bank or other small stages (e.g., the cotyledon stage of *C. obispoensis*). Although limited to only one life history, our results indicate that the DA method can still perform quite well with this limitation. In this regard, our results confirm Meir and Fagan's (2000) finding that observation errors have to be fairly large before influencing DA predictions. Still, our robust findings are probably driven by the high adult survivorship of the particular life history we used. Not including the smallest size class in census counts is certain to have much greater effects on the quality of viability predictions when the larger (adult) size classes are less long-lived. Thus, the method should be used with great caution, if at all, for short-lived perennials or annuals that have seed banks containing a large and vacillating fraction of the population's individuals.

It is important to emphasize that we base our assessment of the "usefulness" of the DA method not on a 0.95 probability of rejecting hypotheses about population growth, but upon a looser standard more appropriate to conservation management. In particular, the ability of the DA approach to more often than not predict the right qualitative dynamics of a population, even with considerable uncertainty (Figs. 7.4, 7.5) indicates that while short periods of census data are not sufficient to make definitive conclusions, they do form a firm enough basis to improve conservation assessment. In this sense, we are highly pragmatic (or even optimistic) in our evaluation of the DA method and its potential to improve PVA and conservation planning.

In summary, the DA method is a useful PVA tool that can inform decisions about the best targets for conservation efforts. Ecologists have developed a number of models in order to understand the interaction of birth and death processes that lead to either a population's growth or its decline. In particular, matrix models and their many elaborations have become standard tools for PVA analysis (Chap. 6, this Vol.). However, these models require large amounts of data. As an increasing number of plant and animal populations have become threatened with extinction, the time available to collect the necessary field data for parameterizing the more complex models is not necessarily available. The DA method employs a relatively simple technique to use count data to estimate population growth and extinction risk. For plants in particular, basic counts of individuals are easy and inexpensive to acquire, making DA methods an especially appealing way to utilize past data as well as current data from ongoing monitoring programs. While it is important to recognize the limitations and uncertainties in the results obtained from the DA method, we believe that it can serve a very useful function in the assessment of viability. Very short time-runs of data will not allow precise analyses of population dynamics, but even modest amounts of data can provide good estimates of qualitative dynamics and especially of relative risk of extinction. With its modest data needs and potential to substantially improve the biological basis of conservation decision-making, the DA approach to viability assessment deserves further use and development as an important conservation tool.

Acknowledgements. We thank Christy Brigham and Mark Schwartz for inviting us to contribute to this volume. Jennie Kluse provided crucial help and support during this project. Conversations with Bill Morris have helped to sharpen many of the ideas presented here. Brian Dennis and one anonymous reviewer provided thoughtful comments that greatly improved this manuscript. We are grateful for NSF awards to Doak, NSF IGERT support for Elderd, and USDE GANN support for Shahani.

References

Beissinger SR, Westphal MI (1998) On the use of demographic models of population viability in endangered species management. J Wildl Manage 62:821–841

Braumann CA (1983) Population growth in random environments. Bull Math Biol 45:635–641

Brook BW, O'Grady JJ, Chapman AP, Burgman MA, Akçakaya HR, Frankham R (2000) Predictive accuracy of population viability analysis in conservation biology. Nature 404:385–387

Capocelli RM, Ricciardi LM (1974) A diffusion model for population growth in random environment. Theor Popul Biol 5:28–41

Cohen JE (1977) Ergodicity of age structure in populations with Markovian vital rates, III: Finite-state moments and growth rate; an illustration. Adv Appl Probab 9:462–475

Cohen JE (1979) Comparative statistics and stochastic dynamics of age-structured populations. Theor Popul Biol 16:159–171

Crouse DT, Crowder LB, Caswell H (1987) A stage-based population model for Loggerhead sea-turtles and implications for conservation. Ecology 68:1412–1423

Dennis B (1989) Stochastic differential equations as insect population models. In: McDonald L, Manly B, Lockwood J, Logan J (eds) Estimation and analysis of insect populations. Proceedings of a Conference, Laramie, Wyoming, 25–29 January 1988. Lecture notes in statistics 55. Springer, Berlin Heidelberg New York, pp 219–238

Dennis B, Taper ML (1994) Density dependence in time series observations of natural populations: estimation and testing. Ecol Monogr 64:205–224

Dennis B, Munholland PL, Scott JM (1991) Estimation of growth and extinction parameters for endangered species. Ecol Monogr 61:115–143
Doak DF, Thomson DM, Jules ES (2002) PVA for plants: understanding the demographic consequences of seed banks for population health. In: Beissinger S, McCullough D (eds) Population viability analysis. University of Chicago Press, Chicago, pp 312–337
Efford M (2001) Environmental stochasticity cannot save declining populations. Trends Ecol Evol 16:177
Engen S, Sæther B (2000) Predicting the time to quasi-extinction for populations far below their carrying capacity. J Theor Biol 205:649–658
Feller W (1939) Die Grundlagen der Volterraschen Theorie des Kampfes ums Dasein in wahrscheinlichkeitstheoretischer Behandlung. Acta Biother 5:11–40
Fieberg J, Ellner SP (2000) When is it meaningful to estimate an extinction probability? Ecology 81:2040–2047
Fiedler PL (1987) Life history and population dynamics of rare and common mariposa lilies (*Calochortus pursh*: Liliaceae). J Ecol 75:977–995
Foley P (1994) Predicting extinction times from environmental stochasticity and carrying capacity. Conserv Biol 8:124–137
Gaston KJ, Nicholls AO (1995) Probable times to extinction of some rare breeding bird species in the United Kingdom. Proc R Soc Lond B 259:119–123
Gerber LR, DeMaster DP, Kareiva PM (1999) Gray whales and the value of monitoring data in implementing the U.S. Endangered Species Act. Conserv Biol 13:1215–1219
Gillespie JH (1972) The effects of stochastic environments on allele frequencies in natural populations. Theor Popul Biol 3:241–248
Goel NS, Richter-Dyn N (1974) Stochastic models in biology. Academic Press, New York
Higgins SI, Pickett STA, Bond WJ (2000) Predicting extinction risks for plants: environmental stochasticity can save declining populations. Trends Ecol Evol 15:516–519
Holmes EE (2001) Estimating risks in declining populations with poor data. Proc Natl Acad Sci USA 98:5072–5077
Karlin S, Levikson B (1974) Temporal fluctuations in selection intensities: case of small population size. Theor Popul Biol 6:383–412
Kendall DG (1949) Stochastic processes and population growth. J R Stat Soc B 11:230–264
Lande R (1993) Risks of population extinction from demographic and environmental stochasticity and random catastrophes. Am Nat 142:911–927
Lande R, Orzack SH (1988) Extinction dynamics of age-structured populations in a fluctuating environment. Proc Natl Acad Sci USA 85:7418–7421
Leigh E (1981) The average lifetime of a population in a varying environment. J Theor Biol 90:213–239
Lima M, Marquet PA, Jaksic FM (1998) Population extinction risks of three neotropical small mammal species. Oecologia 115:120–126
Ludwig D (1996) Uncertainty and the assessment of extinction probabilities. Ecol Appl 6:1067–1076
Ludwig D (1999) Is it meaningful to estimate a probability of extinction? Ecology 80:298–310
Mangel M, Tier C (1993) A simple direct method for finding persistence times of populations and application to conservation problems. Proc Natl Acad Sci USA 90:1083–1086
Mangel M, Tier C (1994) Four facts every conservation biologist should know about persistence. Ecology 75:607–614
May RM (1973) Stability and complexity in model ecosystems. Princeton University Press, Princeton
Meir E, Fagan WF (2000) Will observation error and biases ruin the use of simple extinction models? Conserv Biol 14:148–154

Menges ES (1990) Population viability for an endangered plant. Conserv Biol 4:52–62
Morris W, Doak D, Groom M, Kareiva, P, Fieberg J, Gerber L, Murphy P, Thomson D (1999) A practical handbook for population viability analysis. The Nature Conservancy, Arlington, Virginia
Nicholls AO, Viljoen PC, Knight MH, van Jaarsveld AS (1996) Evaluating population persistence of censused and unmanaged herbivore populations from the Kruger National Park, South Africa. Biol Conserv 76:57–67
Oli MK, Holler NR, Wooten MC (2001) Viability analysis of endangered Gulf Coast beach mice (*Peromyscus polionotus*) populations. Biol Conserv 97:107–118
Oliveira-Pinto F, Conolly BW (1982) Applicable mathematics of non-physical phenomena. Ellis Horwood, Chichester
Possingham HP, Davies I (1995) ALEX – a model for the viability analysis of spatially structured populations. Biol Conserv 73:143–150
Soulé ME (1987) Viable populations for conservation. Cambridge University Press, Cambridge
Tier C, Hanson FB (1981) Persistence in density dependent stochastic populations. Math Biosci 53:89–117
Tuljapurkar SD (1982) Population dynamics in variable environments. II. Correlated environments, sensitivity analysis and dynamics. Theor Popul Biol 21:114–140
Tuljapurkar SD, Orzack SH (1980) Population dynamics in variable environments. I. Long-run growth rates and extinction. Theor Popul Biol 18:314–342
Turchin P (1993) Chaos and stability in rodent population dynamics: evidence from non-linear time-series analysis. Oikos 68:167–172
Turchin P, Taylor AD (1992) Complex dynamics in ecological time series. Ecology 73:289–305
Turelli M (1978) A re-examination of stability in random varying versus deterministic environments with comments on the stochastic theory of limiting similarity. Theor Popul Biol 13:244–267

8 Habitat Models for Population Viability Analysis

J. ELITH and M.A. BURGMAN

8.1 Introduction

Determining what constitutes a suitable habitat for different species is a fundamental part of applied ecology and is the core component of many conservation planning strategies. Information on habitat availability underpins efforts to assess the threats faced by species and to determine the adequacy of conservation reserves (e.g. Margules et al. 1988). Expert rules are often used to categorize the response of a species to changes in available habitat (e.g. Millsap et al. 1990; Master 1991; Lunney et al. 1996; IUCN 2001). For example, the International Union for the Conservation of Nature (IUCN 2001) method uses extent and changes in the amount of habitat to determine the conservation status of species. All these applications assume that, when a model is built and management plans are developed, accurate maps of habitat area and spatial arrangement are available.

Alternatively, threats to species may be evaluated and explored with population viability analysis (PVA, Boyce 1992; Burgman et al. 1993; Possingham et al. 1993, Chap. 6, this Vol.). A PVA will produce a probability of extinction or quasi-extinction over a specified time frame for a given management scenario, providing a direct measure of the expected success of management options for rare, threatened and sensitive species. PVAs can be made spatially explicit and can incorporate information concerning the demographic attributes of a species, habitat preferences, dispersal between habitat patches, and the occurrence of disturbances or catastrophes (Chaps. 6, 7, 11, this Vol.). Data and understanding are never sufficient for precise prediction (Coulson et al. 2001), but PVAs may provide a means of synthesizing what is known, discriminating between management alternatives and establishing an ecological context for interim judgments about preferred spatial configurations of habitat.

Habitat maps are a core component of PVAs for two main reasons. Firstly, before a population model is developed, a species' spatial structure is an important consideration in deciding whether to use a single population model or a metapopulation model (Chaps. 6, 11, this Vol.). The spatial structure also pro-

Ecological Studies, Vol. 165
C.A. Brigham, M.W. Schwartz (Eds.)
Population Viability in Plants

vides the basis for the design of sampling efforts to support the estimation of parameters for disturbance, dispersal, habitat loss, and fragmentation. These parameters may be built into a population model and used to represent trends in habitat quality, extent, and spatial arrangement (Chap. 11, this Vol.). Secondly, management decisions are often spatially explicit. Once the results of a PVA are available, decisions that relate to the protection or enhancement of populations or habitat can only be implemented with a reliable habitat model. These steps require a map that reflects both the current distribution of the species and the potential suitability of the landscape to maintain viable populations and dispersal dynamics.

Habitat maps will be required over different geographic extents and at different resolutions. Maps developed across countries or continents can help to focus attention on issues of broad-scale conservation concern (Olson and Dinerstein 1998; Schwartz 1999). Maps developed with more fine-grained detail can help with conservation planning and management decisions at a more local or regional scale (Ferrier et al. 2002). Mackey and Lindenmayer (2001) define scales (global-, meso-, topo-, micro-, and nano-) that represent natural breaks in the distribution and availability of the primary environmental resources required by a species. The habitat modeling methods described in this chapter are generally applied at the meso- and topo-scales. At the meso-scale, environmental variation is largely caused by the interaction between topography and climate, and by lithology (Mackey and Lindenmayer 2001). At the topo-scale, the local topography modifies the resources arriving at and leaving a site, and soil parent material may also vary considerably (Mackey and Lindenmayer 2001). Regional planning often occurs at the topo-scale (e.g. Ferrier et al., 2002), although data collected at more detailed levels can help inform selection of the most suitable patches (Mackey and Lindenmayer 2001).

This chapter outlines the broad classes of methods available for the development of habitat maps, ranging from statistical models and climate-based envelopes to habitat suitability indices. It describes the data required to develop models and predictions for each approach and the scales at which they are appropriate, and discusses their relative strengths and weaknesses for the development of PVAs and the implementation of management recommendations. The attributes of the various habitat modeling methods are illustrated through the development of models for the shrub species *Leptospermum grandifolium*, an understory component of montane, subalpine and wet forests of southern Australia.

8.2 Methods for Building Habitat Models

The main approaches to habitat modeling can be grouped into seven classes. They are:

- Conceptual models based on expert opinion
- Geographic envelopes and spaces
- Climate envelopes
- Multivariate association methods
- Regression analysis
- Tree-based methods
- Machine learning methods

These methods differ in their data requirements, their assumptions, and the likely bias of their output. Consequently, different methods are likely to predict different habitats for a given species. The following section outlines alternative methods and gives an indication of the considerable differences between the modeling approaches.

8.2.1 Conceptual Models Based on Expert Opinion

Conceptual models are represented by habitat suitability indices (HSIs), which were introduced by the US Fish and Wildlife Service (USFWS 1980) as a means of mapping species habitats (Table 8.1). HSIs are used routinely and extensively, particularly in the United States, for estimating the impacts of management alternatives, ecological assessment, conservation planning, and the identification of steps to avoid habitat losses, or to compensate unavoidable habitat losses (Van Horne and Wiens 1991; Gray et al. 1996; Rand and Newman 1998). The method is based on the judgments of experts who identify critical variables. The HSI for a given species represents a conceptual model that relates each relevant measurable variable of the environment to the suitability of a site for the species, usually scaled between 0 and 1. Each variable is represented by a single suitability index (SI) and SIs are linked by additive, multiplicative or logical functions that reflect relationships among the variables (USFWS 1981). The form of these functions is determined by expert judgment. For example, if experts believe the suitability of land as habitat depends on the presence of all variables, such that they do not compensate for one another, then the geometric mean of the variables may best represent site suitability. If the environmental variables are compensatory (such as the presence of several food types, each of which is equally valuable), then an arithmetic mean or a logical conjunction (AND, OR) may be more appropriate. Indices may be weighted. The functions relating environmental variables to suitability may take any form.

Table 8.1. Modeling methods, data requirements and examples

Method	Key references	Minimum species data	Examples of ecological applications
Habitat suitability indices	Schamberger and O'Neil (1986)	Expert opinion	Birds in the United States (Van Horne and Wiens 1991), bandicoots in Australia (Reading et al. 1996)
Hulls and kernels	Worton (1989), Seaman and Powell (1996), O'Rourke (1998)	Presence	Black howler monkeys in Belize (Ostro et al. 1999), Burgman and Fox in press
Bioclimatic envelopes (e.g. ANUCLIM)	Busby (1991) and CRES (2002)	Presence	Kauri pine in New Zealand (Mitchell 1992), eucalypts in South Africa (Richardson and McMahon 1992)
Simple multivariate distance methods	DOMAIN: CIFOR (1999)	Presence	DOMAIN: marsupials in Australia (Carpenter et al. 1993) Other: wolf distribution in Italy (Corsi et al. 1999)
Ecological niche factor analysis (ENFA)	Hirzel(2001)	Presence	Ibex in Switzerland (Hirzel 2001) and simulated data (Hirzel et al. 2001)
Canonical correspondence analysis (CCA)	ter Braak (1986), Jongman et al. (1995)	Presence-absence	Rock outcrop vegetation, United States (Wiser et al. 1996) Norwegian flora (Birks 1996) and plants in Nevada, USA (Guisan et al. 1999)
Generalized linear models (GLMs)	McCullagh and Nelder (1989) and Agresti (1996)	Presence-absence	Eucalypts in Australia (Austin et al. 1983, 1984, 1990, 1994b), alpine plants in Switzerland (Guisan et al. 1998) and myrtle beech in Australia (Lindenmayer et al. 2000)
Generalized additive models (GAMs)	Hastie and Tibshirani (1990)	Presence-absence	Trees in New Zealand (Yee and Mitchell 1991) eucalypts in Australia (Austin and Meyers 1996) and wetland plants in The Netherlands (Bio et al. 1998)
Decision trees	Breiman et al. (1984)	Presence-absence	Vegetation mapping in Australia (Keith and Bedward 1999), forest types in USA (Lynn et al. 1995)
Neural networks	Aleksander and Morton(1990)	Presence-absence	Fish in France (Mastrorillo et al. 1997) and Himalayan river birds (Manel et al. 1999)
Genetic algorithms	Mitchell(1996)	Presence (but uses pseudo-absences)	GARP algorithm in Australia (Stockwell and Noble 1992; Stockwell 1999; Stockwell and Peters 1999)

In some ways, the general form of the HSI can be thought of as a regression model (see below), but in the case of HSIs the coefficients (or weights) are suggested by the expert rather than estimated from quantitative data. Habitat suitability maps provide a means of integrating expert knowledge and creating habitat maps that synthesize subjective interpretation of biological processes (Burgman et al. 2001). They are explicit attempts to model habitat rather than occupancy. HSIs could suit a range of mapping scales, depending on the level of detail of expert knowledge and the resolution of the predictor variables. One limitation is that any deficiency in expert knowledge is incorporated into predictions and can remain untested, because there are no species or environmental data explicitly behind the model on which to assess "goodness of fit" or any other simple measure of reliability.

8.2.2 Geographic Envelopes and Spaces

Geographic envelopes are models that focus on the geographic distribution of a species. One example is *convex hulls*. These were recommended by the IUCN (1994, 2001), which defined the "extent of occurrence" of a taxon as the area contained within the shortest continuous imaginary boundary that can be drawn to encompass all the known, inferred, or projected sites of present occurrence of the taxon (Fig. 8.1a). This boundary is a minimum convex polygon or convex hull, the smallest polygon containing all known sites in which no internal angle exceeds 180°,- i.e., in which all outer surfaces are convex (Rapoport 1982; O'Rourke 1998). The IUCN definition excludes cases of vagrancy and disjunctions within the overall distributions of taxa. A weakness of the method is that the constraint of convexity on the outer surface yields a hull with a very coarse level of resolution on its outer surface, resulting in a substantial overestimate of the range, particularly for irregularly shaped species ranges (Ostro et al. 1999; Burgman and Fox, in press).

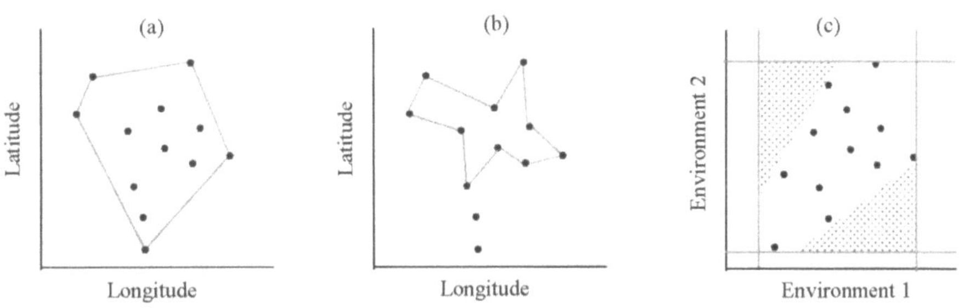

Fig. 8.1a–c. Geographic and climatic envelopes. **a** A convex hull, **b** an α-hull, **c** a two-dimensional box-type envelope. Note in the latter the *shaded areas*, which represent areas in the envelope that are unlikely to support the species

The *α-hull*, a generalization of the convex hull (Edelsbrunner et al. 1983; O'Rourke 1998), is derived from a Delauney triangulation, constructed from lines joining a set of points constrained so that no lines overlap (Fig. 8.1b). α-hulls are constructed by retaining only those vertices of the Delauney triangulation which are shorter in length than a multiple of the chosen value of the parameter α. By varying the value of α, it is possible to manipulate the configuration of the hull and provide a more realistic estimate of habitat extent when the shape of the range is irregular (see also Rapoport 1982; Burgman and Fox, in press). The outer surface of the Delauney triangulation is identical to the convex hull. The α-hull provides a more detailed description of the external shape and is capable of breaking the hull into several discrete areas when it spans an uninhabitable region (Okabe et al. 2000).

Kernel density estimators (Silverman 1986) are an example of a method that defines geograpphic 'space'. These are nonparametric statistical methods that were introduced into ecology for the estimation of home range (Worton 1989; Seaman and Powell 1996). They are also appropriate for other applications such as estimating the range of a species or population (Seaman et al. 1998) or the core area of a species' habitat (Bingham and Noon 1997). They can operate in many dimensions, but are most commonly applied in one or two dimensions (Seaman et al. 1998). The method works by placing a kernel (for example, a normal distribution) over each observation point in a sample. Density is then estimated at evaluation points, which may be each

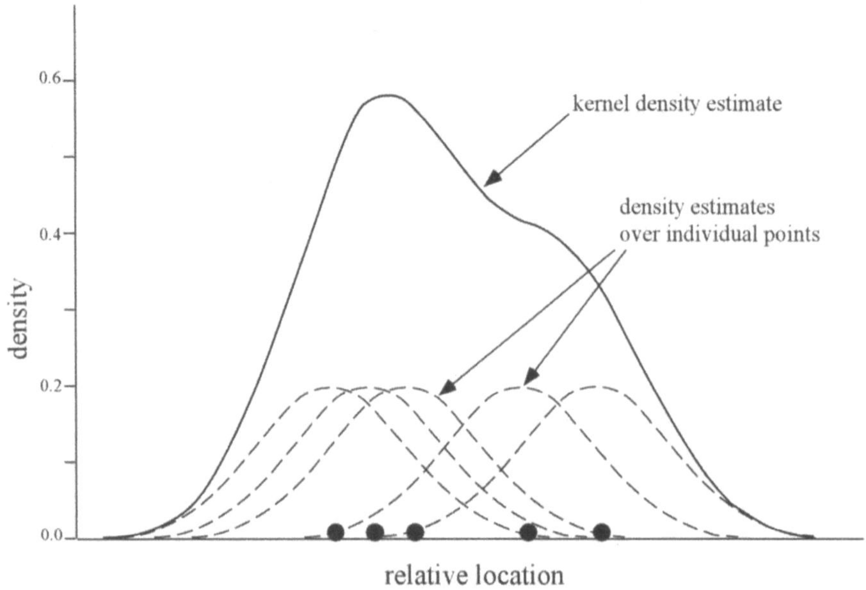

Fig. 8.2. A kernel density estimate calculated from five individual points. (After Levine 2000)

observation point or a regular grid placed over the sample. The density is the sum of all kernel values at the evaluation point (Fig. 8.2). There are different approaches to setting the widths of the kernels (Seaman and Powell 1996), resulting in different degrees of smoothness of the estimate, i.e., different resolutions of spatial scale. Kernel estimators are not substantially biased by irregular shapes (Ostro et al. 1999), and in particular the fixed kernel method generally produces more accurate estimates than the adaptive kernel method (Seaman et al. 1998).

A limitation of all envelope methods is that they are particularly sensitive to missing data and to spatial errors, which result in underestimated and overestimated ranges, respectively (Hansteen et al. 1997; Ostro et al. 1999; Burgman and Fox, in press). They are primarily useful for estimation of ranges but not for more detailed maps of species distributions, because the envelopes will generally encompass many sites that are unsuitable habitat for the species. They are usually applied at smaller cartographic scales (and over larger extents) such as the global- and meso-scales of Mackey and Lindenmayer (2001).

8.2.3 Climate Envelopes

Climate envelopes, like hulls, rely on presence-only records and consequently have the advantage that they can use data such as opportunistic records, sightings, and specimen collections. The intention in using climate rather than geographic location to map habitat is to reflect the underlying ecological processes that affect distribution. ANUCLIM, and specifically its bioclimatic prediction subsystems, BIOCLIM and BIOMAP (Busby 1991; CRES 2002; and see Table 8.1 and Fig. 8.1 c), is an example of a climate-mapping approach to modeling. Presence records are used in conjunction with elevation data and climatic surfaces developed from long-term rainfall, temperature, and radiation records to construct a climate profile for a species. With several climatic parameters the aggregated profile forms a multidimensional space known as a "climatic envelope" which defines the climatic domain of the species (McKenney et al. 1996). A habitat map can then be produced by ranking each location according to its position in the climate profile for the species. Commonly, these maps are grid-based and classify each cell into one of four ranked classes of climatic suitability for the species. A digital elevation model (DEM) is usually the elevation source data used for prediction over regions, and so the cell size of the DEM defines the geographic grain of the predictions (and the precision of the long-term data will affect their numerical precision).

The modeling concept behind ANUCLIM can be applied to any context in which there are environmental data available for sites where the species of interest has been found, and for the sites where prediction is required. Specific use of the ANUCLIM algorithm depends on the development of climate surface coefficient files for the country in question, for example, they have been developed for Australia and for Canada (McKenney et al. 1996).

A limitation of ANUCLIM is that it lacks protocols for selecting relevant climate variables. Its orthogonal geometries (originating from the use of thresholds in each dimension) are ecologically unrealistic because species do not usually occupy extreme combinations of environmental ranges and habitat is often irregularly shaped, even in environmental space (Fig. 8.1 c). Further, the use of a set of equally weighted climate variables implies the unlikely scenario that all variables have equal effect on habitat suitability as determined by their ranges in the region studied. It is unclear in what ways the use of numerous variables, many of which are highly correlated, affect the outcome of the analysis.

HABITAT (Walker and Cocks 1991) is a refinement that uses tree-based modeling (see below) to select a relevant subset of climate variables and a convex hull for defining the envelope. Other refinements of the climatic envelope concept are possible but apparently not commonly applied. For instance, the climate space could be transformed with principal components analysis (PCA, Legendre and Legendre 1998) to remove correlations between the variables. It should also be possible to apply other methods for creating the envelope (such as a kernel density estimator) to the locations in their climatic space. Part of the reason that ANUCLIM tends to be used in its original form is that it is conceptually straightforward and convenient. Typical applications are to define a broad region that is climatically suitable for a species and use this, for example, as a target area to search for a rare species or as an indication of the historical extent of a species (Busby 1986). It is generally used for prediction at global- and meso-scales (e.g. Mackey and Lindenmayer 2001). In common with other envelope methods, it will be sensitive to incomplete samples and spatial errors and will have difficulty in handling irregularly shaped habitat (Burgman and Fox, in press). It will also tend to include, in the envelope, sites where the species is absent (i.e. map many false-positives if the output is considered a habitat map) because there will be other factors affecting the species' distribution.

8.2.4 Multivariate Association Methods

These and all of the following methods are most commonly applied at the topo-scale, although many of them could be applied at other scales if appropriate data are available. The methods that are more constrained by their approach are likely to be less flexible – so, for example, methods that assume certain relationships or that make it difficult or impossible to model interactions (such as canonical correspondence analysis) are unlikely to be applicable at smaller grain sizes (i.e., at fine resolution).

Multivariate association methods are represented most simply by tools such as DOMAIN (Carpenter et al. 1993), which applies a multivariate distance measure to create habitat maps. Like convex hulls and climate envelopes, it requires only presence records. Conceptually, DOMAIN takes the opposite approach to ANUCLIM by defining sites of similarity rather than by determining bounds. It provides a measure of similarity for each site of interest, in which similarity is the environmental similarity between the site of interest and the most similar known record site, as measured by the Gower metric (Legendre and Legendre 1998). The Gower metric scales each parameter by its range to equalize the contributions of all parameters to the final similarity measure. DOMAIN measures range at the presence sites and so candidate sites outside the observed environmental range will be assigned a small (and in some cases a negative) similarity measure. DOMAIN can be used to specify an environmental envelope by selecting a minimum threshold of similarity, or it can be used to map similarities on a continuous scale.

Another example of a multivariate method is ecological niche factor analysis (ENFA). This approach has recently been under theoretical development (Guisan and Zimmerman 2000) and has now been released in the BIOMAPPER package (Hirzel 2001). ENFA uses presence-only species data. It quantifies the niche that the species occupies by comparing its distribution in an ecological space, defined by one or more variables, with the distribution of all cells (the "global distribution") in that space (Fig. 8.3). ENFA focuses on the marginality of the species (how the species mean differs from the global mean) and its specialization (how the species variance compares with the

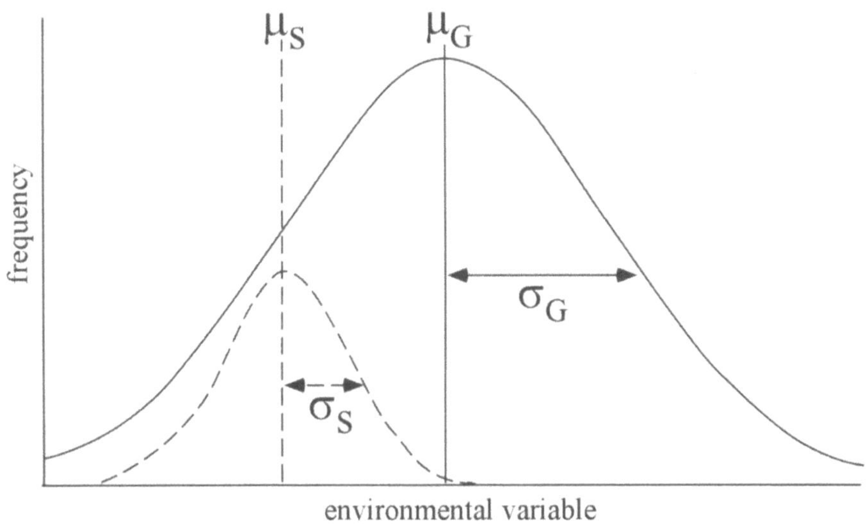

Fig. 8.3. The distribution of a species in a region (*dotted lines*) compared with the global distribution of all cells in the region (*continuous lines*), showing means (μ) and standard deviations (σ). (After Hirzel 2001)

global variance). It uses a factor analysis with orthogonal rotations to (1) transform the predictor variables to a set of uncorrelated factors, and (2) construct axes in a way that accounts for all the marginality of the species in the first axis, and that maximizes specialization in the following axes. Hirzel (2001) presented mathematical details and described methods for predicting to unsampled cells.

Predictions from BIOMAPPER have been tested against those from logistic regression for a set of simulated data (Hirzel et al. 2001). They appear to be robust to data quantity and quality, and able to model an invading species more successfully than a logistic regression model. BIOMAPPER has a number of limitations, discussed by Hirzel (2001); in particular there are as yet no routines implemented for estimating confidence intervals, and the factor analysis assumes that the observed variables are linear combinations of the hypothetical underlying factors (Legendre and Legendre 1998). This second assumption is ecologically unrealistic but can be circumvented to some extent by transforming or combining variables prior to analysis (and see following discussion of CCA).

A popular multivariate method that models presence-absence species records and deals with spaces of reduced dimensionality is canonical correspondence analysis (CCA; see Table 8.1). CCA operates at the plot level by analyzing the relationships between species presence-absence or abundance data and environmental data. It constrains the ordination of the plot data so that the resulting ordination vectors are weighted linear combinations of the environmental variables. CCA is not limited to estimating direct linear combinations, because the constituent variables can be transformed, scaled, or smoothed before being combined in a linear form (ter Braak 1986; Hastie et al. 2001). Unfortunately, it is almost always implemented without such transformations (Austin, 2002), even though linear relationships are difficult to justify ecologically. Key assumptions of CCA are that the species have unimodal responses to the environmental gradients, and that species have equal ecological amplitude (tolerance), equal maxima, and homogeneous spacing (Austin 2002). These assumptions are ecologically unrealistic, and it is not clear whether the method is robust to violations of the assumptions (e.g. Austin et al. 1994a; Jongman et al. 1995). It is worth noting that CCA is based on an ordination of plot data, and therefore one set of environmental predictors combined in one particular way is common to all species in the set (Guisan et al. 1999). In contrast, regression methods allow the selection of species-specific sets of predictors. Guisan et al. (1999) suggested that the advantage of a plot-based approach is that the information from many species helps to define gradients for species that have few presence records, which may be helpful for rare species, particularly where they co-occur with species with similar ecological requirements. Clearly, this could only be true for rare species whose response to the environment is similar to that of the more common co-occurring species. The disadvantage of a plot-based approach is that models for species with adequate data will tend to be less specialized and therefore less likely to be successful for prediction.

8.2.5 Regression Analysis

Regression analysis includes generalized linear models (GLMs) and generalized additive models (GAMs; see Table 8.1). GLMs are a broad class of statistical models that include linear regression and analysis of variance. All GLMs have a response (the species data for models of distribution), a predictor (the explanatory variables, commonly environmental data), and a link function that describes the relationship between the expected value of the response and the predictors. Species distribution models are often constructed from presence-absence species data modeled with logistic regression, i.e., a GLM for data with a binomial distribution, with a logit link function (Guisan et al. 2002). However, a wide variety of data can be accommodated by specifying different distributions for the response and different link functions (e.g. see Austin et al. 1995, for Poisson and quasi-likelihood models applied to simulated abundance data; Guisan 2002, for ordinal, Gaussian and Poisson models applied to Braun-Blanquet data). GLMs are able to model relationships of varying complexity between the response and a predictor variable by specifying linear, β, polynomial, or other functions (Austin et al. 1994b).

GAMs are a nonparametric extension of GLMs, in which at least one of the linear or other parametric functions in a GLM is replaced by a smoothed data-dependent function in a GAM. A variety of smoothers are available (Venables and Ripley 1997; Mathsoft 1999), but smoothing splines (cubic β-splines) are most commonly used in ecological applications. GAMs are considered a useful tool for modeling biological systems because the response is not limited to a parametric function, which means that the fitted response surface may be a more realistic representation of the true response shape. For example, Bio et al. (1998) modeled 156 plant species and found that 77% of them were best fit by models with at least one smoothed function (as compared with the same predictor modeled in a linear or quadratic form).

GLMs and GAMs estimate the probability of presence or the abundance of the species (depending on the type of species data and the model specification), and the models can be used to predict to new sites. Uncertainty can be quantified with confidence intervals derived statistically. When there are sufficient data, regression models provide a sound theoretical framework within which to model species distribution (Guisan et al. 2002) Complex and ecologically realistic relationships can be modelled. However, the requirement for more than presence-only data is limiting, because in many circumstances the only data available are herbarium records and opportunistic collections. Absence data are relatively rare. There has been some exploration of the performance of GLMs and GAMs with pseudo-absence data generated from non-presence sites (see, for example, Ferrier and Watson 1996; Ferrier et al. 2002; Zaniewski et al. 2002). The resulting predictions are expressed as a relative likelihood of occurrence rather than as a probability, because there are no real survey data on the frequency of absence observations. These approaches may be able to produce more realistic models than those built solely on presence data.

8.2.6 Tree-Based Methods

Tree-based methods are based on the concepts of keys, such as those used in botany and medicine, that lead a user through a series of steps to an outcome. For larger problems, such as modeling habitat, they are implemented with automated algorithms (Table 8.1). Breiman et al. (1984) and Clark and Pregibon (1992) were influential in the introduction of automated decision trees. Tree-based methods create a set of decision rules ("splits") that classify presence-absence species data in relation to the environment. They can provide probabilistic predictions for site occupancy. Conceptually, the modeling process is similar to forward selection in regression, without any statistical constraints or assumptions about the data. Tree construction methods vary on how to split each variable, and how to "prune" the tree so the response is not overfitted (Venables and Ripley 1997). A strength of all decision trees is that interactions are modeled within the splitting process and that simple trees are easy to interpret from an ecological perspective. Disadvantages include the need for more than presence-only data, their high variance – meaning that small changes in data can result in a very different series of splits (Hastie et al. 2001), the presence of discontinuities at the nodes, and poor fits for simple linear relationships (Moisen and Frescino 2002).

8.2.7 Machine Learning Methods

Machine learning methods are a broad class of methods that use computers to classify data. Examples include genetic algorithms and neural networks. A *genetic algorithm* is a general-purpose optimization technique based on a set of logical learning rules. An example is GARP (genetic algorithm for rule set production), a program developed by Stockwell and Peters (1999; Scachetti-Pereira 2002) to predict species distributions from environmental variables in a GIS context. The modeling component of the package uses a genetic algorithm to generate, test, and modify rules for predicting distribution. This produces a set of rules that can successfully predict species presence or absence on a training set. For predictions at new unsampled sites, GARP determines which rule to use in any given environmental situation, usually on the basis of prediction accuracy on the training set. In other words, the prediction map will be a grid-based map with many cells, and different rules will have been applied to different cells. Predictions can be viewed as relative likelihoods of the presence of the species. GARP will operate with presence-absence data or presence-only data, but in the latter case pseudo-absences are created. GARP has generated some interest because it is easy to use in Web format (http://biodi.sdsc.edu/). However, the lack of an easily interpretable model, the complex and largely inaccessible background procedure, and few published applications make the wisdom of its adoption in ecological applications questionable. Further assessment of the underlying rules and the performance of the method on independent data would help to clarify its potential contribution.

Neural networks (Table 8.1) learn to predict on the basis of patterns identified in a training set of inputs. They are essentially a nonlinear projection method, using nonlinear, optimally weighted combinations of the original variables to achieve prediction. There are a variety of forms, the most common being a back-propagation network (BPN). In BPNs, the learning sample is passed through the network thousands of times until the network stabilizes. Through this learning procedure, the numerous parameters of the model are estimated. An alternative form is the adaptive resonance theory (ART) network, which has been described as able to achieve fast and stable learning, recognition and prediction with training on only one pass through the data (Carpenter et al. 1997, 1999). Neural networks are most commonly applied to extensive data sets (e.g. in remote sensing), and their application to ecological situations is mostly limited to large data sets such as for vegetation mapping (Fitzgerald and Lees 1992; Carpenter et al. 1999), though some single species examples are given in Table 8.1. Networks are not easily interpreted, i.e., it would be difficult to define the variables influencing a species' distribution from the output of a neural network. Additionally, it is easy to over-fit models or to have problems with local minima in the parameter search, and these need close attention if reasonable predictions are to be achieved (Moisen and Frescino 2002).

Some approaches are intermediate between the machine-learning "black-box" approach and regression methods. One example is the tree-based method described above. Another is Friedman's multivariate adaptive regression splines algorithm (MARS, Friedman 1991, 1993) that fits separate linear splines (i.e., piecewise linear basis functions) to distinct intervals of the predictor variables. Its results are more interpretable than those of a neural network because it uses the predictor variables in their original form (compared with the projections used in neural networks) and reports the contributions of important terms in the model. MARS is promising because it deals with interactions and non-linear responses. Its reported advantages are that it performs predictor subset selection, interaction order selection, and amount of smoothing all automatically (DeVeaux 1995), and it is well suited for problems with a large number of inputs (Hastie et al. 2001). It has been applied to forestry data by Moisen and Frescino (2002). A disadvantage is that confidence intervals can only be derived through cross-validation or bootstrapping, and that there are few published applications. In general, it is easy to over-fit and over-explain the data with these automated approaches (Venables and Ripley 1997), and the methods need to be used with care.

8.3 Issues Affecting Modeling Success

Each modeling method has features that may make it more or less useful in certain circumstances. Some of these were discussed above. The following comments focus on other issues that are particularly important when the methods are used in association with PVAs. The primary aim of a habitat model is to differentiate between suitable and unsuitable habitat, either to define the limits of populations prior to modeling, or to provide a means of making spatially explicit management decisions following modeling. Several issues affect whether a method applied to the available data is able to satisfy this aim.

8.3.1 Comparison of Methods

8.3.1.1 Predictive Performance

Some methods are better than others at producing predictions that discriminate between suitable and unsuitable habitat, as evidenced in the published comparisons of model performance. Broadly, these have shown that the predictive abilities of GLMs and GAMs, when properly applied, exceed those of alternative modeling approaches across a wide range of taxa and environments (Austin and Meyers 1995; Ferrier and Watson 1996; Guisan et al. 1999; Manel et al. 1999; Elith and Burgman 2002). However, some of the emerging methods which have been less frequently tested may have advantages, particularly in automated environments. For example, Moisen and Frescino (2002) found that MARS and a neural network outperformed GAMs and CART on simulated continuous data, although their advantage in real data was reduced. Nevertheless, on real data with a binary response MARS predictions performed best. Moisen and Frescino (2002) argue that, with improvements in abilities to collect and process accurate data, methods such as MARS may have advantages in distribution modeling. Alternatively, Ennis et al. (1998) suggest that the newer methods are unlikely to perform as well as GLMs in situations where there is a low signal-to-noise ratio, such as in ecology. The disadvantages of any of GLMs, GAMS, or MARS are the level of sophistication of the analysis, and the requirement for enough presence-absence records to provide sufficient data for model estimation (see, for example, Austin et al. 1990; Austin and Meyers 1996; Harrell 2001).

Predictions based on presence-only records will often be limited in their discriminatory ability compared with those developed on presence-absence data, because there are no data to help define the conditions leading to absence of the species (e.g. Ferrier and Watson 1996; Elith and Burgman 2002). However, Hirzel et al. (2001) have demonstrated that, when absence records result from unmodeled variation such as suitable sites not yet colonized, models based on presence-only records may out-perform presence-absence models. Under other conditions, methods that use pseudo-absence records may perform better than those based solely on presence records (e.g. see Ferrier and Watson 1996; Ferrier et al. 2002).

8.3.1.2 Understanding the Methods

All methods require some computing and statistical understanding, and in management applications experienced modelers are not always available. Therefore, it is important to recognize the complexities of the methods, as well as the extent and accessibility of literature that explains the methods and demonstrates how to apply them with typical data. Understanding is affected by many things including the complexity of the underlying concepts and the transparency of the procedures. Thus, potential difficulties with the more complex statistical concepts of the GLMs and GAMs are balanced by the extensive documentation and literature associated with them and by the many alternative software platforms for their implementation. The same arguments apply to CCA. MARS is restricted in its availability and published ecological applications. The user-friendly Web interface of GARP is contrasted by the lack of an easily interpretable model, a complex and largely inaccessible background procedure, and few independent published applications. Convex and α-hulls, HSIs, climate envelopes, and DOMAIN are based on simple concepts but there are few readily available software packages to compute them. ANUCLIM has more documentation and published applications than DOMAIN. Nevertheless, it appears that none of the non-statistical methods are accompanied by a comprehensive text or extensive documentation.

8.3.1.3 Estimating Error

GLMs, GAMs, and CCA provide statistical estimates of error and deviance, reflecting the extent to which a model does not fit the modeling data or may not fit new data. HSIs work with expert judgement alone, which makes them broadly applicable but open to errors from subjective judgements. Most of the other methods classify sites without producing a causal model. For many applications, the important issue is whether the predictions are reliable in their classification of new sites. All methods produce predictions that can be tested against new data (e.g. see Elith and Burgman 2002). If the uncertainty of the predictions is also of interest, the only way this can be investigated for methods without statistical error estimates is to use techniques such as bootstrapping (Davison and Hinkley 1997). These can be computationally demanding and have, in the past, been impractical in many applied situations (Ferrier et al. 2002). However, with increases in computer power these methods are now frequently feasible and useful.

8.3.1.4 Model Interpretability

Some ecologists prefer modeling methods that produce an ecologically interpretable model, because the proposed relationships between the species and the environment can be evaluated critically. Others are content with predictions that appear to be valid, even if the model is uninterpretable. It has been argued that good predictions can be obtained from a model that has mechanistic (i.e., conceptual) flaws (Caswell 1976; Rykiel 1996). However, it seems logical that models that are more conceptually adequate will make more robust predictions (Austin 2002). In situations in which it is important that the shape of functional relationships are defined, GLMs or GAMs are most suitable. These are the only methods that allow for explicit, ecologically realistic modeling of the species response to the environment.

8.3.2 Modeling Data

The models are generally computer-based. In such cases the environmental data are commonly stored as themes or "layers" in a geographic information system (GIS). Species data are geo-referenced point locations. Predictions are generally developed by applying the models to grid (raster) representations of the environmental data. Clearly, it is important that these data have sufficient resolution and accuracy to lead to reliable predictions (Elith et al. 2002). Cawsey et al. (2002) have demonstrated the advantages of interfacing a GIS with a relational database to achieve efficient and flexible data management and to keep options open about what maps are produced and what ancillary information is accessible to end-users of the predictions.

8.3.2.1 Species Data

The methods have different requirements for species data, as already discussed (and see Table 8.1). In general, when presence-absence data are available, methods that utilize the information in the absence records are applied. However, Hirzel (2001) makes a case for applying the presence-only methods to species for which the absence data are unreliable (e.g. for cryptic species or those that are poorly understood) or difficult to interpret (e.g. invading species or those in fragmented habitats).

All methods rely on a well-constructed sample of locations. There is relatively little published research on the most effective and efficient means of designing sampling for species modeling, but it tends to favor stratified approaches in which the entire environmental gradient is sampled (Mohler 1983), with modifications to ensure geographic coverage and, in relevant situations, survey efficiency (Austin and Heyligers 1989, 1991; Cawsey et al. 2002).

The envelope methods that construct bounds for the outer limits of the species distributions will be more sensitive to errors in location or incomplete samples than those that build models of the relationship between the species and its environment. Nevertheless, all methods will suffer from biased or incomplete samples, and their output (both models and predictions) need particularly careful analysis when they are developed on non-ideal data.

8.3.2.2 Predictor Variables

Predictors that have a direct causal influence on species distribution (such as temperature or nutrient availability) are more likely to identify important relationships and to reflect a species realized niche than those with indirect effects (such as altitude or soil type) (Austin 1980, 2002; Guisan and Zimmerman 2000). If it is important to identify occupied and unoccupied suitable habitat, as is the case in developing spatial models for PVA, then this aspect is important. However, direct variables are usually not available as GIS layers, or are relatively imprecisely mapped (Guisan and Zimmerman 2000). In this context, climate variables are biologically relevant, direct variables but they usually do not have fine local resolution. Their usefulness depends on the scale of application. The common practice of making do with indirect variables usually results in more locally precise but less transferable (generalizable) models because the indirect variables rarely have a simple linear correlation with the direct gradients (Guisan and Zimmerman 2000; Austin 2002). However, Austin (2002) suggested that this is context-sensitive, because where gradients are steep and environments extreme indirect variables will be as successful as more direct ones in modeling distribution.

A number of issues are important in the collection, manipulation, construction, and modeling of predictor variables. These include scale, relevance to the species (e.g. should the variable be constructed with a neighborhood component, as in the contextual indices reported by Ferrier 2002); amount of correlation between the variables, and so on.

8.3.3 Links Between Occupation and Quality of Habitat

Species distribution models are developed on "static" data, i.e., on records of presence, presence-absence, or abundance that are collected in one or more surveys, without regard to population processes. The models assume that the population is at equilibrium, or at least at quasi-equilibrium where change is slow relative to the life span of the species (Austin 2002). If, instead, the population is increasing (e.g. a species is recovering, colonizing, or invading an area), then many of the observed absence records may be in potentially suitable habitat and from a habitat point of view they will be false-negative observations (Capen et al. 1986). Similarly, if the population is declining (e.g. in oversaturated habitat where the species is present in a habitat that will not support its long-term persistence), a number of the presence records will not reflect good habitat. In these cases it may be possible to model current distribution (especially if variables reflecting geographic location are used as predictors) but the models are unlikely to be good predictors of habitat. Alternatively, it may not even be possible to successfully model current distribution because of the lack of suitable predictors to adequately explain the current observations. In some circumstances, it is possible to include variables reflecting factors such as disturbance, dispersal barriers, competition, population trends, or successional dynamics, and these may assist in modeling non-equilibrium situations (Fewster et al. 2000; Guisan and Zimmerman 2000; Leathwick and Austin 2001). Also, careful analysis of the amount of spatial patterning left unexplained by the environmental predictors may indicate historical disturbance events (Leathwick 1998). Nevertheless, modeling species that are not at equilibrium is a challenging scenario for static models and one that is prone to difficulty (Guisan and Zimmerman 2000). Hirzel et al. (2001) have shown that in these situations the modeling method can have a significant impact on modeling success. In their study, the ENFA models developed on presence-only data were most reliable for colonizing species because the false-negative absence records were excluded, whereas with over-saturated habitat (where it was important to identify the conditions leading to absence) the presence-absence logistic regression models performed better.

It is often assumed that the probability or likelihood of occurrence of a species is positively correlated with the quality of habitat, i.e., the more likely it is that the species is present, the more suitable the habitat (Davey 1989). Similarly, it is assumed that the most suitable habitat supports populations with relatively high fecundities and survivorships (Breininger et al. 1995, 1998). There are plausible and pragmatic reasons for accepting such assumptions (Tyre et al. 2002). However, a number of factors can complicate and obscure the relationships. For example, there are many circumstances in which the species is not at equilibrium (see above, and Clark 1991); there are species which, even at equilibrium, do not saturate all good habitat (Kareiva 1990; McKenna and Houle 2000); the likelihood that a patch is occupied varies with its proximity to good habitat, and bad habitat close to good habitat may be more likely to be occupied than good habitat remote from any other good patch (Hanski 1994; Tyre et al. 2002); individuals of a species may not congregate in the most suitable locations because of behavior, intraspecific competitive exclusion, or dispersal dynamics (Van Horne 1983); if mortality is high in a species, a survey is likely to observe absences in good habitat because of recent death without recolonization (Tyre et al. 2002). As a result of such processes, there is considerable noise in the relationship between occupancy (even if well-modeled) and habitat quality. Buckland and Elston (1993) suggested that different models may be needed to estimate occupancy and suitability. Tyre et al. (2002) concluded that species with high survival rates will make better targets for statistical habitat modeling. Franklin (1995) and Austin (2002) emphasized the role of direct as compared with indirect variables in producing more generalizable models.

Despite this, models of species distribution are usually the best available representation of habitat, and they are frequently called habitat maps and frequently used as such. One of the methods reviewed here, HSIs, explicitly attempts to model habitat rather than occupancy, and does so by quantifying expert opinion. However, it is likely that experts are relying on their observations of occurrence, because habitat is not something that can be observed with any certainty in the field. To do so would require that the observer could capture in his or her assessment of a site all conditions that lead to suitable habitat and recognize any conditions that would make the site unsuitable for the species. This is the reason that none of the data-based methods can easily be used to directly model habitat (i.e., with data on the presence or absence of habitat as the response). The main alternative is to develop spatially explicit dynamic models that include information on population processes such as survival, dispersal, and succession, but these data are costly to collect (Guisan and Zimmerman 2000; Tyre et al. 2002). Another option is illustrated in Fewster et al.'s inclusion of population trends in a GAM (Fewster et al. 2000). However, in general, models of species distribution will continue to be used as surrogates for models of habitat, and assessments of likely uncertainties and shortcomings of the predictions when used in this way would be helpful.

8.3.4 Habitat and Patches

Spatially explicit PVA usually involves abstracting a habitat map into a representation of patches and dispersal pathways. The process of amalgamating regions of high habitat suitability into patches also involves a decision threshold. At what spatial scale should individual pixels in a habitat map be coalesced into patches? A lower limit is generally set by the scale of resolution of the maps. Akçakaya (1994) solved this problem by finding clusters of nearby cells (cells within a neighborhood distance) that have suitability values greater than the specified threshold. Patches were formed from sets of interconnected high-value cells. The average habitat suitability and area of the patch were used to calculate the carrying capacity of the patch.

8.4 Assessing the Reliability of a Habitat Model

All habitat maps, regardless of how they are constructed, make spatially explicit predictions. Usually, the prediction is expressed relative to a threshold of habitat suitability such that, above the threshold, a species is more likely to be present than absent. A difficulty in interpreting habitat maps for PVA is that it is unclear how to define an ecologically relevant threshold for predicting habitat. Fielding and Bell (1997) suggested several approaches. However, there will always be some places above the threshold from which the species is absent and places below the threshold where the species is present. Such prediction errors may be the result of vagrants, error in the choice or measurement of the predictor variables, observational error, errors in spatial location, inter and intra-specific social interactions, the effects of historical events, etc. Often the "prediction error" is a reflection of the fact that any threshold that converts a probability or a likelihood into a binary outcome will classify some cases as positive that were originally given some chance of being negative and vice versa.

Prediction errors may be characterized as false-positives (predicting a presence when the species is in fact absent) and false-negatives (predicting an absence when the species is in fact present). The threshold for habitat suitability at which the decision is made may be adjusted. However, any decrease in the false-positive rate will increase the false-negative rate, and vice versa. The number of false-positives, false-negatives, true-positives and true-negatives (Table 8.2) give an indication of how well a habitat model is working in predicting the presence and absence of a species.

Each number in the classification table (Table 8.2) represents a field sample. The model makes a prediction for the presence of a species, and field validation assesses whether or not each prediction is true. If a map is a perfect

Table 8.2. Classification table (confusion or error matrix) for predictions from a habitat model

		Model prediction Species present	Species absent
Outcome of validation:	Species present	True-positive	False-negative
	Species absent	False-positive	True-negativ

representation of species distribution, then all values would fall on the diagonal of Table 8.2. Predictions may be made for the presence of a species, or for the presence of suitable but unoccupied habitat.

If the threshold for prediction of presence were to be lowered, so that the model predicts more presences, then the true-positive count would increase. The false-positive count would also increase. The person interpreting the habitat model has to make a decision about the threshold that is best for the current context. This decision is essentially a trade-off between the costs and benefits of false-positives versus false-negatives. For example, if a species is rare and it is important not to miss potential occurrences, then the decision may be taken to make a conservative interpretation of the habitat model and lower the decision threshold. The result will be an increased likelihood of capturing occurrences of the species within the limits of the mapped habitat, but at the cost of increasing the number of places that are not, in fact, suitable.

The converse situation may also arise. For example, if a species is to be the focus of a survey to establish its distribution, planners may wish to stratify the area and conduct searches in areas that have the highest chances of success. In these circumstances, it is desirable to exclude as much unsuitable habitat as possible. This may be achieved by increasing the threshold to map only those areas that have a very high probability of containing the species.

Data in the classification table can be summarized in numerous ways (Fielding and Bell 1997). One useful measure is Cohen's κ (Cohen 1960), which measures the proportion of agreement after chance agreement is removed from consideration. That is, $\kappa=(p_o-p_c)/(1-p_c)$ where p_o is the proportion of units in which the predictions and observations agree, and p_c is the proportion of units in which agreement is expected by chance. κ is preferable to some of the more common alternatives (such as percent correctly classified) because it is less sensitive to prevalence of the species (Fielding and Bell 1997; Manel et al. 2001). It measures aspects of both calibration and discrimination (Pearce and Ferrier 2000; Elith and Guisan, pers. comm.).

More generally, modeled predictions can be tested before they are classified as binary outcomes. This is useful because it allows evaluation of the predictive capability of a model for the full range of thresholds over which predictions may be made. The Mann-Whitney statistic (Hanley and McNeil 1982) reports the probability that, for a randomly selected pair of presence/absence observations, the prediction for presence will be greater than the prediction for absence. It measures discrimination (Harrell 2001) and is a nonparametric summary of the area under a receiver operating characteristic (ROC) curve. [See Swets et al. (2000) for an accessible introduction to ROC curves, and Pearce and Ferrier (2000) for an ecological application.] The ROC curve is a plot of the true-positive rate against the false-positive rate over a range of thresholds, and thus there is a connection between the Mann-Whitney statistic and the classification table.

8.5 Application to *Leptospermum grandifolium*

Leptospermum grandifolium is a shrub or small tree, abundant in subalpine regions of southeastern Australia. In surveys undertaken between 1976 and 1996, it was recorded in 179 of 3,522 visited locations (Fig. 8.4). It is primarily associated with drainage lines and commonly occurs in dense vegetation in montane areas. It occurs close to gullies and springs in the sub-alpine zones and grows occasionally in lower altitude, wet forests, particularly along rocky slopes beside drainage lines. At higher altitudes, above about 1,200 m, it is not restricted to streamside locations but occurs more broadly across the landscape (D. Frood, pers. comm.).

Models for *L. grandifolium* have been constructed. A digital terrain model for the region at a 9-s scale (cell size ~250 m) provided the basis for the construction of variables related to topography and climate. Other available data included a wetness index (a proxy for drainage patterns), vegetation classes, and geology (Elith 2000; Elith and Burgman 2002). These data and the ecological characteristics described above were used to construct eight different habitat models, one each for the HSI, CONVEX and α-HULLS, ANUCLIM, DOMAIN, GLM, GAM, and GARP (Figs. 8.5, 8.6). The particular protocols used for several of the methods are detailed in Elith et al. (1998) and were based on existing approaches to modeling that could be applied to many

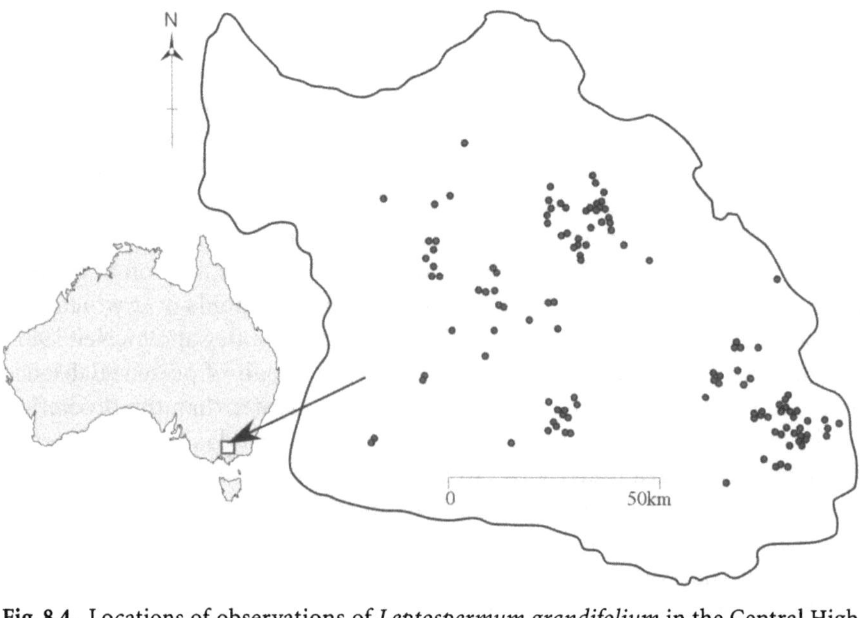

Fig. 8.4. Locations of observations of *Leptospermum grandifolium* in the Central Highlands region of Victoria, Australia. The *black outline* around the locations is the boundary of the Central Highlands RFA area

species in a relatively automated framework. The model predictions were subsequently validated using an additional 161 sample sites that were selected to test a range of the predictions within the known geographic extent of the species. The extent of agreement between observations and predictions was calculated as the maximum kappa across the range of possible thresholds (Guisan et al. 1998; Fig. 8.7), and as a Mann-Whitney statistic on the predictions with no threshold applied (Fig. 8.8).

In this particular example, all modeling methods produced models with at least some discriminatory ability, but methods which used presence-absence

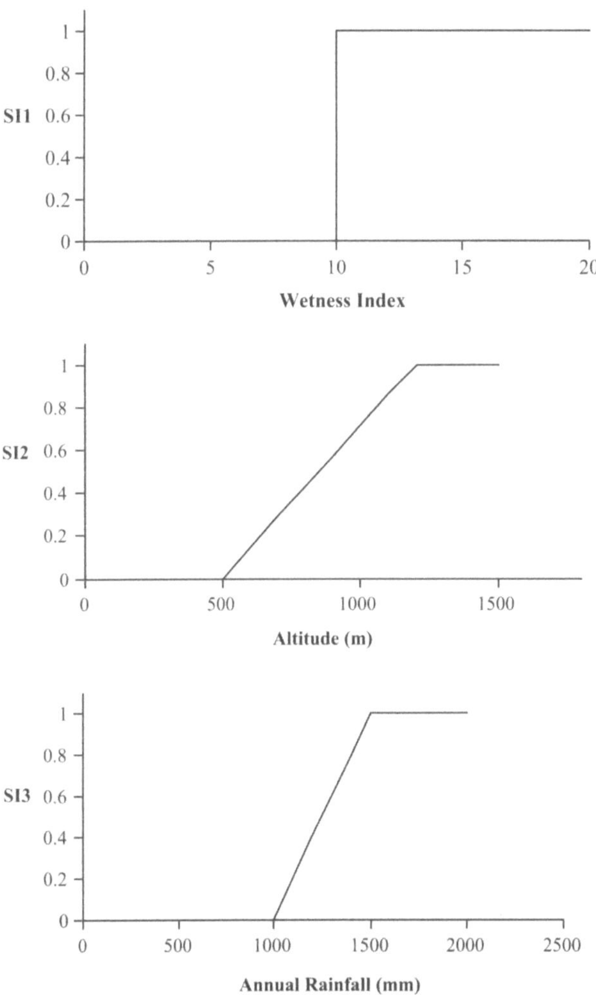

$$HSI = \sqrt{\mathbf{SI2} \bullet max\,(\mathbf{SI1,SI3})}$$

Fig. 8.5. Suitability indices (SI1–SI3) and the combined habitat suitability index (HSI) function for *Leptospermum grandifolium.* (wetness index is based on flow accumulation and indicates the volume of water draining to the site)

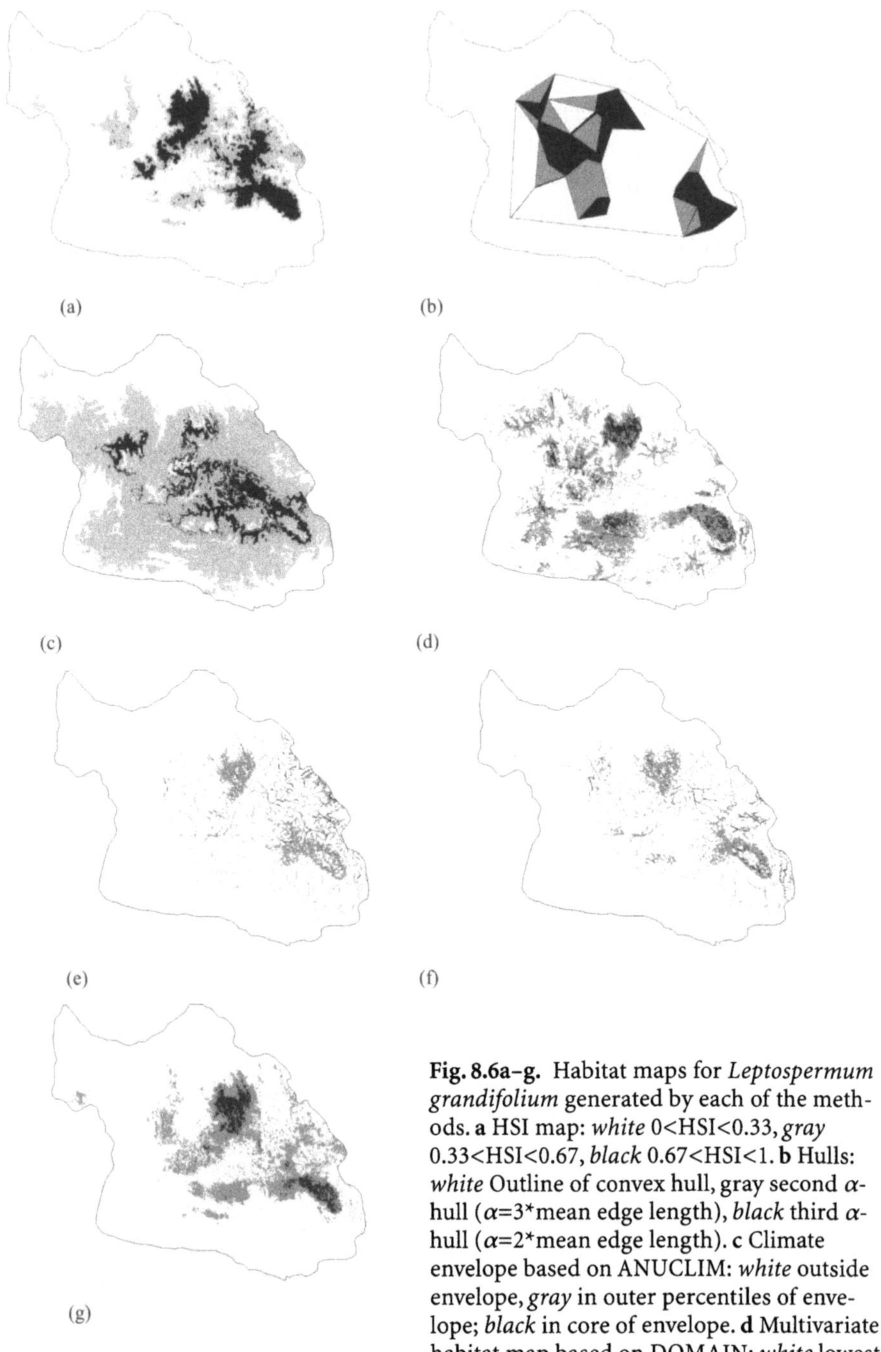

Fig. 8.6a–g. Habitat maps for *Leptospermum grandifolium* generated by each of the methods. **a** HSI map: *white* 0<HSI<0.33, *gray* 0.33<HSI<0.67, *black* 0.67<HSI<1. **b** Hulls: *white* Outline of convex hull, gray second α-hull (α=3*mean edge length), *black* third α-hull (α=2*mean edge length). **c** Climate envelope based on ANUCLIM: *white* outside envelope, *gray* in outer percentiles of envelope; *black* in core of envelope. **d** Multivariate habitat map based on DOMAIN: *white* lowest similarity to occupied sites (35–90), *gray* moderate similarities (90+-95), *black* similarities 95+ to 100. **e** GLM predictions: *white* probability≤0.1, *gray* 0.1<probability≤0.5, *black* probability >0.5. **f** GAM predictions, same scale as for GLM. **g** Likelihood of species presence based on GARP: *white* 0–100, *gray* 101–200, *black* 201–255

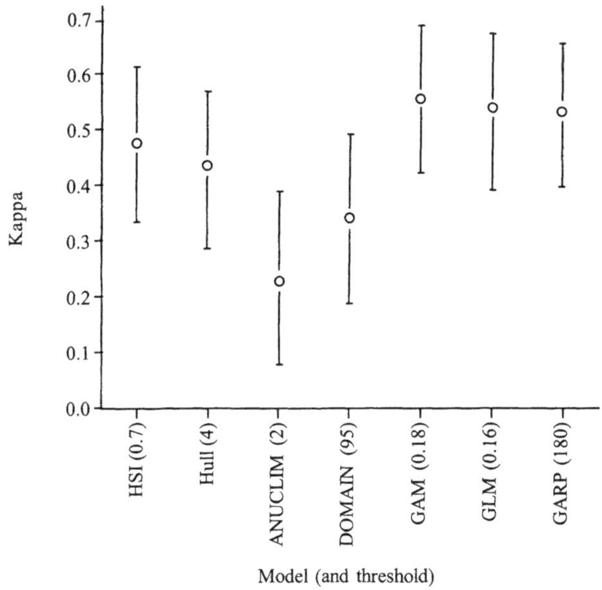

Fig. 8.7. The κ statistics and 95 % confidence intervals for each of the habitat models. Thresholds at which κ was calculated (all predictions greater than or equal to this threshold become predictions for presence) are in *parentheses* after the name of the method

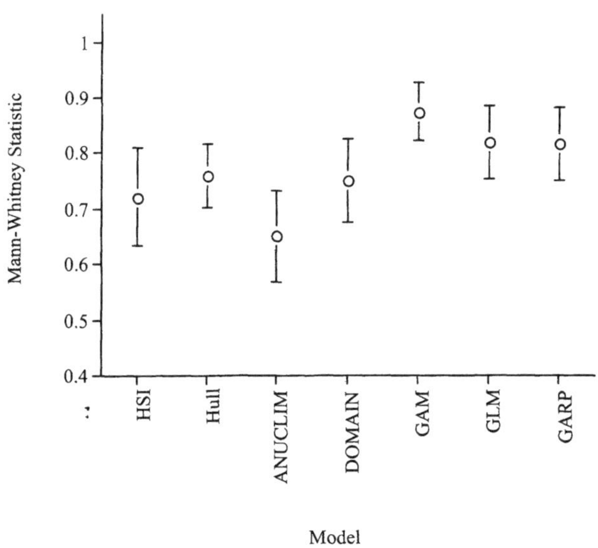

Fig. 8.8. Mann-Whitney statistics and 95 % confidence intervals for each of the habitat models

data tended to perform better than those based on presence-only data. For example, without a threshold, the GAM predictions had significantly better discriminatory ability than the three simplest methods (HSIs, hulls and ANUCLIM; Fig. 8.8), and with a threshold, GAMs were substantially better than ANUCLIM (Fig. 8.7). As noted above, there is generally broad correspondence between different methods for predicting habitat, and the methods applied here have generated habitat maps that are roughly equivalent, although they differ in important detail. The benefits of using the technique that performs best (GAMs) need to be weighed against the technical overhead involved in running it. There is a significant investment in software tools and statistical skills necessary to implement a GAM, and the benefit of improved prediction should be worth the cost.

8.6 Conclusions

The choice of a method for developing habitat models usually is determined by precedent, institutional culture, the kind and quality of available data, and the availability of relevant expertise. For PVAs, the required level of detail and reliability of a habitat map should be determined primarily by the management context. If management options include the protection of alternative pieces of habitat or the creation of dispersal corridors, then the habitat should be resolved at a scale that differentiates reliably between patches and corridors of different quality. If the options relate to the protection of a proportion of habitat at much coarser scales, then the habitat model need only discriminate at coarser levels of resolution at the scale of the landscape within which the habitat is to be allocated.

As outlined above, the results of habitat mapping exercises are used in several ways to support conservation planning decisions. Many decision support systems include thresholds for extent of occurrence such that if the estimated area is lower than the threshold, the species is classified into an associated threat category. For example, the IUCN (2001) rule that classifies a species as critically endangered includes the criterion that the extent of occurrence should be estimated to be less than 100 km^2. Such data are also used to estimate parameters in PVA models, such as the carrying capacity and boundaries of habitat patches (Akçakaya et al. 1995).

Habitat models are also used for inferring trends in range. For example, a species will be considered critically endangered under the IUCN (2001) rules if data suggest a reduction in extent of occurrence of more than 80 % in the last 10 years. Trends in carrying capacity and habitat quality are also used in PVAs to model the consequences of threatening processes.

Area and spatial arrangement of habitat are used to set wildlife management priorities and to constrain activities within a landscape. Habitat maps provide the basis for estimating the relative impacts of management alternatives and for planning the mitigation of their impacts. The methods have different utilities depending on the amount, quality, and type of available data. They are also prone to different types of uncertainties, including simplifying assumptions, inaccuracies in the data, poor implementation of the method, and lack of an interpretable model. These should be borne in mind when the choice of habitat modeling technique is being made.

All habitat maps are uncertain, but levels of uncertainty are rarely quantified, and uncertainty is not factored into the decision-making processes to which the maps contribute. Furthermore, error and uncertainty exist, not only because modeling methods are biased, but because occupancy does not always indicate habitat that is suitable for long-term persistence. Van Horne (1983) pointed out that density and demographic success are not necessarily closely related. The simulations of Tyre et al. (2002) further support the concept that habitat models, while indicating ability to disperse to an area, do not provide information about survival rates. These considerations need to be part of the understanding of the uncertainty of conservation decisions based on habitat models.

Simple and intuitive methods for habitat modeling such as climate envelopes and α-hulls are known to be biased (Elith 2000; Burgman and Fox, in press). However, these biases may be tolerable if the magnitudes or errors are not too large relative to the scales at which predictions are made. Despite their theoretical flaws, they will continue to be used for the foreseeable future to set conservation priorities and to form the basis for spatially explicit management decisions. This situation creates an imperative that the kinds and magnitudes of uncertainties associated with each of the methods for habitat mapping be explored and their implications for management decisions be documented in a way that makes their flaws apparent to managers and planners.

Acknowledgements. We are grateful to Doug Frood for his insights into the biology of *L. grandifolium* and for conducting much of the fieldwork for the validation exercise. We thank Julian Fox for creating the convex hull and α-hull, and for generating the HSI map for the species.

References

Agresti A (1996) An introduction to categorical data analysis. Wiley, New York

Akçakaya HR (1994) RAMAS/GIS: linking landscape data with population viability analysis (version 1.0). Applied Biomathematics, New York

Akçakaya HR, McCarthy MA, Pearce JL (1995) Linking landscape data with population viability analysis - management options for the Helmeted Honeyeater *Lichenostomus melanops cassidix*. Biol Conserv 73:169–176

Aleksander I, Morton H (1990) An introduction to neural computing. Chapman and Hall, London

Austin MP (1980) Searching for a model for use in vegetation analysis. Vegetatio 42:11–21

Austin MP (2002) Spatial prediction of species distribution: an interface between ecological theory and statistical modelling. Ecol Model 157:101–118

Austin MP, Heyligers PC (1989) Vegetation survey design for conservation: gradsect sampling of forests in north-eastern NSW. Biol Conserv 50:13–32

Austin MP, Heyligers PC (1991) New approach to vegetation survey design: gradsect sampling. In: Margules CR, Austin MP (eds) Nature conservation: cost effective biological surveys and data analysis. CSIRO, Canberra, Australia, pp 31–36

Austin MP, Meyers JA (1995) Modelling of landscape patterns and processes using biological data, subproject 4: real data case study. Division of Wildlife and Ecology, CSIRO, Canberra

Austin MP, Meyers JA (1996) Current approaches to modelling the environmental niche of eucalypts: implications for management of forest biodiversity. For Ecol Manage 85:95–106

Austin MP, Cunningham RB, Good RB (1983) Altitudinal distribution in relation to other environmental factors of several eucalypt species in southern New South Wales. Aust J Ecol 8:169–80

Austin MP, Cunningham RB, Fleming PM (1984) New approaches to direct gradient analysis using environmental scalars and statistical curve-fitting procedures. Vegetatio 55:11–27

Austin MP, Nicholls AO, Margules CR (1990) Measurement of the realized qualitative niche: environmental niches of five eucalypt species. Ecol Monogr 60:161–177

Austin MP, Meyers JA, Doherty MD (1994a) Predictive models for landscape patterns and processes, sub-project 2, modelling of landscape patterns and processes using biological data. Division of Wildlife and Ecology, CSIRO, Canberra

Austin MP, Nicholls AO, Doherty MD, Meyers JA (1994b) Determining species response functions to an environmental gradient by means of a beta-function. J Veg Sci 5:215–228

Austin MP, Meyers JA, Belbin L, Doherty MD (1995) Modelling of landscape patterns and processes using biological data. subproject 5: simulated data case study. Division of Wildlife and Ecology, CSIRO, Canberra

Bingham BB, Noon BR (1997) Mitigation of habitat "take": application to habitat conservation planning. Conserv Biol 11:127–139

Bio AMF, Alkemande R, Barendregt A (1998) Determining alternative models for vegetation response analysis - a non-parametric approach. J Veg Sci 9:5–16

Birks HJB (1996) Statistical approaches to interpreting diversity patterns in the Norwegian mountain flora. Ecography 19:332–340

Boyce MS (1992) Population viability analysis. Annu Rev Ecol Syst 23:481–506

Breininger DR, Larson VL, Duncan BW, Smith RB, Oddy DM, Goodchild MF (1995) Landscape patterns of Florida scrub-jay habitat use and demographic success. Conserv Biol 9:1442–1453

Breininger DR, Larson VL, Duncan BW, Smith RB (1998) Linking habitat suitability to demographic success in Florida scrub-jays. Wildl Soc Bull 26:118–128
Breiman L, Friedman JH, Olshen RA, Stone CJ (1984) Classification and regression trees. Wadsworth International Group, Belmont, CA
Buckland ST, Elston DA (1993) Empirical models for the spatial distribution of wildlife. J Appl Ecol 30:478–495
Burgman MA, Ferson S, Akçakaya HR (1993) Risk assessment in conservation biology. Chapman and Hall, London
Burgman MA, Breininger DR, Duncan BW, Ferson S (2001) Setting reliability bounds on habitat suitability indices. Ecol Appl 11:70–78
Burgman MA, Fox JC (in press) Bias in species range estimates from minimum convex polygones. Animal Conservation
Busby JR (1986) A biogeographic analysis of *Nothofagus cunninghamii* (Hook.) Oerst. in south-eastern Australia. Aust J Ecol 11:1–7
Busby JR (1991) BIOCLIM – a bioclimate analysis and prediction system. In: Margules CR, Austin MP (eds) Nature conservation: cost effective biological surveys and data analysis. CSIRO, Canberra, Australia, pp 64–68
Capen DE, Fenwick JW, Inkley DB, Boynton AC (1986) Multivariate models of songbird habitat in New England forests. In: Verner J, Morrison ML, Ralph CJ (eds) Wildlife 2000: modeling habitat relationships of terrestrial vertebrates. Based on an International Symposium held at Stanford Sierra Camp, Fallen Leaf Lake, California. The University of Wisconsin Press, Madison, Wisconsin, pp 171–175
Carpenter G, Gillison AN, Winter J (1993) DOMAIN: a flexible modelling procedure for mapping potential distributions of plants and animals. Biodiv Conserv 2:667–680
Carpenter GA, Gjaja MN, Gopal S, Woodcock CE (1997) ART neural networks for remote sensing: vegetation classification from Landsat TM and terrain data. IEEE Trans Geosci Remote Sensing 35:308–325
Carpenter GA, Gopal S, Macomber S, Martens S, Woodcock CE, Franklin J (1999) A neural network method for efficient vegetation mapping. Remote Sensing Environ 70:326–338
Caswell H (1976) The validation problem. In: Patten B (ed) Systems analysis and simulation in ecology. Academic Press, New York, pp 313–325
Cawsey EM, Austin MP, Baker BL (2002) Regional vegetation mapping in Australia: a case study in the practical use of statistical modelling. Biodiv Conserv (in press)
CIFOR (1999) http://www.cifor.cgiar.org/domain/index.htm. (last visited May 2002) Centre for International Forestry Research, Bogor, Indonesia
Clark JS (1991) Disturbance and tree life history on the shifting mosaic landscape. Ecology 72:1102–1118
Clark LA, Pregibon D (1992) Tree-based models. In: Chambers JM, Hastie TJ (eds) Statistical models. S. Wadsworth and Brooks/Cole Advanced Books and Software, Pacific Grove, California
Cohen J (1960) A coefficient of agreement for nominal scales. Educ Psychol Meas 20:37–46
Corsi F, Duprè E, Boitani L (1999) A large-scale model of wolf distribution in Italy for conservation planning. Conserv Biol 13:150–159
Coulson T, Mace GM, Hudson E, Possingham H (2001) The use and abuse of population viability analysis. Trends Ecol Evol 16:219–221
CRES (2002) http://cres.anu.edu.au/outputs/anuclim.html. (last visited May 2002)
Davey SM (1989) Thoughts towards a forest management strategy. Aust For 52:56–67
Davison AC, Hinkley DV (1997) Bootstrap methods and their application. Cambridge University Press, Cambridge
DeVeaux RD (1995) A guided tour of modern regresssion methods. Proceedings of the section on physical and engineering sciences. In: Invited talk given at the 1995 Fall

Technical Conference, St Louis, MO, http://www.williams.edu/Mathematics/rdeveaux/pubs.html (last visited May 2002)
Edelsbrunner H, Kirkpatrick DG, Seidel R (1983) On the shape of a set of points in the plane. IEEE Trans Inform Theor IT-29:551–559
Elith J (2000) Quantitative methods for modeling species habitat: comparative performance and an application to Australian plants. In: Ferson S, Burgman MA (eds) Quantitative methods in conservation biology. Springer, Berlin Heidelberg New York, pp 39–58
Elith J, Burgman MA (2002) Predictions and their validation: rare plants in the Central Highlands, Victoria, Australia. In: Scott JM, Heglund PJ, Morrison M (eds) Predicting species occurrences: issues of scale and accuracy. Island Press, Covelo, California
Elith J, Burgman MA, Minchin P (1998) Improved protection strategies for rare plants: consultancy report for environment Australia. Environment Australia, Canberra, Australia
Elith J, Burgman MA, Regan HM (2002) Mapping epistemic uncertainties and vague concepts in predictions of species distribution. Ecol Model Ecol Model 157:313–329Ferrier S, Watson G (1996) An evaluation of the effectiveness of environmental surrogates and modelling techniques in predicting the distribution of biological diversity. Consultancy report prepared by the NSW National Parks and Wildlife Service for Department of Environment, Sport and Territories, Canberra, Australia
Ennis M, Hinton G, Naylor D, Revow M, Tibshirani R (1998) A comparison ot statistical learning methods on the GUSTO database. Statistics in Medicine 17:2501–2508
Ferrier S, Watson G, Pearce J, Drielsma M (2002) Extended statistical approaches to modelling spatial pattern in biodiversity: the north-east New South Wales experience. I. Species-level modelling. Biodiv Conserv (in press)
Fewster RM, Buckland ST, Siriwardena GM, Baillie SR, Wilson JD (2000) Analysis of population trends for farmland birds using generalized additive models. Ecology 81:1970–1984
Fielding AH, Bell JF (1997) A review of methods for the assessment of prediction errors in conservation presence/absence models. Environ Conserv 24:38–49
Fitzgerald RW, Lees BG (1992) The application of neural networks to the floristic classification of remote sensing and GIS data in complex terrain. In: Proceedings of the XVII Congress of the International Society for Photogrammetry and Remote Sensing, Washington, USA
Franklin J (1995) Predictive vegetation mapping: geographic modelling of biospatial patterns in relation to environmental gradients. Prog Phys Geogr 4:474–499
Friedman JH (1991) Mutlivariate adaptive regression splines (with discussion). Ann Stat 19:1–141
Friedman JH (1993) Estimating functions of mixed, ordinal and categorical variables using adaptive splines. In: Morgenthaler S, Ronchetti E, Stahel WA (eds) New directions in statistical data analysis and robustness. Birkhäuser, Basel
Gray PA, Cameron D, Kirkham I (1996) Wildlife habitat evaluation in forested ecosystems: some examples from Canada and the United States. In: DeGraaf RM, Miller RI (eds) Conservation of faunal diversity in forested landscapes. Chapman and Hall, New York, pp 406–533
Guisan A (2002) A semi-quantitative response model for predicting the spatial distribution of plant species. In: Scott JM, Heglund PJ, Morrison M (eds) Predicting species occurrences: issues of scale and accuracy. Island Press, Covelo, California
Guisan A, Zimmerman NE (2000) Predictive habitat distribution models in ecology. Ecol Model 135:147–186
Guisan A, Theurillat JP, Kienast F (1998) Predicting the potential distribution of plant species in an alpine environment. J Veg Sci 9:65–74

Guisan A, Weiss SB, Weiss AD (1999) GLM versus CCA spatial modeling of plant species distribution. Plant Ecol 143:107–122
Guisan A, Edwards Jr TC, Hastie T (2002) Generalized regression in predictive modeling of species distribution: setting the scene. Ecol Model 157:89–100
Hanley JA, McNeil BJ (1982) The meaning and use of the area under a Receiver Operating Characteristic (ROC) curve. Radiology 143:29-36
Hanski I (1994) Patch occupancy dynamics in fragmented landscapes. Trends Ecol Evol 9:131–134
Hansteen TL, Andreassen HP, Ms RA (1997) Effects of spatiotemporal scale on autocorrelation and home range estimators. J Wildlife Manage 61:280–290
Harrell FE (2001) Regression modeling strategies with applications to linear models, logistic regression and survival analysis. Springer, Berlin Heidelberg New York
Hastie T, Tibshirani R (1990) Generalized additive models. Chapman and Hall, London
Hastie T, Tibshirani R, Friedman JH (2001) The elements of statistical learning: data mining, inference, and prediction. Springer, Berlin Heidelberg New York
Hirzel AH (2001) When GIS come to life. Linking landscape- and population ecology for large population management modelling: the case of Ibex (*Capra ibex*) in Switzerland. In: Science faculty. The University of Lausanne, Lausanne, http://www.unil.ch/biomapper/bibliography.html (last visited Jan 2002), Switzerland
Hirzel AH, Helfer V, Metral F (2001) Assessing habitat-suitability models with a virtual species. Ecol Model 145:111–121
IUCN (1994) IUCN red list categories. As approved by the 40th Meeting of the IUCN Council. Prepared by the International Union for the Conservation of Nature Species Survival Commission, Gland, Switzerland
IUCN (2001) International union for the conservation of nature. Red list categories, version 3.1. IUCN Species Survival Commission, Gland, Switzerland
Jongman RH, ter Braak CJF, van Tongeren OFR (1995) Data analysis in community and landscape ecology, 2nd edn. Cambridge University Press, Wageningen
Kareiva P (1990) Population dynamics in spatially complex environments: theory and data. Philos Trans R Soc Lond B Biol Sci 330:175–190
Keith DA, Bedward M (1999) Native vegetation of the South East Forests region, Eden New South Wales. Cunninghamia 6:1–218
Leathwick JR (1998) Are New Zealand's *Nothofagus* species in equilibrium with their environment? J Veg Sci 9:719–732
Leathwick JR, Austin MP (2001) Competitive interactions between tree species in New Zealand's old-growth indigenous forests. Ecology 82:2560–2573
Legendre L, Legendre P (1998) Numerical ecology, 2nd edn. Elsevier, New York
Levine N (2000) CrimeStat: a spatial statistics program for the analysis of crime incident locations (v 1.1). Ned Levine and Associates, Annandale, VA, and the National Institute of Justice, Washington, DC. http://www.icpsr.umich.edu/ NACJD/crimestat/Crime StatManual.Ch7.pdf (last visited Jan 2002)
Lindenmayer DB, Mackey BG, Cunningham RB, et al. (2000) Factors affecting the presence of the cool temperate rain forest tree myrtle beech (*Nothofagus cunninghamii*) in southern Australia: integrating climatic, terrain and disturbance predictors of distribution patterns. J Biogeogr 27:1001–1009
Lunney D, Curtin A, Ayers D, Cogger HG, Dickman CR (1996) An ecological approach to identifying the endangered fauna of New South Wales. Pacific Conserv Biol 2:212:231
Lynn H, Mohler CL, DeGloria SD, McCulloch CE (1995) Error assessment in decision-tree models applied to vegetation analysis. Landscape Ecol 10:323–335
Mackey BG, Lindenmayer DB (2001) Towards a hierarchical framework for modelling the spatial distribution of animals. J Biogeogr 28:1147–1166
Manel S, Dias JM, Ormerod SJ (1999) Comparing discriminant analysis, neural networks

and logistic regression for predicting species distributions: a case study with a Himalayan river bird. Ecol Model 120:337–347

Manel S, Ceri Williams H, Ormerod SJ (2001) Evaluating presence-absence models in ecology: the need to account for prevalence. J Appl Ecol 38:921–931

Margules CR, Nicholls AO, Pressey RL (1988) Selecting networks of reserves to maximise biological diversity. Biol Conserv 43:663–676

Master LL (1991) Assessing threats and setting priorities for conservation. Conserv Biol 5:559–563

Mastrorillo S, Lek S, Dauba F (1997) Predicting the abundance of minnow *Phoxinus phoxinus* (Cyprinidae) in the River Ariege (France) using artificial neural networks. Aquat Living Resour 10:169–176

Mathsoft (1999) S-PLUS 2000 guide to statistics. Data Analysis Products Division, Mathsoft, Seattle, Washington

McCullagh P, Nelder JA (1989) Generalized linear models, 2nd edn. Chapman and Hall, London

McKenna MF, Houle G (2000) Under-saturated distribution of *Floerkea proserpinacoides* Willd. (Limnanthaceae) at the northern limit of its distribution. Ecoscience 7:466–473

McKenney DW, Mackey BG, Hutchinson MF, Sims RA (1996) An accuracy assessment of a spatial bioclimatic model. In: Mowrer HT, Czaplewski RL, Hamre RH (eds) Spatial accuracy assessment in natural resources and environmental sciences: second International Symposium. USDA Forest Service Report RM-GTR-277, Fort Collins, Colorado, pp 291–300

Millsap BA, Gore JA, Runde DE, Cerulean SI (1990) Setting the priorities for the conservation of fish and wildlife species in Florida. Wildl Monogr Suppl J Wildl Manage 54:5–57

Mitchell M (1996) An introduction to genetic algorithms. MIT Press, Cambridge, MA

Mitchell ND (1992) The derivation of climate surfaces for New Zealand, and their application to the bioclimatic analysis of the distribution of Kauri (*Agathis australis*). J R Soc N Z 21:13–24

Mohler CL (1983) Effect of sampling pattern on estimation of species distributions along gradients. Vegetatio 54:97–102

Moisen GG, Frescino TS (2002) Comparing five modeling techniques for predicting forest characteristics. Ecol Model 157:209–225

Okabe A, Boots B, Sugihara K, Chiu SN, Kendall DG (2000) Spatial tessellations: concepts and applications of voronoi diagrams, 2nd edn. Wiley, Chichester

Olson DM, Dinerstein E (1998) The global 200: a representation approach to conserving the Earth's most biologically valuable ecoregions. Conserv Biol 12:502–515

O'Rourke J (1998) Computational geometry in C. Cambridge University Press, Cambridge

Ostro LET, Young TP, Silver SC, Koontz FW (1999) A geographic information system method for estimating home range size. J Wildl Manage 63:748–755

Pearce J, Ferrier S (2000) Evaluating the predictive performance of habitat models developed using logistic regression. Ecol Model 133:225–245

Possingham HP, Lindenmayer DB, Norton TW (1993) A framework for the improved management of threatened species based on population viability analysis. Pac Conserv Biol 1:39–45

Rand GM, Newman JR (1998) The applicability of habitat evaluation methodologies in ecological risk assessment. Human Ecol Risk Assess 4:905–929

Rapoport EH (1982) Aerography. Pergamon Press, Oxford

Reading RP, Clark TA, Seebeck JH, Pearce J (1996) Habitat suitability index model for the eastern barred bandicoot, *Perameles gunnii*. Wildl Res 23:221–235

Richardson DM, McMahon JP (1992) A bioclimatic analysis of *Eucalyptus nitens* to identify potential planting regions in southern Africa. S Afr J Sci 88:380–387

Rykiel EJJ (1996) Testing ecological models: the meaning of validation. Ecol Model 90:229–244

Scachetti-Pereira R (2002) DesktopGARP. http://beta.lifemapper.org/desktopgarp/ (last visited May 2002), University of Kansas Biodiversity Research Center

Schamberger ML, O'Neil LJ (1986) Concepts and constraints of habitat-model testing. In: Verner J, Morrison ML, Ralph CJ (eds) Wildlife 2000: modeling habitat relationships of terrestrial vertebrates. Based on an International Symposium held at Stanford Sierra Camp, Fallen Leaf Lake, California. The University of Wisconsin Press, Madison, WI, pp 5–10

Schwartz MW (1999) Choosing the appropriate scale of reserves for conservation. Annu Rev Ecol Syst 30:83–108

Seaman DE, Powell RA (1996) An evaluation of the accuracy of kernel density estimators for home range analysis. Ecology 77:2075–1088

Seaman DE, Griffith B, Powell RA (1998) KERNELHR: a program for estimating animal home ranges. Wildl Soc Bull 26:95–100

Silverman BW (1986) Density estimation for statistics and data analysis. Chapman and Hall, London

Stockwell DRB (1999) Genetic algorithms II: species distribution modelling. In: Fielding A (ed) Machine learning methods for ecological applications. Kluwer, Dordrecht, pp 123–144

Stockwell DRB, Noble IR (1992) Induction of sets of rules from animal distribution data: a robust and informative method of data analysis. Math Comput Simulat 33:385–390

Stockwell D, Peters D (1999) The GARP modelling system: problems and solutions to automated spatial prediction. Int J Geogr Inform Sci 13:143–158

Swets JA, Dawes RM, Monahan J (2000) Better decisions through science. Sci Am 283:82–88

ter Braak CJF (1986) Canonical correspondence analysis: a new eigenvector technique for multivariate direct gradient analysis. Ecology 67:1167–1179

Tyre AJ, Possingham HP, Lindenmayer DB (2002) Matching observed pattern with ecological process: can territory occupancy provide information about life history parameters? Ecol Appl 11:1722–1738

USFWS (1980) Habitat evaluation procedures. United States Fish and Wildlife Service, Department of the Interior, Washington, DC

USFWS (1981) Standards for the development of habitat suitability index models. United States Fish and Wildlife Service, Department of the Interior, Washington, DC

Van Horne B (1983) Density as a misleading indicator of habitat quality. J Wildlife Manage 47:893–901

Van Horne B, Wiens JA (1991) Forest bird habitat suitability models and the development of general habitat models. United States Department of the Interior Fish and Wildlife Service, Washington, DC

Venables WN, Ripley BD (1997) Modern applied statistics with S-PLUS, 2nd edn. Springer, Berlin Heidelberg New York

Walker PA, Cocks KD (1991) HABITAT: a procedure for modelling a disjoint environmental envelope for a plant or animal species. Global Ecol Biogeogr Lett 1:108–118

Wiser SK, Peet RK, White PS (1996) High-elevation rock outcrop vegetation of the Southern Appalachian Mountains. J Veg Sci 7:703–722

Worton BJ (1989) Kernel methods for estimating the utilization distribution in home-range studies. Ecology 70:164–168

Yee TW, Mitchell ND (1991) Generalized additive models in plant ecology. J Veg Sci 2:587–602

Zaniewski AE, Lehmann A, Overton JM (2002) Predicting species distribution using presence-only data: a case study of native New Zealand ferns. Ecol Model 157:261–280

III. Addressing Plant Life Histories in Population Viability Analysis

9 Assessing Population Viability in Long-Lived Plants

M.W. Schwartz

9.1 Introduction

Understanding how to best estimate viability in long-lived plants is a problem that has not been given sufficient study in the plant population viability analysis (PVA) literature. In fact, annual plants are the focus of population studies to such an extent that one might conclude that most plant species are annuals (but see Chap. 8, this Vol. for habitat modeling with a perennial and Chaps. 11, 12, this Vol., for two PVAs of perennial plants). Quite the contrary, a perennial lifespan is the norm in plants. It is estimated that approximately 75 % of the flora of the United States and Canada are perennial species (Kartesz 1994). The overrepresentation of annuals in population studies is, no doubt, owing to their greater tractability for short-term studies.

Perennials are also well represented on lists of plants of conservation concern. For example, Stout (2001) summarized the rare plants of Florida Scrub habitats and found that 81 % of 48 rare and threatened taxa are perennial. Similarly, at least 94 % of the nearly 300 rare wetland-associated taxa are perennial (Edwards and Weakley 2001). Thus, it is critical that plant conservation biologists develop a new understanding of perennials so as to better estimate their viability for effective plant conservation strategies.

Perennials, by contrast to annuals, carry several significant problems for assessing the status of plant populations. Indeterminate, variable and clonal growth, along with overlapping generations and complex stage transitions (e.g., seed banks, adult dormancy (Shefferson et al. 2001)) make population projection in perennials both difficult and uncertain. As an example of variable lifespan, Fair et al. (1999) analyzed census data on *Bouteloua gracilis* across a 38-year interval and found that while the average lifespan of an individual of this short-lived perennial grass was less than 4 years, some individuals lived through the entire 38 years. These complex and variable life histories of perennial plants make estimating viability of long-lived plants difficult and extremely data-intensive. This chapter addresses these issues in attempt to summarize methods and suggest new, underutilized alternatives. Further-

Ecological Studies, Vol. 165
C.A. Brigham, M.W. Schwartz (Eds.)
Population Viability in Plants

more, the purpose of this chapter is to review basic strategies that can be used to assess viability in long-lived plants. A general difficulty in assessing viability in long-lived plants has resulted in a variety of creative approaches to the problem. Here, traditional viability concepts are expanded on to include examples of analyzing data often collected on long-lived plants but not typically applied to the specific problem of assessing viability.

There are two obvious topics not covered by this chapter that deserve special mention. First, seed dormancy adds a level of difficulty to plant viability analysis for all plants. Seed dormancy is not discussed here because long-lived plants tend to have lower seed dormancy rates than short-lived plants (but see Chap. 8, this Vol., for one approach to PVA approach that can be used for plants with an unknown seed bank). The second area not covered in this chapter is the issue of spatial ecology and metapopulation dynamics with a viability assessment (see Chaps. 6, 11, this Vol. for discussion of spatial issues in PVAs). Using spatial models and patch dynamics, viability analyses can be constructed within the context of patch incidence functions (Hanski 1991). Since species with long generation times are the focus of this chapter, there are few instances in which data are sufficient to assess population turnover and metapopulation dynamics. Thus incidence models are not considered in this chapter despite an increasing appreciation for their importance in viability assessment in general (Sjogren-Gulve and Ebenhard 2000).

9.2 Strategies for Population Viability Analysis

Although demographic data with which viability parameters may be assessed are relatively available in the ecological literature (Silvertown et al. 1996), very few viability assessments of rare plants have been published (Menges 2000). Viability studies of rare plants generally use one of two common viability assessment methods. First, a numerically simple means for assessing viability is to use simple surveys to examine population trend data (Dennis et al. 1991, expanded in Chap. 8, this Vol.). Interannual variation in population size provides an estimate of λ, the change in population size from one year to the next. A second, and more common approach is that of transition matrix population projection (Caswell 2001, Chap. 6, this Vol.). Population projection using transition matrices allows one to both estimate probability of persistence and to determine how different stage classes contribute to the probability of persistence of a population (Caswell 2001, Chap. 6, this Vol.). This specificity, however, may serve to promote unrealistic confidence in the population forecasts (Taylor 1995; Ludwig 1999; Fieberg and Ellner 2000; Thompson et al. 2000).

A third approach for predicting extinction vulnerability in long-lived plants is the use of life history attributes and historical reconstruction. Both of these have been used to understand the performance of tree species within a forest community context, but not to assess viability per se. Nonetheless, the principle is general; an understanding of life history can be used to predict the future distribution of size classes of individuals. Size class structure can be used to assert inadequate population replacement (Alvarez-Buylla and Martinez-Ramos 1992; Rentch et al. 2000). For example, in a species in which disturbance is considered to play a minor role in recruitment, deviation from a stable stage distribution may be a sign of a population persistence problem. Similarly, transition matrix modeling may be augmented by using tree rings to enhance estimates of transition rates of individuals across stage classes (e.g., Abrams et al. 1997).

A fourth approach for assessing population viability may be to use community-level models of community dynamics, which are particularly well developed for forest trees [e.g., JABOWA, FORET, ZELIG (Shugart 1984; Urban and Shugart 1992; Botkin 1993)], to provide an estimate of viability of a population within its community. Again, this has not, to date, been used to assess viability per se. Single species models of population persistence have been criticized as lacking biological reality [e.g., density dependence, interactions among species within their habitats (Groom and Pascual 1998)]. Models such as JABOWA, FORET and ZELIG (Shugart 1984; Urban and Shugart 1992; Botkin 1993) have been used to project transitions in local abundance of multiple species through time. These models can also be used to assess the likelihood that individual species will persist within a community. Similar models may be used to predict when seed recruitment is insufficient to maintain a population (e.g., SIMSEED, Rogers and Johnson 1998).

The common usage of PVA is to make a quantitative prediction of the likelihood of persistence of a population into the future. A quantitative prediction of extinction likelihood, however, is perhaps not the best use of a PVA (Chaps. 6, 7, this Vol.). Beissinger and Westphal (1998) asserted that, owing to the large error associated with viability estimates, exact probabilities of persistence are less practical than examining comparative persistence probabilities among populations under different management strategies or different environmental conditions. Burgman et al. (2001) proposed a set of criteria in order to circumvent a lengthy viability analysis in order to make conservation recommendations with respect to land area required for a species. Numerous other authors have offered other methods for rapid estimation of viability. These will be reviewed below and another decision guideline framework to categorize relative threat will be provided.

9.2.1 Population Trend Assessment

Past change in population size can be a powerful predictor of the future. Population biologists sometimes represent the net change of population size across a time step as a proportion (N_{t+1}/N_t) and call this value λ_t, the discrete population growth rate. The geometric mean of many such inter-annual changes is an estimate of the long-term growth rate (λ_s). However, this simple measure of population change is misleading of possible future outcomes if it lacks a measure of the variance in λ_s. Variance estimates are necessary in order to make a statement regarding extinction likelihood (Dennis et al. 1991; Ferson and Burgman 1995; Morris and Doak 2002). Consider two populations where population A is observed to have λ_s=1.1 and population B has λ_s=1.2. Population A, with no variance, would grow at a slower rate than B, but both would grow. With variance, either or both populations may have a high vulnerability to extinction. Population A would be less vulnerable to extinction than B if the variance in population A is sufficiently less than that of B. The biggest advantage of using census data to assess viability is that the data needs are minimal: simple population size survey data are adequate.

An assessment of viability using census data often includes a statement regarding the probability that λ_s is significantly larger than 1.0 (Alvarez-Buylla and Slatkin 1991). Clearly, if λ_s is significantly greater than 1.0, then persistence is likely. There are, however, two problems with this approach. First, documenting λ_s sufficiently well to identify a significant difference from 1.0 is a challenge because of the long time series likely to be required for long-lived plants. Second, asserting that a population growth rate is significantly greater than 1.0 is somewhat artificial. We expect the long-term value of λ_s to equal 1.0 when populations are stable and have saturated their potential environment. We expect high mean λ_s if a species had declined to low numbers and then undergoes a robust recovery. The only time we expect to document that λ_s is significantly greater than 1.0 is when a species is expanding rapidly or invading new habitats, not the sort of species that are likely to be subjected to a viability assessment.

If λ_s is not significantly greater than 1.0, then a more detailed analysis of persistence is required. Likelihood of persistence can be estimated using either of two methods. First, an analytic method using a regression technique developed by Dennis et al. (1991) and elaborated upon by Morris and Doak (2002) and discussed in Chapter 7 (this Vol.), uses the slope of the regression of individual estimates of λ and the time interval, run through the origin, to estimate the long-term mean population growth rate and confidence intervals. From these measures, an estimate of extinction probability over a given time span can be calculated.

Second, one can project population size into the future through a stochastic simulation model in order to predict extinction probabilities. Simulations begin with the current estimated population size and randomly draw a change in the population size for that time step from the observed distribution of λ_t. This observed change is used to project the population size for a single time step. Random draws from the observed distribution of λ_t are repeated through a time series, often 100 years. The simulated population is scored as having gone extinct if it passes through some minimum population threshold size (quasi-extinction; e.g., 2, 10, 20, 100). Replication of this simulation (e.g., 1,000 times) provides a distribution of extinction probabilities as a function of time.

Both the regression and simulation methods should result in the same qualitative predictions and the choice is really a matter of preference for analysis. Both methods allow a sensitivity analysis of the extinction probability based on the variance of λ_t as well as allowing the incorporation of biological complexities such as density-dependent effects is possible. Both methods are improved with increased sample size in population size transition from year to year. One way to increase this estimate is through sampling multiple populations in a single year. The cost of this is that sites may have correlated responses, and treating them as independent estimators of λ_t would lead to an undue inflation of confidence in the estimate. Remarkably, however, very few studies have been published that use these methods despite their ease of use and simple data requirements.

9.2.2 Transition Matrix Modeling

Bierzychudek (1982, 1999) provided insight into the robustness of viability assessment by examining the population status of *Arisaema triphyllum*, a perennial herb, 15 years after her initial study. Observing a failure of the early study to predict subsequent dynamics accurately, she highlighted three reasons why transition matrix studies may yield inaccurate population projections (insufficient duration of initial study, inaccurate transition probabilities, complex transitions). In order to graphically depict the magnitude of these problems, I created a simple matrix model using a hypothetical tree species with plausible life history attributes (Harcombe 1987) and deterministic recruitment and mortality. This hypothetical tree has four stage classes (seedlings, saplings, sub-adults and adults). For simplicity, I specify: (1) only adults reproduce; (2) survivorship is low for seedlings (10 %), intermediate for saplings (45 %); (3) subadults and adults have high survivorship (95 %); (4) transitions to larger size classes are infrequent (5 % in all cases); (5) transitions to smaller size classes are impossible; and (6) seed production is constant (24 seedlings per adult per year; Table 9.1). I also assume 100 % germination of seeds (no seed mortality and no seed dormancy). When these values are fixed, the demography is deterministic and λ=1.0 and a stable stage distribution yields a population of 77 % seedlings and 3 % adults (Table 9.1).

Using this simple model, I explore the potential magnitude of aforementioned problems with transition matrix models in each of the next two sections. This is done by adding stochastic behavior to the simple model by randomly selecting from a distribution of transition probabilities among size classes to explore model sensitivity. Error owing to episodic events in the demography was estimated by repeating simulations with 1,000 runs. For

Table 9.1. Transition values for a simple population matrix for a hypothetical tree that results in the stable stage distribution (*SSD*) and a constant population (λ=1.0)

	Seedling	Sapling	Subadult	Adult
Seedling	0	0	0	24
Sapling	0.1	0.4	0	0
Subadult	0	0.05	0.9	0
Adult	0	0	0.05	0.9
Proportion at SSD	0.77	0.13	0.065	0.032

simplicity, stochasticity in seed production and in adult survival were independent. In reality, these are likely to co-vary.

9.2.2.1 Insufficient Sampling Interval

Bierzychudek (1999) suggested that most studies are not of a sufficient duration to assess a sufficient breadth of environmental conditions that may affect population performance. The lack of sufficiently long study duration is most significantly a problem in species that respond to episodic infrequent events such as disturbances or anomalous rainfall years.

I varied initial model conditions to incorporate stochastic behavior in adult and seedling survival. Mimicking episodic events, stochasticity was modeled with variation drawn from an exponential distribution in which populations conform to the mean 90 % of the time and have drastically different outcomes in 10 % of years. The distribution of λ values that result are very strongly leptokurtic (i.e., more peaked than expected; Fig. 9.1). With this amount of stochasticity, 10 years of sampling are required in order for 95 % of estimates of λ to be within 10 % of the true value (Table 9.2). I used the same exponential distribution to generate similar amounts of stochasticity for each of the subsequent model variants.

Next, I varied demographic parameters to simulate a declining population and then again estimated the number of years of sampling required to accurately detect a decline. First, seed production was reduced tenfold to 2.4 seeds/adult/year. The resultant model yields λ=0.93 in the deterministic version, a rapid decline for a species with long-lived adults. However, sampling the population, without measurement error, for 2 years would result in a 35 % probability of estimating a growing (λ>1) population despite the actual population experiencing an order of magnitude reduction in seed production. It would require between 10 and 20 years of sampling, without measurement error, in order to detect that λ<1.0 with 95 % accuracy (Table 9.2).

Reducing adult survivorship by 50 %, similarly, yields λ=0.92 in the deterministic model. Adding stochasticity via episodic events on a decadal scale, as above, I found that it requires 5 years of sampling to detect λ<1.0 in 95 % of

Table 9.2. Results of a population projection simulation estimating the effect of the length of a study on the ability to accurately predict the long term change in population size

Run conditions	Number of years of data	1	2	3	5	10	20
Baseline	Estimated λ	1	1	1	1	1	1
	95 % CI	0.51–1.64	0.65–1.36	0.75–1.29	0.84–1.21	0.92–1.13	0.96–1.09
	Estimated probability λ>1	0.5	0.5	0.5	0.5	0.5	0.5
10 % Seed prod	Estimated λ	0.94	0.94	0.94	0.94	0.94	0.94
	95 % CI	0.52–1.26	0.64–1.15	0.69–1.10	0.86–1.08	0.84–1.02	0.88–1.00
	Estimated probability λ>1	0.35	0.35	0.33	0.29	0.12	0.02
50 % adult survival	Estimated λ	0.92	0.92	0.92	0.92	0.92	0.92
	95 % CI	0.66–1.35	0.77–1.18	0.82–1.10	0.89–0.99	0.90–0.99	0.91–0.97
	Estimated probability λ>1	0.21	0.21	0.17	0.01	0.01	<0.001
Skewed distribution of years of high growth or mortality	Estimated λ	1.03057	1.030459	1.030542	1.033862	1.030253	1.030099
	95 % CI	0.43–1.62	0.66–1.37	0.77–1.27	0.94–1.15	0.93–1.13	0.97–1.09
	Estimated probability λ>1	0.60	0.60	0.61	0.72	0.76	0.82

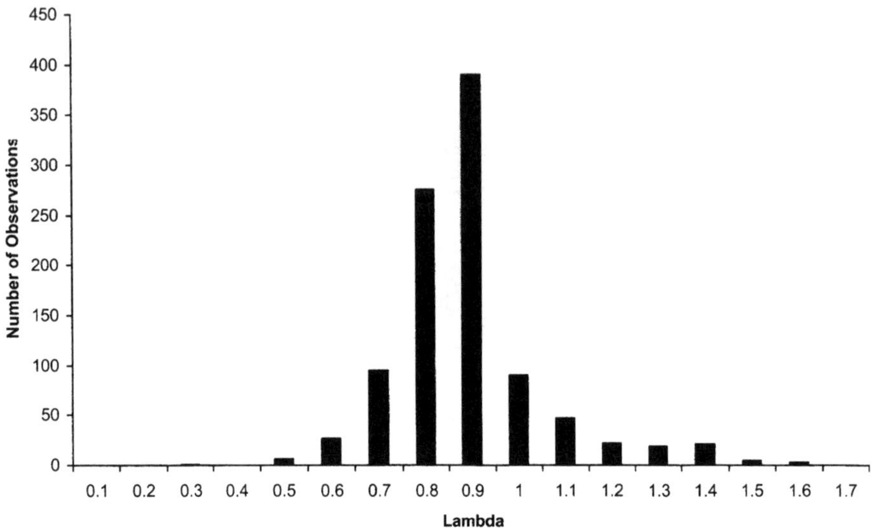

Fig. 9.1. The distribution of λ values for simulation models resulting from a exponential distribution of random numbers in which normative conditions apply in approximately 90 % of simulation draws. The result is a strongly humped distribution centered with a mean of 0.89 as a result of a 50 % reduction in adult survival

replications (Table 9.2). Adding measurement error such that λ is estimated from incomplete sampling of the population would increase the estimate of time required to accurately detect decline. To place this in perspective, Menges (2000) found that few of the 95 PVA studies reviewed lasted more than 5 years.

A frequently attempted solution to this sampling problem is to increase the number of populations sampled. Ten years' worth of estimates of λ from ten different populations each year will increase the overall estimate of λ. Multiple populations, however, may have correlated responses to interannual environmental variability. As a result, 10 years of sampling ten populations would provide fewer than 100 independent estimates of λ. For example, if regional differences in rainfall dictate seed production and all populations are affected by inter-annual variation in rainfall, then samples are not independent. Treating populations as independent inflates confidence in the estimate of λ and leads to inaccurate recommendations.

9.2.2.2 Inaccurate Transition Probabilities and Complex Transitions

A related data insufficiency problem is that the sample sizes found in most studies are insufficient to accurately estimate low frequency state transitions. This problem is likely to be acute in long-lived species with persistent above-ground biomass (e.g., trees and shrubs) where residence time in stage classes is often long. Demographic models of trees tend to classify populations into relatively few size classes (e.g., canopy adults) and average residency time in a single stage class may be on the order of decades to centuries (e.g., Schwartz et al. 2000). As a result, hundreds of trees must be sampled in order to estimate accurately these transition probabilities. For example, a survey population of 100 individuals in which 10 % of the individuals are in the sub-adult size class would yield only ten observations of a potential transition per year of survey. If the probability of a particular transition is low (i.e., <10 %) we might only encounter it once per year, on average. Thus, it would require many years of survey data and several hundred individuals in the population to accurately estimate the transition and its variance.

An example of this problem is provided by Schwartz et al. (2000), who estimated the time to extinction for *Torreya taxifolia*. This species is a coniferous tree endemic to the Apalachicola River basin in the southeastern USA, and the study was based on a 9-year survey of over 100 individuals. The largest stem of individual trees occasionally dies back, leaving a smaller resprout as the primary stem. This does not happen often for larger trees, which are few. Despite nearly 1,000 tree observations, downward transition probabilities among individuals in the largest size class were based on three observations of primary stem loss. The resulting transition probabilities are low (1.4 %) and likely contain a substantial measurement error, being based on three observations. An unobtainably large sample of trees across decades would be necessary in order to accurately detect the transition probability for these rarely encountered transitions and account for their impact on λ.

Similarly, transitions among stage classes of herbaceous species may be complex where individuals may skip stages, in both directions, and grow either larger or smaller. For example, individuals of A. *triphyllum* can either leap forward stage classes or regress (Bierzychudek 1982). The resulting transition matrices contain many additional non-zero elements that must be estimated. If these transition elements are common, then one might reasonably estimate them from population sampling. Unfortunately, life histories that create matrices with many non-zero elements typically contain many transitions that are infrequent and, thus, difficult to estimate in rare plants with small population sizes.

Clonal reproduction is a final example of complex transitions in perennial plants that add non-zero elements that must be estimated in a transition matrix model. If clonal reproduction is integral to estimating population persistence, then accurate transitions must be estimated. Projecting populations with clonal reproduction is made difficult, but not impossible, by the need to distinguish sexual from clonal reproduction in population matrices. Damman and Cain (1998) assessed viability in the clonal herb *Asarum canadense*. They found that adult survivorship was more important than clonal reproduction for persistence of the population. However, one can envision clonal plants for which clonal reproduction is integrally linked to population persistence. A review of the 298 wetland-associated rare plant species along the southeastern coastal plain of the USA found that 94 % spread through vegetative reproduction (Edwards and Weakley 2001).

Small population size from which to sample is an acute problem for rare plants. For example, 63 % of the 207 endangered and threatened plants of Hawaii covered in nine multi-species recovery plans were represented by fewer than 200 individuals (USFWS 1994; USFWS 1995a, b; USFWS 1996a, b; USFWS 1997; USFWS 1998a–c). For many species most in need of some estimate of viability, there are simply too few individuals to estimate accurate transition probabilities in less time than decades. Estimating transition probabilities for a demographic model of these species may not be worth the effort, given that most of these species have relatively straightforward problems relating to seed predation by exotic vertebrates (USFWS 1994; USFWS 1995a, b; USFWS 1996a, b; USFWS 1997; USFWS 1998a,b,c).

9.2.2.3 Transition Matrix Element Elasticity

One of the cited benefits of transition matrix modeling is the ability to conduct a sensitivity analysis on matrix element elasticities (Lesica 1995; Oostermeijer 1995; Groom and Pascual 1997; Menges 1998; Grant and Benton 2000; Heppell et al. 2000a, b; Kaye et al. 2001; Chap. 6, this Vol.). In short, an elasticity is a measure of the change in the overall estimate of population growth (λ) given small changes in the transition probability of a single element (Tuljapurkar and Caswell 1997; Caswell 2001, also see definition in Chap. 6, this Vol.). As a result, matrix elements with a high elasticity contribute the most to overall population growth (e.g., Silvertown et al. 1996). The argument has been made that management efforts to achieve population persistence may be most effectively applied to increasing matrix elements with high elasticities.

This approach suffers from two obvious shortcomings. First, matrix elements are not often independent (Menges and Dolan 1998; Chaps. 6, 7, this Vol.). Thus, a simple sensitivity analysis may mask total effect on population viability unless it incorporates the correlation between matrix elements. By including correlations among matrix elements we could, in theory, find regions of the transition matrix (sensu Silvertown et al. 1996) that have more impact on population growth than others and that individual elasticities may not accurately reflect predict this. This has not been adequately modeled.

Second, management of attributes that affect particular matrix elements (particular transitions at particular life stages) are neither equally possible nor effective. For example, Silvertown et al. (1996) show that long-lived plants tend to have high elasticities in life stages associated with adult survivorship.

For trees that are harvested, the management directive is relatively straightforward: reduce adult mortality (i.e., reduce logging). For many other trees this is simply not the problem. Several endangered Hawaiian trees are known from fewer than ten individuals (USFWS 1995a,b; USFWS 1996a,b; USFWS 1997; USFWS 1998a,b,c) – hardly a sufficient number upon which to base a quantitative PVA. How these populations came to be so small is not clear. Lacking the ability to conduct a full viability analysis because of small population size, one might heed to patterns observed by Silvertown et al. (1996) and suggest that management should focus on facilitating persistence of the adults. Yet, there seems to be no problem associated with adult survivorship. Any amount of management to help preserve these four remaining adults is not likely to extend these individuals past a normal lifespan for this species. This lifespan is not likely to exceed a few hundred years. The problem, biologically, is straightforward: seeds that are produced are reliably consumed by non-native herbivores (i.e., pigs and goats) and are not germinating. Since there is no regeneration, managing for increased reproduction would quite clearly help this species. A similar pattern emerges from the viability assessment of *Torreya taxifolia* (Schwartz et al. 2000). The real problem for *T. taxifolia* is a lack of seed production, not survivorship. As of 1999, there had been no seed production observed in this dioecious tree for more than two decades.

Simply using analysis of elasticities to guide conservation action may lead managers down the wrong path. A better approach would be to estimate the degree to which management may be able to increase a particular transition matrix element. This approach would lead managers to determine which matrix elements under realistic management would have the greatest impact on population growth. Alternatively, one might estimate the degree to which an observed matrix element varies from what is presumed to be a probability in an unimpacted population. Some combination of matrix elements with high elasticities and that are most different from an unimpacted population may afford the biggest potential impact on population persistence. Again, this approach has not been adequately addressed in the literature.

9.2.3 Reconstructing Performance from Population Size Structure

Long-lived plants have physical attributes that provide evidence of past performance and thus assist in assessing population viability into the future. This evidence can take at least three forms: woody growth rings, plant size class structure, and the rate of spread of clonal ramets.

Trees in seasonal environments leave a record of their past growth in tree rings (Cook and Kairiukstis 1990). For a subset of trees we expect continual recruitment within a population in order to maintain long-term persistence. Thus, assessing the distribution of ages among individuals should provide information on the rate at which new individuals recruit into a population and allow an estimate of the number of small trees that would need to be recruited each year in order to balance mortality across all size classes. Assessing the distribution of tree ages within a population, along with growth rates, suggests whether the species is maintaining a stable population size through time (e.g., Kullman 1996; Abrams et al. 1997). For example, Abrams et al. (1997) used tree rings to show that episodic recruitment among oaks in the Blue Ridge Mountains has been the norm over the past 200 years, and that the current stand structure is a relic of the high oak recruitment rates that followed the mid-twentieth century chestnut blight. Hence, the current low recruitment rate might indicate a return to historically normal abundance and not be a sign of population inviability. In northern Japan, Abrams et al. (1999) used tree rings to describe a nearly 400-year growth record of oaks to document long-term variability in recruitment. These researchers found that the observed on-going recruitment failure is consistent with a prognosis of long-term oak decline. Thus, tree ring records have proven to be an effective means to predict future dynamics of a population based on a better understanding of historical population dynamics.

A second and less definite means by which a plant population may provide evidence of recruitment failure in a population is through its size structure. For woody plants we anticipate continual increases with size, usually measured by the girth of stems, with age. Despite indeterminate growth in plants, many species demonstrate significant correlations between the size and age of individuals. In this case, a plausible way to make an initial assessment of viability for a long-lived species is to examine the distribution of individuals within size classes. Tree populations are characterized by predictable size distributions in populations at a stable size class distribution (Harcombe 1987; Barbour et al. 1998). In habitats that lack periodic disturbance events that

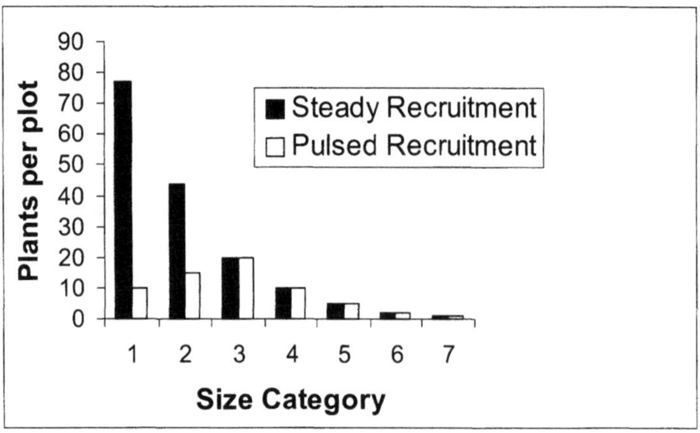

Fig. 9.2. Hypothetical abundance distributions of a plant by size class. *Dark bar* A typical self-replacing population at a stable stage distribution. In contrast, the *open bars* represents a species undergoing pulsed recruitment or recruitment failure

focus mortality or recruitment (e.g., catastrophic fire), we expect a constant mortality rate within size or age classes leading to attrition of individuals within cohorts through time. This distribution is also known as the Deevey type III survivorship curve, and is frequently observed among trees. Barnes et al. (1998) referred to these tree species as "gap species." As a result, populations in these habitats should be characterized by decreasing densities of individuals from small to large size classes (Fig. 9.2). Even species that are characterized by episodic events of recruitment may appear to exhibit this classic inverse J-shaped curve if the episodes of recruitment are at short intervals relative to the life span of adults.

In contrast, however, many species are characterized by very infrequent recruitment, or recruitment following disturbances that kill the previous generation of adults (e.g., fires in jack pine, sand pine). For these cases, we expect populations to be characterized by one or a few size classes of individuals, a situation that ordinarily suggests a lack of viability. In this case, any viability assessment, even an informal one, requires a suite of populations at various stages of maturity and an assessment of the continuation of processes that give rise to new stand formation (e.g., natural disturbance regimes). In both cases, a sign of likely population inviability is an under-representation of individuals in the smallest size classes (Fig. 9.2).

Rentch et al. (2000) used a lack of recruitment into smaller size classes to predict a population decline of Carolina hemlock (*Tsuga caroliniana*) during the coming decades. These authors interpreted the high current density of Carolina hemlock in the overstory as a response to chestnut decline. Similarly, Smith and Nicholas (2000) used size class structure of Fraser fir (*Abies fraseri*) to assess recruitment after mortality due to wooly adelgid infestations. These authors concluded, through analysis of recruitment patterns in infested stands, that Fraser fir is likely to decline through time as repeated infestations slowly decrease its dominance.

As discussed earlier, many long-lived plants spread through clonal growth. Clonal growth creates difficulties for assessing viability because physically isolated individuals may be genetically identical. Thus, defining the unit for assessing a population, the genetic or physical individual, creates challenges with clonal plants. Nonetheless, clonal growth can provide an historical record of growth that can be used to help assess past population performance as a guideline to determine whether current performance is poorer than required to maintain a population. For example, the age and spread of individuals of *Symphoricarpus occidentalis* was reconstructed through rhizomes (Pelton 1953). Similarly, ages of clones have been determined through growth rings on rhizomes or following ramet expansion for a variety of shrubs and herbs (see Harper 1977). Historical data derived from woody rhizomes can be used to help parameterize a model to predict viability by estimating an age distribution of a population, the likely survivorship of individuals, the spread rate of individuals, or the rate at which clonal growth gives rise to new individuals.

9.2.4 Predicting Community Dynamics

There is a well-established precedent for assessing the viability of populations of trees within the context of forest community dynamics (e.g., Horn 1975). These assessments, however, are an outgrowth of studies of forest succession and have been conducted either within the context of community transition matrix models (Horn 1975) or forest gap models (Shugart 1984; Botkin 1993). At the present time, forest gap models are frequently used to assess changes in forest composition through time (e.g., Hansen et al. 1993; Szwagrzyk 1994; Mladenoff and Baker 1999; Zolbrod and Peterson 1999). These models use physiological attributes of trees (e.g., maximum growth rate, size-specific mortality rate) to assess their relative probabilities of gap capture. Replication across thousands of potential gaps predicts the abundance of a species within a forest. Perturbation of the model parameters (e.g., changing climate) results in alternative predictions of forest composition. These models, although not traditionally framed as viability assessments, predict the probability of the continued existence of a species within a landscape and assess potential drivers of vegetation change as mechanisms that affect the likelihood of persistence.

For example, Weinstein et al. (2001) used forest community simulators (ZELIG and TREGRO) to predict the future declines in population size of yellow-poplar (*Liriodendron tulipifera*). In this case, the physiological response model TREGRO was used to predict responses of yellow-poplar to ozone pollution. Because of the greater sensitivity of yellow-poplar to pollution, it was predicted that other species would increase at the expense of yellow-poplar and that the density of yellow-poplar was likely to decline. In this case, the authors were not arguing an issue of population viability, but the models provide a framework under which one could estimate viability among populations. Similarly, Zolbrod et al. (1999) used ZELIG to predict that *Tsuga mertensiana* populations are not viable under climate change scenarios for the Olympic Mountains in Washington. They predicted that *T. mertensiana* would be replaced by *Abies amabilis* if climate warms as predicted.

9.2.5 Sample Variance

A fundamental problem associated with all viability assessments is sample variance adding uncertainty to estimates of viability. For long-lived species this problem is severe because transitions among size classes may be infrequent (e.g., trees) and can run in both directions (e.g., perennial herbs that can get larger and smaller through time). Further, recruitment and mortality among long-lived species are often episodic and difficult to measure without a very long time series of data. Finally, because perennial individuals are likely to persist through multiple census intervals, a study of long duration is needed to get a good estimate of mortality, recruitment, and population growth rates. Quite simply, populations of long-lived species may change little from year to year despite a consistent trend. In fact, the variance in estimates of most ecological parameters increases with the length of sample period (Pimm and Redfearn 1988).

Thus, collecting large data sets may be required to produce a believable viability assessment using any method. Unfortunately, acquiring a long-term data set is often unrealistic for a viability assessment of endangered species because sample sizes are, by definition, small and time is limited. For less endangered species, a large sample size may help alleviate the problem of sparse data, although large sample size over a short time period does not alleviate the problem of variable transitions (episodic seed production, recruitment or mortality) that depend on edaphic conditions (e.g., Menges 1990).

One method for reducing the problem of poor estimates of population transitions or population growth rate is to collect data on several populations within each sample year. Variance across sites within years, however, should be interpreted with caution. Inter-site estimation of the variance in performance parameters is valid only to the extent that sampled populations are independent with respect to environmental variation (e.g., climate, disease). Thus, the power of the estimate must be reduced to the extent to which these variables may co-vary – an attribute that is typically not known.

9.3 Expressing Viability: Extinction Likelihood/Generation Time

The appropriate time frame of viability for long-lived species is an important issue to consider. Expressions of viability for long-lived species should vary from those of short-lived species under many circumstances. Barring severe diebacks, the process of population failure in long-lived species may be relatively slow (e.g., Schwartz et al. 2000). Many stands of blue oak (*Quercus douglassii*) appear to have complete reproductive failure (Gordon and Rice 2000). Stand extirpations are guaranteed. *Torreya taxifolia*, with no seed production in the wild, is clearly moving steadfastly toward extinction (Schwartz et al. 2000). Nonetheless, the process of extinction for both of these species is likely to take well over 100 years. Standard threat categories used to classify endangerment of species use values expressed over periods of 10–100 years (e.g., Mace and Lande 1991). With long-lived species, a viability assessment that predicts a 90 % chance of survival for 100 years may not distinguish between a healthy population and one declining toward extinction. As a result, expression of viability for long-lived species must be in terms of generation time. A prediction of extinction within one to two generations should be alarming no matter how long the generation time for the species.

9.4 Nonquantitative Assessments of Viability

A growing literature argues that quantitative predictions of extinction probabilities are often not very accurate (Taylor 1995; Brook et al. 1997; Bierzychudek 1999; Chapman et al. 2001; Chap. 6, this Vol.) and that the effort to construct them is misplaced (Ludwig 1999; Fieberg and Ellner 2000; Coulson et al. 2001). These concerns are usually derived from the extensive data required to make population forecasts and the unrealistic assumptions one must often make in projecting populations (e.g., density independence, constant demographic rates). Nonetheless, conservation management often requires an estimate of vulnerability among species in order to prioritize protection (e.g., Millsap et al. 1990). Lacking demographic data on all species under consideration of potential vulnerability, expert opinion is often used to place species on conservation watch lists (e.g., Lunney et al. 1996; Carter et al. 2000).

A variety of methods for making rapid, non-quantitative predictions of species extinction vulnerability have emerged as an alternative to PVAs. Most of these methods set conservation priorities based on perceived relative extinction vulnerability by tabulating points toward extinction vulnerability across a variety of categories (e.g., Millsap et al. 1990; Lunney et al. 1996; Todd and Burgman 1998; Carter et al. 2000). Categories for scoring vulnerability can include habitat factors, distribution factors, species characteristics, and life history attributes. Qualitative point-based rankings are somewhat controversial, but often necessary to set a priority list of species (Burgman et al. 2001). An alternative method is to use multivariate analyses of species attributes to cluster species that share vulnerability attributes (e.g., Given and Norton 1993). The value of using a multivariate approach is that species may be similarly vulnerable for different sets of reasons, and the approach of Given and Norton (1993) purports to capture these instances better than linear methods. All of these ranking methods for species vulnerability have been developed for, and applied to, vertebrates.

A primary point of this chapter has been that long-lived perennial plants suffer from high uncertainty in quantitative population forecasts. Further, there are very few plants for which sufficient data allows a robust PVA. Thus, developing a decision support structure for rendering informed expert opinion on the degree of extinction risk for plants is critical. I propose a decision tree structure to assist structuring an expert opinion in a preliminary "relative population viability assessment "(RPVA, Fig. 9.3). This decision support structure borrows concepts developed by previous efforts of linear, point-scoring systems for prioritizing species for protection. I use readily detectable attributes of species and their environments that are demonstrably linked to species endangerment (e.g., habitat loss, exotic species (Wilcove et al. 1998)). The particular decision tree presented here is general. Specific applications require altering the structure of the decision tree to accommodate environ-

AREA	SCORE
Habitat Loss: Multiple protected populations = 0 points Few protected populations = 5 points Moderate loss, little protected habitat = 10 points Severe loss, no protected habitat= 15 points Nearly complete habitat loss = 20 points	______
Disturbance regime disruption Natural disturbance regime maintained = 0 points Severe disruption of disturbance regime = 20 points Example: fire suppression with obvious indication of community change.	______
Habitat Degradation Habitat Intact = 0 points Severe habitat degradation = 20 points Example: invasive weed usurping habitat	______
Population Performance Apparently normal = 0 points Poor performance = 20 points Example: disease killing all seedlings Loss of pollinator reducing seed set	______
TOTAL	______

Fig. 9.3. Example score sheet for a relative population viability assessment (RPVA). Score categories are broken into four major categories of habitat loss, disruption of natural disturbance cycles, degradation of the biotic community, and evidence of population failure. Several potential drivers of population failure are possible under each principal heading. Investigators must evaluate each criterion for each species and score them from no degradation (0 points) to degradation sufficient to render the population inviable (20 points). Points are then summed to provide a relative risk of extinction when compared to other such assessments

mental conditions and species attributes. I propose using environmental attributes and conditions to make a rapid assessment of the likely status of a species. It is assumed that no demographic data exist with which to make this assessment, but that the expert has examined the population(s) in question during the reproductive season.

A potential weakness of any point-scoring system is that it presumes that the user can accurately scale the points across categories with respect to how each contributes to potential extinction likelihood. Similarly, the strength of point-scoring methods is that they allow the user to weight different factors different amounts and does not require them to be on the same scale. Thus, the method is flexible, but also prone to errors owing to its subjective nature.

The general notion is to tabulate point-scoring toward increasing concern regarding potential viability. Tabulating points is meant to help track cumulative problems across categories of the decision tree in assessing viability. The particular number of points awarded across each area is subjective and relegated to expert opinion. In this example I describe point-scoring across a viable (0 points) to inviable (20 points) spectrum in each of four general areas (habitat loss, disturbance regime alteration, habitat degradation, and population performance). For each case, points are to be scaled, using expert judgment, based on the degree to which that particular threat may lead to concern for the ability of the species to persist.

9.4.1 Habitat Loss

The first step in the decision tree is to assess habitat loss. Habitat loss is the leading cause of endangerment (Wilcove et al. 1998). Ongoing habitat loss for any species with a limited distribution is likely to be a clear indication of potential viability problems (for a discussion of how to determine habitat for a species see Chap. 8, this Vol.). Nonetheless, a RPVA should clearly identify both the role of historic and likely future habitat loss. Past habitat loss may or may not place a species within a precarious position because populations have already been lost. Nonetheless, since management of the population for persistence will focus on remaining habitat, the focus of the RPVA must be to evaluate remaining extant habitat to predict what may be required in order for the species to remain viable. In order to estimate the contribution of likely future habitat losses toward threat, points are scored that indicate the degree of ongoing habitat loss and how those losses would contribute to further losses in populations or population size. If habitat loss rates are high and complete habitat destruction seems likely, and there are neither protected sites nor any mechanisms to create them, then the answer is simple; viability is unlikely under the current status and protection should be the top priority. In this case, award 20 points, indicating that habitat loss, in and of itself, is rendering viability unlikely.

Alternatively, if ongoing habitat loss is negligible, then award 0 points and continue to step 2. Intermediate conditions of habitat loss receive intermediate numbers of points depending on the perceived risk of threat owing to potential future habitat loss. For example, one might assign 5 points if there are a some protected areas and low rates of habitat loss, 10 points if habitat loss rates are moderate, and 15 points if either habitat loss rate are high or there is virtually no protected habitat.

Historical habitat loss creates a potential different type of problem. For example, even if historical habitat loss were severe, if there are a significant number of protected areas such that there should remain sufficient habitat for persistence, then no points would be assigned. Points in this case represent the degree to which the expert is uncertain of viability owing to low numbers of protected populations.

9.4.2 Disturbance Regime

The second class of RPVA decision tree criteria requires classifying a species' relationship to disturbance (Fig. 9.3). Long-lived plants can generally be classified into one of three groups based on their relationship to disturbance: (1) species that are self-maintaining in the absence of disturbance; (2) species that tolerate and thrive under more frequent low intensity disturbance; and (3) species maintained through recruitment following chronic disturbance. Although there are likely to be some species that do not fit these categories well, these disturbance categories are used because they are common across a wide range of species and disturbance types. For assessing the role of disturbance in viability, the logical progression is to assess the nature of the historical disturbance regime for the region, the status of the habitat condition with respect to maintenance of the historical disturbance regime, and then predicted plant performance within extant populations under current versus historical disturbance regimes. Thus, points may be awarded either for species accustomed to a frequent disturbance regime when that disturbance is removed from the system (e.g., fire suppression), or when a disturbance is introduced to a historically undisturbed system (e.g., increasing fire frequency).

Historical range of variation (Haufler et al. 1999) concepts are useful for classifying current disturbance regimes relative to historic disturbance regimes. If protected habitats fall outside of the estimated historic range of variability for disturbance, then assigning inviability points for populations under these altered disturbance regimes in a RPVA is warranted. This may also suggest targeting management and research toward understanding the effects of the altered disturbance regime.

Late-successional species that recruit in stable communities (category a) are exemplified by shade-tolerant trees that recruit into canopy gaps. If disturbance rates are not unduly enhanced, reducing the likelihood of recruitment, then no points are awarded for alteration of disturbance regimes. If disturbance has been artificially increased, then points should be assigned depending upon the severity of the disturbance and the perceived threat of these disturbances to the population. Species in this class are unlikely to be severely impacted by disturbance rates since the examples of habitats where abiotic disturbance rates, such as fire intervals, have increased are rare.

Species in category b are those that are maintained by frequent low-intensity disturbance (e.g., ground fires). For these species, disturbance may provide recruitment opportunities and may reduce competition from disturbance intolerant species. A classic example of a fire-maintained ecosystem is eastern tallgrass prairie, where fire suppression results in hardwood invasion and succession to forest. Assessing species viability in tallgrass prairie using RPVA would require knowledge of the fire regime and an estimate of the species response to fire. Once again, an assessment of disturbance regimes is required to assess viability. If the landscape in question deviates from estimated historical levels of disturbance, then awarding inviability points is warranted. An assessment of recruitment in the absence of disturbance assists rendering an opinion regarding inviability points. Although this assessment requires long-term data, size structure of the extant population may provide clues as to continuous reproductive success for woody species. Long-lived herbaceous species and clonal herbs may present fewer signals with which to assess continued reproductive success.

For species that recruit primarily in response to chronic disturbance (category c), it may be impossible to assess viability from the recruitment performance of a single population and a multi-population approach is required. Menges (1990) study of Furbish's lousewort (*Pedicularis furbishiae*), the first widely cited study on plant population viability analysis, fits this category. Furbish's lousewort is a species in which populations are expected to slowly decline toward local extirpation even when the system is intact and the species performing well as a whole. Continual re-establishment of new populations through flood scouring is required. As such, assessing species viability requires a multi-population assessment of region-wide disturbance rates and the likelihood of new population establishment relative to the average longevity of populations. In this case, the species requires landscapes that contain patches of habitat at various stages of maturity in order to maintain a disturbance cycle (*sensu* Pickett and Thompson 1978). This life history is likely to require a very large area that includes potential, as well as actual, population sites. As a result, habitat loss attributes should be more heavily scrutinized.

A full analysis of a disturbance-established species would require information on time to maturity for populations, seed production, and dispersal among likely establishment sites and disturbance rates (e.g., Harper 1977). In lieu of this information we might make a rapid assessment based on the number of populations at various stages. For example, if we have no reason to believe that the disturbance cycle is disrupted and a sample of populations can demonstrate that newly established, mature and senescent populations exist, then we can assert that the large-scale environment is not limiting viability. If the assessment fails any of these criteria, then award points toward an assessment of inviability.

9.4.3 Habitat Degradation

The third step in the decision tree is to assess the degree to which existing habitats remain intact. Habitat degradation through alien species, as plant competitors, herbivores, or diseases, is a leading cause of species endangerment (Menges 1990; Wilcove et al. 1998). A visual assessment of habitat conditions could provide information with which to assert that a species is likely placed in a vulnerable position as a result of the invasion of weeds in its habitat. This may or may not be a simple process. The mere presence of non-native species does not warrant undue concern. Most communities have some non-native plants. Nonetheless, strong dominance of a plant community by non-natives is an indication of potential trouble and warrants caution. The evidence that exotic pest species create problems by reducing population abundances of native species abounds (e.g., Lambrinos 2000; Mooney and Hobbs 2000). The point scale, in this case, is used to assess the degree to which non-native species appear to detract from the potential of the target species to reach perceived presettlement densities.

9.4.4 Population Performance

The fourth step for the RPVA is to assess whether there is any reason to believe that population performance is below expectations. In this case, an examination of the reproduction and dispersal of the target species is warranted. Demonstration that seed production or dispersal is disrupted would again warrant adding inviability points and caution against a determination of probable viability. This is likely to be a difficult attribute to assess lacking data. Seed bank attributes, for example, may not be known and yet may be very important to estimates of viability.

Summing all inviability points allows a general ranking of a species relative to other species considered for more formal viability assessment. Clearly, this sort of rough classification is subjective and can lead to erroneous decisions. Nonetheless, it is a way to formalize consideration of a set of criteria that relate directly to viability. As such, it provides a defensible decision structure in the absence of data. A RPVA should conclude with an expert opinion on what protection or management steps may be taken to best enhance the likelihood of viability and what data needs, if filled, would most likely enhance our ability to accurately assess viability.

9.4.5 Obtaining Expert Opinion

Many of the vertebrate point-scoring methods have utilized methods to try to minimize personal bias and errors by surveying the opinions of numerous authors in order to gauge many experts' opinions. One formal method for surveying experts to obtain relatively robust standard opinions is the Delphi method (Linstone and Turoff 1975; Mendoza and Prabhu 2000). In short, the Delphi method suggests that expert opinion be polled, and the results of this survey tabulated and then returned to the original experts. These experts can then either stick with their original answer or change in accordance with the results from the first round. The theory is that those that are sure of their opinions are less likely to change than those that have low certainty on particular questions. This process is iterated until the responses settle down and do not change in distribution. Empirical tests of the Delphi method demonstrate that it can be a very accurate form of prediction. To the extent possible, adopting Delphi methods for obtaining a RPVA is likely to improve accuracy and reduce bias inherent in individual expert opinion.

9.4.6 Estimating Habitat Area Requirements

Burgman et al. (2001) outline a detailed a method for setting habitat protection area targets for species of conservation concern. They then applied this technique to estimate land area requirements for three long-lived shrubs (*Banksia cuneata, Boronia keysii,* and *Parasonia dorrigoensis*). Their method uses twelve steps to assess the likely extinction probabilities for populations under different habitat conditions. The authors then suggested combining the presumed extinction likelihood for individual populations into an overall extinction estimate for a given habitat area (Table 9.3). The fundamentally difficult portion of their method is to provide an estimate of the population size that is likely to persist. Once this is completed, all other steps detail factors that would require increasing target population size in order to accommodate for factors that would cause increased extinction probabilities (e.g., disturbance or deterministic anthropogenic factors causing population loss).

Table 9.3. A brief synopsis of the 12-step program for estimating habitat area requirements for rare and endangered plant species of Burgman et al. (2001)

Step	Task
1	Estimate the population size likely to persist under demographic and environmental uncertainty assuming no human disturbance (e.g., 500)
2	Identify populations experiencing common disturbance regimes and treat them independently for all remaining steps
3	Identify and map the area of potential habitat within each disturbance regime class
4	Outline the area of potential habitat that has been surveyed
5	Estimate the population size within surveyed potential habitats
6	Estimate target area for protection based on background disturbance processes
7	Identify small-scale disturbances affecting the species from which a population may recover within the management time frame (e.g., 50 years)
8	Adjust the target area for deterministic trends that affect available habitat for the target species (i.e., land development)
9	Adjust the target area for anthropogenic factors that would permanently reduce the density within populations (e.g., grazing)
10	Identify catastrophes that may eliminate populations
11	Sum target areas across disturbance regimes and choose the area the results in a low quasi-extinction probability
12	Evaluate habitat maps and set strategies for patch selection and protection

9.5 Conclusions

There are numerous tools that remain under-exploited which can, in some measure, provide insight into the likely persistence of populations of long-lived plants. From using simple census data, to historical reconstructions, more data than strict demographic stage-transition data can be collected for many species. Despite the many options for assessing viability of long-lived plants, there are likely to be severe constraints on the confidence we place in most viability assessments. Few resources to initiate studies results in few species for which existing data are sufficient to provide even a rough viability assessment. When data are collected, small population sizes, short study intervals, complex life histories, and long lifespan all conspire to inflate expected error rates in these assessments. Thus, comparing relative viability among patches that differ in habitat management or habitat condition may result in more robust assertions.

Given the many tools available, it is prudent to employ as many as possible in any viability assessment of long-lived plants. I know of no studies that use community or historical data in order to augment assessment of viability using mathematical projections of population size as is typical of a standard PVA. No matter what tools are used to express viability estimates, these should be expressed in terms of generation time. A population that is destined for extinction within a few generations merits conservation attention whether or not these generations encompass hundreds of years.

Formal PVA will be restricted in its application by limits of resources to capture sufficient data. In recognition of the large job of assessing viability for many species (e.g., Edwards and Weakley 2001; Stout 2001), I propose using a decision structure to formalize expert opinion, such as in a RPVA, in order to provide preliminary assessments of the relative vulnerabilities of species or habitat area requirements of species. These methods are considerably less costly and rely on data that are available for species. These methods are not, however, simple and cost-free. They require expert opinion, and gaining multiple opinions is likely to lead to a better overall assessment.

Acknowledgements. I thank Christy Brigham, Dan Doak, Sharon Hermann, Kelly Lyons, Eric Menges, and Phil van Mantgem for conversations over the years that have helped me to formalize these thoughts. I thank Steve Beissinger and one anonymous reviewer who helped improve this manuscript.

References

Abrams MD, Orwig DA et al. (1997) Dendroecology and successional status of two contrasting old-growth oak forests in the Blue Ridge Mountains, U.S.A. Can J For Res 27(7):994–1002

Abrams MD, Copenheaver CA, Copenheaver CA, Terazawa K, et al. (1999) A 370-year dendroecological history of an old-growth *Abies-Acer-Quercus* forest in Hokkaido, northern Japan. Can J For Res 29:1891–1899

Alvarez-Buylla ER, Martinez-Ramos M (1992) Demography and allometry of *Cecropia obtusifolia*, a neotropical pioneer tree: An evaluation of the climax-pioneer paradigm for tropical rain forests. J Ecol 80:275–290

Alvarez-Buylla ER, Slatkin M (1991) Finding confidence limits on population growth rates. Trends Ecol Evol 6:221–224

Barbour MG, Burk JH, Pitts WD, Gilliam FS, Schwartz MW (1998) Terrestrial plant ecology. Cummings, Menlo Park, CA

Barnes BV, Zak DR, Denton SR, Spurr SH (1998) Forest ecology. Wiley, New York

Beissinger SR, Westphal MI (1998) On the use of demographic models of population viability in endangered species management. J Wildl Manage 62(3):821–841

Bierzychudek P (1982) The demography of jack-in-the-pulpit, a forest perennial that changes sex. Ecol Monogr 52:335–351

Bierzychudek P (1999) Looking backwards: assessing the projections of a transition matrix model. Ecol Appl 9:1278–1287

Botkin DB (1993) Forest dynamics: an ecological model. Oxford University Press, New York

Brook BW, Lim L, Harden R, Frankham R (1997) Does population viability analysis software predict the behaviour of real populations? A retrospective study on the Lord Howe Island Woodhen (*Tricholimnas sylvestris* (Sclater). Biol Conserv 82(2):119–128

Burgman MA, Possingham HP, Lynch AJJ et al. (2001) A method for setting the size of plant conservation target areas. Conserv Biol 15(3):603–616

Carter MF, Hunter WC, Pashley DN, Rosenberg KV 2000) Setting conservation priorities for landbirds in the United States: the Partners in Flight approach. The Auk 117:541–548

Caswell H (2001) Matrix population models : construction, analysis, and interpretation. Sinauer, Sunderland, MA

Chapman AP, Brook BW, Clutton-Brock TH, et al. (2001) Population viability analyses on a cycling population: a cautionary tale. Biol Conserv 97(1):61–69

Cook ER, Kairiukstis LA (1990) Methods of dendrochronology : applications in the environmental sciences. Kluwer, Dordrecht

Coulson T, Mace GM, Hudson E, Possingham H (2001) The use and abuse of population viability analysis. Trends Ecol Evol 16:219–221

Damman H, Cain ML (1998) Population growth and viability analyses of the clonal woodland herb, *Asarum canadense*. J Ecol 86:13–26

Dennis B, Munholland PL, Scott JM (1991) Estimation of growth and extinction parameters for endangered species. Ecol Monogr 64:205–224

Edwards AL, Weakley AS (2001) Population biology and management of rare plants in depression wetlands of the Southeastern Coastal Plain, USA. Nat Area J 21:12–35

Fair J, Lauenroth WK, Coffin DP (1999) Demography of *Bouteloua gracilis* in a mixed prairie: analysis of genets and individuals. J Ecol 87:233–243

Ferson S, Burgman MA (1995) Correlations, dependency bounds and extinction risks. Biol Conserv 73(2):101–105

Fieberg J, Ellner SP (2000) When is it meaningful to estimate an extinction probability? Ecology (Washington, DC) 81(7):2040–2047

Given DR, Norton DA (1993) A multivariate approach to assessing threat and for priority setting in threatened species conservation. Biol Conserv 64:57–66

Gordon DR, Rice KJ (2000) Competitive suppression of *Quercus douglasii* (Fagaceae) seedling emergence and growth. Am J Bot 87(7):986–994

Grant A, Benton TG (2000) Elasticity analysis for density-dependent populations in stochastic environments. Ecology (Washington DC) 81(3):680–693

Groom M, Pascual M (1997) Elasticity analyses and viability models: are we taking the right approach? Bull Ecol Soc Am 78(Suppl 4):99

Groom MJ, Pascual MA (1998) The analysis of population persistence: an outlook on the practice of viability analysis. In: Fiedler PL, Kareiva PM (eds) Conservation biology. Chapman and Hall, New York, pp 4–27

Hansen AJ, Garman SL, Marks B (1993) An approach for managing vertebrate diversity across multiple-use landscapes. Ecol Appl 3(3):481–496

Hanski I (1991) Single-species metapopulation dynamics: concepts models and observations. Biol J Linn Soc 42:17–38

Harcombe PA (1987) Tree life tables. Bioscience 37:557–568

Harper JL (1977) Population biology of plants. Academic Press, London

Haufler JB, Mehl CA, Roloff GJ (1999) Conserving biological diversity using a coarse-filter approach with a species assessment. In: Baydack RK, Campa H III, Haufler JB (eds) Practical approaches to the conservation of biological diversity. Island Press, Washington, DC, pp 107–126

Heppell S, Pfister C, de Kroon H (2000a) Elasticity analysis in population biology: methods and applications. Ecology 81:605–606

Heppell SS, Crouse DT, Crowder LB (2000b) Using matrix models to focus research and management efforts in conservation. In. Kaye TN, Pendergrass K, Finley K, Kauffman JB (eds) The effect of fire on the population viability of an endangered prairie plant. Ecol Appl 11:1366–1380

Horn HS (1975). Markovian properties of forest succession. In: Cody ML, Diamond JM (eds) Ecology and evolution of communities. Harvard University Press, Cambridge, pp 196–211

Kartesz JT (1994) A synonymized checklist of the vascular flora of the United States, Canada, and Greenland. Timber Press, Portland, OR

Kaye TN, Pendergrass KL, Finley K, Kauffman JB (2001) The effect of fire on the population viability of an endangered prairie plant. Ecol Appl 11:1366–1380

Kullman L (1996) Rise and demise of cold-climate *Picea abies* forest in Sweden. New Phytol 134(2):243–256

Lambrinos JG (2000) The impact of the invasive alien grass *Cortaderia jubata* (Lemoine) Stapf on an endangered mediterranean-type shrubland in California. Div Distrib 6(5):217–231

Lesica P (1995) Demography of *Astragalus scaphoides* and effects of herbivory on population growth. Great Basin Nat 55(2):142–150

Linstone HA, Turoff M (1975) The Delphi method, techniques and applications. Addison-Wesley, Reading, MA

Ludwig D (1999) Is it meaningful to estimate a probability of extinction? Ecology (Washington, DC) 80(1):298–310

Lunney A, Curtin A, Ayers D, (1996) An ecological approach to identifying the endangered fauna of New South Wales. Pac Conserv Biol 2:212–231

Mace GM, Lande R (1991) Assessing extinction threats: toward a reevaluation of IUCN threatened species categories. Conserv Biol 5(2):148–157

Mendoza GA, Prabhu R (2000) Development of a methodology for selecting criteria and indicators of sustainable forest management: a case study of participatory assessment. Environ Manage 26:659–673

Menges ES (1990) Population viability analysis for an endangered plant. Conserv Biol 4(1):52–62

Menges ES (1998) Evaluating extinction risks in plant populations. In: Fiedler PL, Kareiva PM (eds) Conservation biology, 2nd edn. Chapman and Hall, New York

Menges ES (2000) Population viability analyses in plants: challenges and opportunities. Trends Ecol Evol 15(2):51–56

Menges ES, Dolan RW (1998) Demographic viability of populations of Silene regia in midwestern prairies: relationships with fire management, genetic variation, geographic location, population size and isolation. J Ecol 86(1):63–78

Millsap BA, Gore JA, Runde DE, Cerulean SI (1990) Setting priorities for the conservation of fish and wildlife species in Florida. Wildl Monogr 111

Mladenoff DJ, Baker WJ (eds) (1999) Spatial modeling of forest landscape change : approaches and applications. Cambridge University Press, Cambridge

Mooney HA, Hobbs RJ (eds) (2000) Invasive species in a changing world. Island Press Washington, DC

Morris W, Doak D (2002) Quantitative conservation biology : theory and practice of population viability analysis. Sinauer, Sunderland, MA

Oostermeijer G (1995) The viability and management of small plant populations: a reaction. Levende Nat 96(6):223–227

Pelton J (1953) Studies on the life history of *Symphoricarpus occidentalis* Hook. in Minnesota. Ecol Monogr 23:17–84

Pickett STA, Thompson J (1978) Patch dynamics and the design of nature reserves. Biol Conserv 13:27–37

Pimm SL, Redfearn L (1988) The variability of animal populations. Nature (Lond) 334:613–614

Rentch JS, Adams HS, Coxe Rb, Stephenson SL 2000) En ecological study of a Carolina hemlock (*Tsuga caroliniana*) community in southwestern Virginia. Castanea 65:108

Rogers R, Johnson PS (1998) Approaches to modeling natural regeneration in oak-dominated forests. For Ecol Manage 106:45–54

Schwartz MW, Hermann SM, van Mantgem PJ (2000) Population persistence in Florida torreya: comparing modeled projections of a declining coniferous tree. Conserv Biol 14:1023–1033

Shefferson RP, Sandercock BK, Proper J, Beissinger SR (2001) Estimating dormancy and survival of a rare herbaceous perennial using mark-recapture models. Ecology (Washington, DC) 82(1):145–156

Shugart HH (1984) A theory of forest dynamics. Springer, Berlin Heidelberg New York

Silvertown J, Franco M,, Menges E et al. (1996) Interpretation of elasticity matrices as an aid to the management of plant populations for conservation. Conserv Biol 10(2):591–597

Sjogren-Gulve P, Ebenhard T (2000) The use of population viability analyses in conservation planning. Munksgaard, Copenhagen

Smith GF, Nicholas NS (2000) Size- and age-class distributions of Fraser fir following balsam woolly adelgid infestation. Can J For Res 30:948–957

Stout IJ (2001) Rare plants of the Florida Scrub, USA. Nat Area J 21:50–60

Szwagrzyk J (1994) Simulation models of forest dynamics based upon the concept of tree stand regeneration in gaps. Wiadomosci Ekologiczne 40(2):57–75

Taylor BL (1995) The reliability of using population viability analysis of risk classification of species. Conserv Biol 9(3):551–558

Thompson PM, Wilson B, Grellier K, Hammond PS (2000) Combining power analysis and population viability analysis to compare traditional and precautionary approaches to conservation of coastal cetaceans. Conserv Biol 14(5):1253–1263

Todd CR, Burgman MA (1998) Assessment of threat and conservation priorities under realistic levels of uncertainty and reliability. Conserv Biol 12(5):966–974

Tuljapurkar S, Caswell H (1997). Structured-population models in marine, terrestrial, and freshwater systems. Chapman and Hall, New York

Urban DL, Shugart HH (1992). Individual-based models of forest succession. In: Glenn-Lewin DC, Peet RK, Veblen TT (eds) Plant succession: theory and prediction. Chapman and Hall, New York, pp 249–292

USFWS (1994) Recovery plan for the Wahiawa plant cluster: *Cyanea undulata*, *Dubautia pauciflorula*, *Herperomannia lydgatei*, *Labordia lydgatei* and *Viola helenae*. US Fish and Wildlife Service, Portland, Oregon, Appendix B

USFWS (1995a) Lanai plant cluster recovery plan: *Abutilon eremitopetalum*, *Abutilon menziesii*, *Cyanea macrostegia* ssp. gibsonii, *Cyrtandra munroi*, *Gannia lanaiensis*, *Phyllostegia glabra* var. lanaiensis, *Santalum freycinetianum* var. lanaiense, *Tetramolopium remyi*, and *Viola lanaiensis*. US Fish and Wildlife Service, Portland, OR

USFWS (1995b) Recovery plan for the Kauai Plant Cluster. US Fish and Wildlife Service, Portland, OR

USFWS (1996a) Big Island plant cluster recovery plan. US Fish and Wildlife Service, Portland, OR

USFWS (1996b) Recovery plan for the Molokai plant cluster. US Fish and Wildlife Service, Portland, OR, 143 pp

USFWS (1997) Recovery plan for the Maui plant cluster. U.S. Fish and Wildlife Service, Portland, OR

USFWS (1998a) Big Island II: addendum to the recovery plan for the Big Island plant cluster. US Fish and Wildlife Service Portland, OR

USFWS (1998b) Kauai II: addendum to the recovery plan for the Kauai plant cluster. US Fish and Wildlife Service, Portland, OR

USFWS (1998 c) Recovery plan for Oahu plants. US Fish and Wildlife Service, Portland, OR

Weinstein DA, Gollands B, Retzlaff A (2001) The effects of ozone on a lower slope forest of the Great Smoky Mountain National Park: Simulations linking an individual tree model to a stand model. For Sci 47:29–42

Wilcove DS, Rothstein D, Dubow J, et al. (1998) Quantifying threats to imperiled species in the United States. Bioscience 48(8):607–615

Zolbrod AN, Peterson DL (1999) Response of high-elevation forests in the Olympic Mountains to climatic change. Can J For Res 29(12):1966–1978

10 Considering Interactions: Incorporating Biotic Interactions into Viability Assessment

M.A. Morales, D.W. Inouye, M.J. Leigh, and G. Lowe

10.1 Introduction

Most analyses of population viability focus on changes in numbers of the focal species independent of other members of their community or any other biotic interactions. It is difficult to incorporate all relevant factors into a viability analysis, but leaving out biotic interactions may be a critical flaw in some analyses. While single-species population viability analyses (PVAs) implicitly incorporate the effect of species interactions on population growth rate parameters (i.e., vital rates), models that explicitly consider changes in species dynamics as ecological conditions change may be needed. Unfortunately, these models will significantly increase data requirements. Because robust data sets are notoriously difficult to acquire even for single-species PVAs, it is important to evaluate the relative importance of species interactions before considering development of a PVA model that incorporates them explicitly. In this chapter, we discuss the various kinds of interactions that plants are involved in, evaluate when species interactions are likely to matter, consider strategies for deciding when to incorporate these interactions into PVA models, and discuss relevant modeling approaches.

10.2 What Kinds of Interactions Are Plants Involved in?

The major categories of species interactions that plants are involved in include typically negative interactions such as competition, parasitism, seed predation, and herbivory – and typically positive interactions such as pollination, seed dispersal, protection (e.g., ant guards), and nutrient exchange (e.g., nitrogen-fixing bacteria and mycorrhizal fungi; Boucher et al. 1982).

Ecological Studies, Vol. 165
C.A. Brigham, M.W. Schwartz (Eds.)
Population Viability in Plants

Although we can define major categories of interactions, the outcome of species interactions is often dynamic (Thompson 1988). Thus, while negative and positive effects may seem like clear-cut distinctions, interactions are not always easily categorized (Thompson 1988). For example, the interaction between *Lasius niger* ants and *Aphis fabae* aphids can range from mutualism to predation depending on the abundance of alternative sugar sources (Offenberg 2001). Similarly, relationships between plants and mycorrhizal fungi (Kretzer et al. 2000) can range from mutualism to parasitism due to differential costs and benefits (Johnson et al. 1997; Lapointe and Molard 1997).

Although usually a strictly negative interaction, herbivory can also have positive effects on plant growth (e.g., Mattson and Addy 1975). For example, Simberloff et al. (1978) reported that branching induced in aerial roots by root-boring isopods and insects might increase the stability of mangrove plants and their resistance to being pushed over. Such observations have created interest in discovering examples of overcompensation by plants in response to herbivory, which may increase their fitness (e.g., Inouye 1982; Paige and Whitham 1987; see also the review of ecology of tolerance to consumer damage by Stowe et al. 2000). Another plant-related example of an interaction with a range of outcomes from positive to negative is that between seed predation and successful seed dispersal by the same seed predator (e.g., Levey and Byrne 1993; Norconk 1998). Unfortunately, our understanding of how these effects balance out is limited, because most attention has been given to situations in which animals act primarily as only predator or disperser. An analogous situation occurs in brood-site pollination mutualisms, such as between yuccas and yucca moths, in which effects can span a continuum from positive to negative depending on the fraction of seeds consumed (Addicott 1986). Other seemingly negative events such as nectar robbing can also have a range of effects, because many "robbers" may also act as pollinators (Maloof and Inouye 2000). These studies emphasize that the distinction between positive and negative interactions is not always obvious or even static.

Where variation in the outcome of species interactions is context dependent, the term "conditionality" has been used (Cushman and Addicott 1991; Bronstein 1994). Conditionality in species interactions can qualitatively change the predictions of PVAs if the ecological context is changing. Recent research dealing with species interactions in plants emphasizes this spatiotemporal variability. For example, plant herbivory can induce plant defenses, which in turn influence the population dynamics of herbivores (Karban and Myers 1989; Karban and Kuc 1999; Underwood 1999). In cotton, herbivory can induce production of extrafloral nectar that in turn attracts a defensive ant guard (Wackers and Wunderlin 1999). In addition to the better-known induction of chemical defenses, morphological defenses like spines can also be induced (Young and Okello 1998), and although best known for herbaceous species, induced defenses have also been reported in trees (Wold and Marquis 1997). These plant-herbivore relationships and their effects on plant fitness can be quite intricate, involving multiple herbivore species (e.g., Agrawal 1999), temporal variation (Underwood 1998), and even transgenerational effects (Agrawal and Laforsch 1999). Such intricacies suggest that it may be complicated to incorporate the effects of herbivory - and species interactions in general - into PVA models.

On the other hand, there is no doubt that species interactions can potentially influence both population and community dynamics. For example, the effects of mycorrhizal mutualisms can influence tri-trophic interactions with herbivores (Borowicz 1997; Gange and Nice 1997; Gehring et al. 1997), and recent work has implicated mycorrhizal fungi as a potentially important agent of community structure by creating linkages among different plant species (Zelmer and Currah 1995). Thus, an important preliminary question becomes whether including species interactions in PVAs is likely to influence estimates of extinction risk or the development of management strategies.

10.3 When Are Species Interactions Likely to Matter?

Probably 90 % of the estimated 250,000 species of flowering plants (Heywood 1993) require the services of pollinators for sexual reproduction (Buchmann and Nabhan 1996), and as many as 300,000 species of animals visit flowers as pollinators (Nabhan and Buchmann 1997). Multiple species of pollinators may visit a single plant species, with widely varying degrees of pollinating effectiveness (e.g., Kearns and Inouye 1994). The details of pollination relationships are well known for only a small number of plants, and the threats facing many plant-pollinator relationships (Kearns et al. 1998; Chaps. 2, 3, this Vol.) suggest that these relationships are changing and that we may have a limited amount of time to learn about some of them. Although plant-pollinator interactions can involve many physiological, genetic, and behavioral factors for both pollinator and plant, these intricacies can sometimes be ignored in PVAs as long as the bottom line - successful production of seeds - is known (Chap. 3, this Vol.).

Ecologists distinguish between realized and intrinsic growth rates. Whereas the intrinsic growth rate represents the theoretical rate of increase for a population as it approaches zero density, the realized growth rate represents the observed rate of increase for a given population. Similarly, all realized parameters are the value of their corresponding intrinsic parameters after modification by environmental factors– including modification by species interactions. For example, estimates of seed production will incorporate the degree of pre-dispersal seed predation. Most PVAs implicitly include the effects of species interactions in realized parameter estimates of vital rates, and for some systems such inclusion is sufficient. For example, community effects (e.g., competition among plant species, Chap. 3, this Vol.) may be adequately accounted for by the ordinary population growth rate parameters used in a PVA. However, such will *not* be the case if the identities and/or distributions of neighboring plants – or of other interacting organisms – are themselves changing significantly. In this latter case, it may be necessary to explicitly incorporate these changes in the PVA to produce a model with predictive value.

In general, we suggest that incorporating species interactions into PVAs will be especially important if: the population dynamics of the focal plant species are strongly influenced by the species interaction; the outcome of the species interaction is strongly dependent on ecological conditions; and the ecological conditions are likely to change. We begin by addressing the following conditions: (1) strong community effects; (2) density-dependent species interactions, including the Allee effect; and (3) critical interactions that might break down.

10.3.1 Community Effects

If a focal plant species is strongly affected by its neighborhood composition – and neighborhood composition is changing – any PVA of the focal species will be unreliable unless neighborhood effects and changes in neighborhood are taken into account. One obvious example of this is in plant communities undergoing succession, where failure to model the effects of vegetative succession will result in unreliable PVAs for affected species (e.g., Oostermeijer 2000).

In many cases it may be necessary to know the focal species' competitive position vis-à-vis prevalent neighboring heterospecifics– and the distribution of those heterospecifics – to predict the focal species' likelihood of persistence. If large portions of the habitat suitable for and accessible to the focal species are occupied by dominant competitors, or populations are threatened by invasive species, failure to recognize this can result in unrealistic predictions. Interestingly, although the invasion potential of exotic plants has been modeled extensively (Reeves and Usher 1989; Goodwin et al. 1999; Parker 2000; Zalba et al. 2000), and a few studies have quantified the impact of exotic species on the population dynamics of threatened plants (e.g., Lesica and Shelly 1996; Carlsen et al. 2000; Chaps. 2, 3, this Vol.), we are unaware of any studies that have incorporated competition from exotics into PVAs.

The surrounding plant community may also affect plant species persistence in a number of less obvious ways, by affecting interactions between the focal species and its herbivores, pollinators, seed predators, and seed dispersers. For example, even though fragmentation may result in lower numbers of a focal frugivore-dependent plant species, the presence of neighboring fruiting plants may enable the rarer focal species to still receive regular visits (Whelan et al. 1998; see also Sect. 10.3.2). Similarly, structural or chemical properties of neighboring heterospecifics may make it easier or more difficult for interacting species to find the focal species, or may affect the survivorship or behavior of species interacting with the focal species. Such effects have been reported with neighboring heterospecifics impacting focal plant species by decreasing herbivory (Holmes and Jepson-Innes 1989; Hambäck et al. 2000; Chap. 3, this Vol.), increasing herbivory (Karban 1997; White and Whitham 2000), and increasing pollination (Laverty 1992). Seed predation can also be affected by the identity of neighboring plants and the interactions in which those plants are involved. For example, rodent seed predators have shown neighborhood-dependent seed preferences (Thompson 1985), as have granivorous birds (Willson and Harmeson 1973). Further, primates may serve as either seed predators or dispersers depending on the abundance of fruit (Gautier-Hion et al. 1993; Kaplan 1998).

In another example of the effects of plant community on species interactions, infestation by a shared pre-dispersal dipteran seed predator has been found to increase in one plant species with proximity to a heterospecific ant-defended plant. While prior research showed that seed predation on the montane sunflower *Helianthella quinquenervis* was decreased for focal plants by the activities of its ant partners (Inouye and Taylor 1979), recent research suggests that this displacement of flies causes them to oviposit in nearby non-ant-tended host plants. As a result, individuals of one alternate host plant species experienced a nine-fold increase in fly infestation rates when growing near ant-tended plants compared to those growing near plants from which ants had been excluded (Leigh, unpubl. data). These higher rates of fly infestation correlate with increased amounts of seed predation, and suggest that the identities and species interactions of heterospecific neighbors may have a significant effect on focal plants.

10.3.2 Density-Dependent Species Interactions

Incorporating species interactions such as competition and mutualism into PVA models increases in importance where the per-capita effect of the interaction depends on the density of either species. Density dependence in species interactions could affect estimates of population growth rate parameters in a variety of ways [e.g., pollination rates are a combined function of the density and effectiveness of pollinators as plant density changes (Jennersten and Nilsson 1993)], and ignoring this density dependence could significantly bias estimates of mean vital rates and their variance (Chap. 7, this Vol.).

In general, the influence of density-dependent interactions on estimates of mean vital rates will depend on the functional form of the interaction – that is, the per-capita effect of species A on focal plant species B as a function of B's density (Sih and Baltus 1987; Morales 2000; Holland and DeAngelis 2001). For example, where interactions are characterized by a monotonically saturating curve (type II, Fig. 10.1), the mean interaction effect will increase relative to a model that assumes density-independent interaction (Fig. 10.1 and legend). Where interactions are characterized by an S-shaped curve (type III, Fig. 10.1), the direction of change in mean interaction effect will depend on the density of the focal population relative to the range of the functional form (Fig. 10.1 and legend). Both types II and III functional forms are common in species interactions, including examples from pollination and seed dispersal

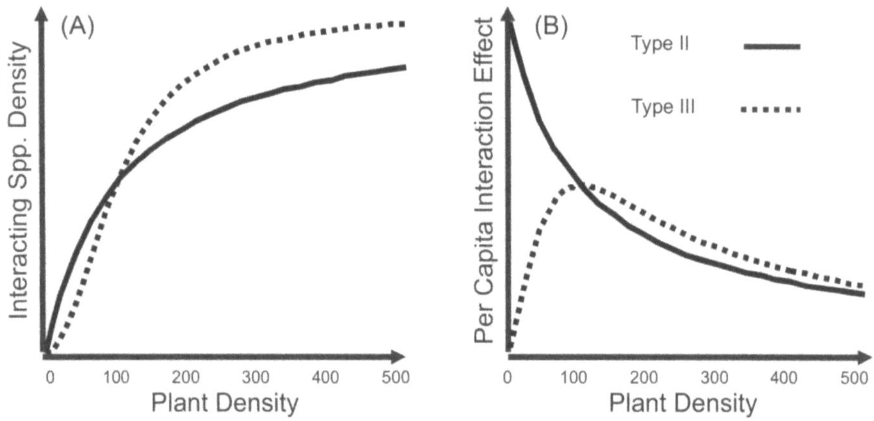

Fig. 10.1. Recruitment/functional response (**A**) and per-capita benefit to plants (**B**) as a function of plant density (*P*) for type II and III species "interactors" (e.g., seed predators, pollinators):

$$\frac{\alpha_i P}{P+\beta} \quad \text{Type II}$$

$$\frac{\alpha_i P^2}{P^2+\beta^2} \quad \text{Type III}$$

where α_i is the interaction coefficient and β is the density of plants at which the effect of interacting is one-half the maximum (i.e., the half-saturation constant). The maximum per-capita interaction effect is at low or intermediate plant densities for a type II or III response, respectively. Note that small differences in the total density of interacting species generate large differences in patterns of per-capita benefit (α=1, β=100 in the example above). Because a type II response is a decelerating function of plant density, the mean interaction effect is greater than the interaction effect at the mean plant density (assuming symmetric variation around the mean plant density). In other words, at any point along the type II curve in **B** (call this point the mean interaction effect), reducing plant density by some amount will cause a greater change in the interaction effect than augmenting plant density by the same amount. For a type III response, the change in mean interaction effect will depend on whether plant density is at the accelerating or decelerating portion of the response curve (i.e., in **A**)

systems (Howe and Kerckove 1979; Sih and Baltus 1987; Sargent 1990), although other functional forms are possible.

A special case of density dependence is the Allee effect, defined as a decrease in growth rate at low densities (i.e., positive density dependence). For plant species, Allee effects can result from species interactions in at least two ways. First, the response of pollinators (or other mutualists) to low plant densities can result in declines or even complete losses of seed production (Lamont et al. 1993; Widén 1993; Groom 1998; Cunningham 2000a, b; Hackney and McGraw 2001). In particular, where the effectiveness or density of mutualists (i.e., pollinators, seed dispersers, or ant defenders) is an S-shaped function of plant population density (Fig. 10.1), plant populations will show an Allee effect in the vital rates affected by those mutualists (Morales 2000). Whether this translates into an Allee effect in overall growth rate depends on the relative importance of the mutualism to the population dynamics of the focal plant population (Stephens et al. 1999; Chaps. 2, 3, this Vol.). Accordingly, a decrease in mutualist visitation rate or effectiveness in fragmented host populations is increasingly cited as a possible mechanism that may contribute to the decline of these populations (Rathcke and Jules 1993; Aizen and Feinsinger 1994a; Kunin and Gaston 1997; Groom 1998; Cunningham 2000b). For example, Sargent (1990) demonstrated that the amount of fruit around a fruiting plant affects fruit removal, such that fragmentation could depress seed dispersal rates. Unfortunately, while empirical support is increasing for the hypothesis that habitat fragmentation may disrupt host-visitor mutualisms (Aizen and Feinsinger 1994b; Groom 1998; Steffan-Dewenter and Tscharntke 1999; Cunningham 2000b), the impact of this disruption on the persistence of host plant populations is largely unknown.

Allee effects will also be seen in populations of plant species that support and require the services of obligate mutualists. A consistent prediction arising from models of obligate-obligate mutualism is the existence of a threshold density (defined as the density of species A required to offset the extinction rate of species B), below which species will show a deterministic decline to extinction (Wolin 1985). For obligate plant species that interact with facultative mutualists and have a monotonically saturating benefit function (e.g., type II), there may or may not be a threshold density, depending on the values of the parameters. On the other hand, if the benefit function is S-shaped (e.g., type III), there will always be a threshold density for biologically reasonable parameter values (Morales 1999). This results in part because a threshold density of mutualists is required to offset the extinction rate of plants; however, where visitors show an S-shaped recruitment response, a threshold density of plants is also required to attract this threshold density of mutualists (Fig. 10.2). A potential example of this was found in Groom's study of the annual herb *Clarkia concinna* (Groom 1998), in which she documented complete reproductive failure of small and isolated populations of the herb due to their inability to attract pollinators.

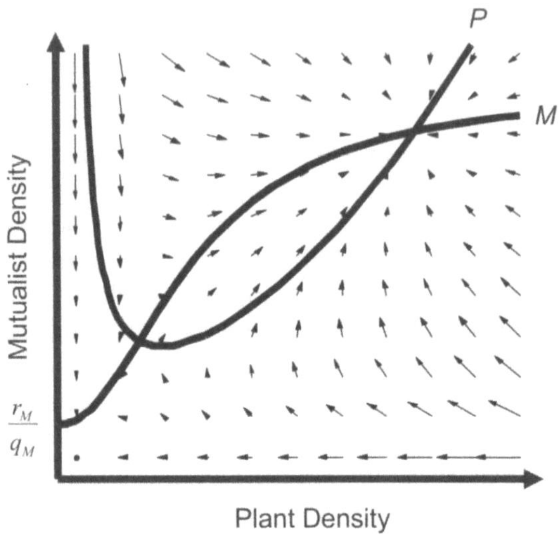

Fig. 10.2. Phase-plane graph of obligate-facultative mutualism with a type III benefit response, where plants (*P*) are obligate and mutualists (*M*) are facultative. Model equations are:

$$\frac{dP}{dt} = r_P P - q_P P^2 + \frac{\alpha_P P^2 M}{P^2 + \beta^2} \quad \text{Plant}$$

$$\frac{dM}{dt} = r_M M - q_M M^2 + \frac{\alpha_M P^2 M}{P^2 + \beta^2} \quad \text{Visitor}$$

where r is the intrinsic growth rate in the absence of the mutualist, and q is the slope of the growth rate as density increases (α and β are defined in the legend for Fig. 10.1). Because the benefit function has an inflection point at the half-saturation constant β, the host isocline curves upwards at this host density. Thus, there is always a low-density unstable equilibrium if a biologically reasonable solution to these equations exists (*arrows* indicate the population growth vectors for a given density combination)

10.3.3 Critical Interactions and Feedback Dynamics

Even where the per-capita effect of species interactions is density-independent, the *total* interaction effect could change if the density of either species changes. For generalist species, changes in interaction effect will be buffered to the extent that alternate species compensate for changes in the densities of each other. For example, reproductive success of *Dinizia excelsa* (Mimosaceae) in Amazonian forest fragments is maintained by introduced Africanized honeybees in the absence of its native pollinators (Dick 2001). On the other hand, plants that are involved in highly-specific, obligate interactions will be sensitive to changes in the density of their partner (Aizen and Feinsinger 1994a; Kearns and Inouye 1997; Chap. 3, this Vol.). For example, fig trees depend exclusively on highly species-specific fig wasps for pollination, while fig wasps require figs for brood development (Janzen 1979). Because fig wasps are short-lived, they require continuous production of figs to maintain their populations. Recent droughts in Borneo associated with El Niño resulted in the almost complete absence of flowering fig trees from January to March 1998, leading to the local extinction of fig wasps associated with eight species of dioecious fig trees. As of April 1999 - over 1 year after the drought - fig wasps still had not recolonized four species of fig trees. Given the high level of endemism in Borneo and the species-specific nature of fig pollination systems, these fig wasp extinctions may ultimately result in the extinction of their associated fig trees (Harrison 2000).

For seed dispersal, the importance of feedback dynamics on plant population viability seems to exist primarily at the guild level (Fleming 1991). Except for a few examples, like phainopepla (*Phainopepla nitens*) and mistletoe (*Phoradendron californicum*) (e.g., Walsburg 1975; Larson 1996), most animal-dispersed plants rely on a variety of species for dispersal, although the relative efficacy of the various partners is rarely known (Livingston 1972; Howe and Primack 1975; Herrera and Joradano 1981; Murray 1998). Given this pattern of generalization, it may not be necessary to model separately each of the species involved in dispersal of a particular plant (an exception to this may be differences in seed dispersal agents that swallow seeds and then defecate them; see Traveset et al. 2001). On the other hand, it *will* be important to consider how disperser services change at the guild level. For example, habitat fragmentation may alter overall rates of fruit removal (see Sect. 10.3.2). Similarly, invasive species may change dispersal effectiveness by altering community composition (see Sect. 10.3.1). In the extremely diverse South African fynbos, up to 30 % of plants rely on ants for seed dispersal - a dispersal system known as myrmecochory. Ant dispersal is essential to these plants because they protect the seeds from rodent seed predators and fire. The recent invasion by Argentine ants (*Linepithema humile*) in these areas has displaced native ant species that preferentially disperse large-seeded plants, which may be causing a shift in plant composition away from these large-seeded plants following fire disturbances (Christian 2001).

10.4 Strategies for Evaluating the Importance of Species Interactions

In the sections above, we outlined characteristics of species interactions that suggest those interactions will be important when modeling PVAs. Below, we present approaches to evaluating the importance of species interactions for specific PVAs.

Although it might seem as though seed production, for example, would be a crucial aspect of plant population biology to model, there have been few experimental studies comparing seed input with seedling establishment or other stages of the life cycle to determine effects on plant population size and structure (however, see Louda and Potvin 1995; Ackerman et al. 1996; Maron and Simms 1997). Seed predation can certainly be substantial, as evidenced by reported predation levels of 60 % (Inouye and Taylor 1979), 80 % (Snow and Snow 1986), and as high as 100 % of total seed production (Crawley 1992, and references therein). However, simply measuring the percentage of seeds consumed by seed predators may not be relevant to predictions of extinction risk, because recruitment may not be affected by seed predation (e.g., Sousa and Mitchell 1999; Alcantara et al. 2000). Rather, populations of some species may be limited by suitable microsites for seedling establishment (Turnbull et al. 2000), or may be able to offset seed predation, at least temporarily, by relying on seed banks for recruitment (Crawley 1992).

One approach to evaluating the importance of seed predation is through a series of experiments designed to assess the effect of seed predators relative to autecological factors (Schemske et al. 1994). Consideration of spatiotemporal effects will also be important, because different seed predators show different patterns of abundance and may handle food items in different ways (Holthuijzen et al. 1993; Diaz et al. 1999). For example, migratory granivorous birds may only have brief access to a plant population's seeds in the fall (Diaz et al. 1999); rodents, on the other hand, may exert more constant pressure, with a high amount of annual variation due to population cycles. Furthermore, different predators may search for food items in different microhabitats, and some seed predators may act as seed dispersers if they cache seeds and fail to recover them (Crawley 1992; Diaz et al. 1999).

The dispersal shadow (where the majority of seeds land) is also an important element to consider for at least two reasons. First, some studies have shown that distance from the parent plant or a conspecific adult may be an important variable affecting seed predation pressure (Janzen et al. 1976; Augspurger 1984; Holthuijzen et al. 1993; Terborgh et al. 1993). Second, microhabitats may differ significantly in their rates of seed mortality (Schupp et al. 1989; Terborgh et al. 1993; Diaz et al. 1999; however. see Whelan and Willson 1991). Optimally, one would measure how far seeds travel from the parent plant and how many seeds fall into different microhabitats (e.g., gap, forest interior, edge) when calculating dispersal success.

Finally, multiyear investigations are preferable, because the effects of seed predators vary annually (Schupp 1990; Whelan and Willson 1991). The need for multi-year investigations is even greater with herbaceous perennials, which may produce highly variable numbers of seeds each year (Crawley 1992). Similarly, multiyear investigations are essential for masting plants, since masting may affect the degree of seed predation during years of extensive fruit production (Curran and Leighton 2000; Kelly et al. 2000; see also Sect. 10.3.2).

10.4.1 Evaluating the Importance of Species Interactions: Matrix Modeling Approaches

A complementary approach for evaluating the importance of species interactions is based on matrix models. This is illustrated by the example of seed dispersal of wild ginger (*Asarum canadense*). Like many spring-flowering herbs in eastern North America, wild ginger is a myrmecochore, i.e., its seeds are dispersed by ants (Beattie 1985). The interaction between ants and wild ginger satisfies many of the properties outlined in the opening of this chapter. First, seed dispersal of wild ginger by ants increases as its relative and absolute density decreases (Smith et al. 1989; i.e., significant density dependence). Second, as has been shown in an ecologically similar system of myrmecochory (Morales and Heithaus 1998), the food reward to ants from wild ginger (elaiosomes) may increase the output of queen ants (i.e., positive feedback on ant populations). Finally, habitat fragmentation has been shown to decrease significantly the density of those ant species that are the most effective seed dispersers (Pudlo et al. 1980: i.e., significant community effects).

Even though the properties outlined above are satisfied, it may not be necessary to explicitly model this species interaction in a PVA. Development of matrix models followed by elasticity or sensitivity analysis can provide insight into the importance of species interactions to the persistence of a given population by examining the relative contributions of stage-specific transitions to overall growth rate (Caswell 2000; Chap. 6, this Vol.). For example, ants significantly benefit wild ginger primarily by reducing post-dispersal seed predation by mice (Heithaus 1981). Using elasticity analysis, Damman and Cain (1998) showed that seed germination success has relatively little impact on the population dynamics of this species – a result that is supported by simulation models of the same system (Heithaus 1986). These results suggest that even if ants were included in a dynamic model with wild ginger, predictions of population persistence would remain relatively unchanged.

Within a matrix modeling framework, life table response experiments (LTREs) provide the best approach to assess the relative importance of species interactions (Caswell 2000). For example, one could manipulate seed dispersal by ants to evaluate its effect on stage-specific vital rates. LTREs are important because the stage-specific effect of species interactions must be well established for elasticity analysis to identify accurately the relative importance of species interactions.

LTREs that have examined the stage-specific contribution of ants for other species of myrmecochorous plants have found that ants can increase seedling survivorship by dispersing seeds to suitable microsites (Hanzawa et al. 1988). In the case of wild ginger, there is no obvious indication of microsite enhancement (Heithaus 1986), although this has not been tested experimentally. However, because elasticity analysis identified seedling limitation as an important predictor for the long-term persistence of wild ginger (Damman and Cain 1998), future studies should examine the role of ants in seedling performance for this species. This example illustrates the utility of matrix modeling approaches combined with sensitivity analysis to identify the potential importance of species interactions, and in focusing future research on stage-specific effects of species interactions.

The use of matrix modeling approaches to investigate the effect of species interactions on plant population viability can also include human impacts such as harvesting or trampling (Chap. 6, this Vol.). For example, Nantel et al. (1996) used stochastic matrix projection models (see below) to evaluate the impact of various harvesting regimes on the population persistence of American ginseng (*Panax quinquefolius*) and wild leek (*Allium tricoccum*). This approach could be extended to species interactions more broadly.

10.5 Modeling Species Interactions in PVAs

Much of the information in this chapter emphasizes the variable or conditional nature of species interactions. Nevertheless, this variation presents a larger problem to understanding the effects of species interactions on plants than it does to modeling those effects. One approach to handling conditionality is to have multiple transition matrices, each reflecting a different set of conditions, or by using Markov chain approaches (Caswell 2000; Chaps. 6, Chap. 11, this Vol.). Stochasticity can be included by using bootstrap approaches or by sampling parameter estimates from a probability distribution (Caswell 2000; Chaps. 6, 11, this Vol.). For dynamic models (see below), stochastic and conditional variation can be explicitly modeled by sampling parameters from an underlying process or probability distribution (Hilborn and Mangel 1997).

Although matrix modeling approaches provide valuable insight into the influence of species interactions for the persistence of plant populations in many cases, these approaches do not come without limitations. In particular, matrix modeling approaches analyze the *current* properties of a given population (Caswell 2000). Where state variables such as population size are likely to change significantly, these analyses may no longer apply. In addition to quantitative changes, qualitative differences may also arise, especially if interactions are strongly density-dependent or where obligate interactions are involved (see above). For example, wild ginger reproduces both sexually and vegetatively, so loss of its seed disperser is not necessarily catastrophic, at least in the short term. In contrast, fig trees depend exclusively on species-specific fig wasps for their pollination. Models of obligate mutualism predict a threshold density below which populations go extinct, and for these systems it will be necessary at a minimum to evaluate the risk of partner loss. In general, matrix modeling approaches will not work well where the effect of species interactions is likely to strongly influence the dynamics of both species.

One approach to modeling the persistence of strongly interacting populations is to consider the interaction while ignoring internal dynamics. An example of this can be found in two-species metapopulation models, used to examine minimum viable patch densities (Nee et al. 1996). Although these models are probably not useful for PVA per se, these analyses have generated useful predictions. For example, metapopulation models of obligate mutualists predict a threshold density of patches below which the metapopulation will go extinct (Nee et al. 1996). Interestingly, this prediction is supported by phenology-based models developed for the interaction between figs and fig wasps (Bronstein et al. 1990; Anstett et al. 1995, 1997).

In general, monoecious fig trees show within-tree synchrony and between-tree asynchrony in flowering. This phenological pattern maximizes outcrossing and increases the likelihood that brood-site resources are available to fig wasps throughout the year (Fig. 10.3, and discussion above – Sect. 10.3.3). However, as the number of fig trees decreases, the probability of a temporal

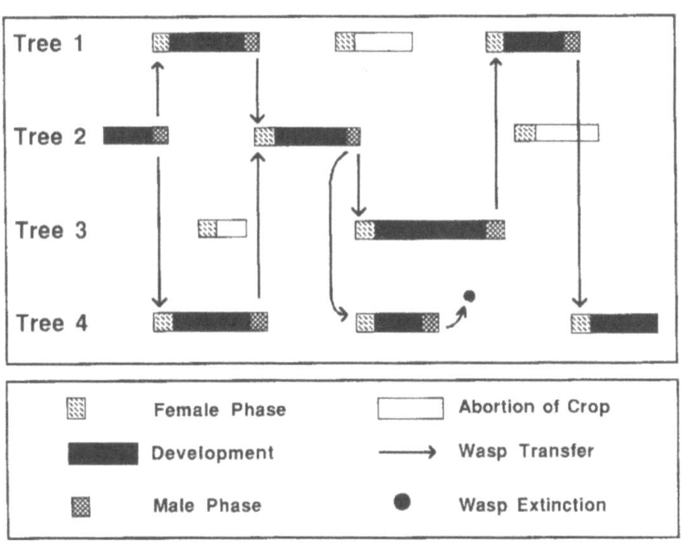

Fig. 10.3. Phenology of flowering and pollination for fig trees and fig wasps. Female wasps oviposit in figs during female-flowering phase. Wasp broods emerge, mate within figs, and female wasps exit after collecting pollen during the fig's male-flowering phase. For fig trees that are in female-flowering phase during periods when no adult wasps are available, fruit crops are aborted (e.g., tree 3). If female wasps do not find trees in female-flowering phase, that brood goes extinct (e.g., tree 1). In this example, the population of wasps persists through the sequence 4→2→5→2→3. A temporal break in this sequence would result in total wasp extinction, followed by local extinction of the fig tree population in the absence of re-colonization by wasps. (Bronstein et al. 1990)

gap in flowering increases – which can cause a local extinction of fig wasps if the temporal gap is longer than the short life span of ovipositing fig wasps (Bronstein et al. 1990).

Several models have been developed that predict the critical or minimum viable population size of fig trees required to sustain a local population of fig wasps (Bronstein et al. 1990; Anstett et al. 1995, 1997). These analyses are based on the density of trees necessary to produce continuously overlapping periods of wasp emergence and receptive figs. Parameterized solely from observed phenology, these models generate estimates for the minimum population density of fig trees required for the long-term persistence of both fig tree and fig-wasp populations. Even though these models do not incorporate any information on the population dynamics of figs or fig wasps (e.g., the models assume 100% colonization success), such simplifications may produce fairly accurate predictions. In particular, leaving out the population dynamics of fig wasps from these models is probably appropriate, since the population dynamics of fig wasps operate on a time scale that is orders of magnitude faster than the population dynamics of fig trees. Consistent with this assumption, recent work has documented extreme resilience following disturbance for the interaction between figs and fig wasps in southern Florida (Bronstein and Hossaert-McKey 1995).

On the other hand, if species interactions are likely to affect the population dynamics of both populations, more complicated models will be needed. While we are unaware of any PVA models that incorporate dynamic species interactions for plant focal populations, analogous approaches are available for animal systems. For example, Hochberg et al. (1992, 1994) have examined the persistence of big blue butterflies (*Maculinea arion*) as a function of their interaction with host plants and ants (on which the larvae sequentially feed). Unfortunately, there are no available computer programs that allow researchers to model these systems; rather, individualized models must be developed on a case-by-case basis. Given the complexity required for dynamic interaction models, theoretical studies that consider the importance of species interactions in PVA are needed. For example, empirical and theoretical work suggests that mutualisms characterized by a type III benefit function merit special consideration, because these systems are likely to show Allee effects and associated increased risks of extinction (see discussion of Allee effects, Sect. 10.3.2).

10.5.1 Genetic Consequences of Species Interactions

So far, we have only considered the effect of species interactions on plant population dynamics. However, pollination and seed dispersal interactions also may have important consequences for the *genetic* structure of plant populations (Young 1996; however, see Holsinger and Gottlieb 1991; Chaps. 2, 3, this

Vol.). Currently available PVA packages can incorporate inbreeding as a function of population size for animal systems, and analogous approaches are possible for plant systems – but unlike animals, there is no simple relationship between inbreeding and population size in plants. In pollination systems, the degree of self-pollination (including geitonogamy) will vary as a function of pollinator behavior, pollen carryover (the pattern of self vs. outcross pollen deposition), and breeding system (Snow et al. 1996). Furthermore, pollinator communities may vary temporally, such that there may be no consistent relationship between plant density and inbreeding even for a single plant population. Finally, the consequences of inbreeding are likely to be extremely variable among plant species due to inherent differences in breeding systems (Chap. 3, this Vol.). For example, as plant population density declines, self-incompatible species may experience reduced seed-set, while most obligately selfing species will remain genetically unaffected.

Currently, few PVA models have explicitly considered the genetic consequences of interactions with pollinators or seed dispersers on the persistence of plant populations. One exception is a PVA of *Gentiana pneumonanthe* (Gentianaceae) (Oostermeijer 2000). Oostermeijer used a stochastic matrix modeling approach to examine the effects of ecological and genetic factors on the viability of restored *G. pneumonanthe* populations in "late" successional stages. Small populations of *G. pneumonanthe* experience increased selfing (and thus inbreeding depression) due to a decrease in insect pollination, and adding inbreeding depression to the PVA significantly decreased the predicted time to extinction in populations as large as 250 individuals. More models of this kind are needed.

10.6 Conclusions

Currently, species interactions are rarely included in models used to evaluate the risk of extinction for plant populations. The information summarized in this chapter underscores the complex ways in which species interactions may influence plant populations. Unfortunately, knowing when to explicitly include these dynamics may not always be obvious. In many cases, while we may suspect that such interactions exist and are important, they have not been studied enough to permit quantitative estimates to be incorporated into models. The kinds of models currently in use may not be able to incorporate our growing knowledge of such interactions, and more sophisticated models may be needed in the future.

References

Ackerman JD, Sabat A, Zimmerman JK (1996) Seedling establishment in an epiphytic orchid: an experimental study of seed limitation. Oecologia 106:192–198
Addicott JF (1986) Variation in the costs and benefits of mutualism: the interaction between yucca and yucca moths. Oecologia 70:486–494
Agrawal AA (1999) Induced responses to herbivory in wild radish: effects on several herbivores and plant fitness. Ecology 80:1713–1723
Agrawal AA, Laforsch C, Tollrian R (1999) Transgenerational induction of defences in animals and plants. Nature 401:60–63
Aizen MA, Feinsinger P (1994a) Forest fragmentation, pollination, and plant reproduction in a Chaco dry forest, Argentina. Ecology 75:330–351
Aizen MA, Feinsinger P (1994b) Habitat fragmentation, native insect pollinators, and feral honey bees in Argentine "Chaco Serrano". Ecol Appl 4:378–392
Alcantara JM, Rey PJ, Sanchez-Lafuente AM, Valera F (2000) Early effects of rodent post-dispersal seed predation on the outcome of the plant-seed disperser interaction. Oikos 88:362–370
Anstett MC, Michaloud G, Kjellberg F (1995) Critical population size for fig/wasp mutualism in a seasonal environment: effect and evolution of the duration of female receptivity. Oecologia 103:453–461
Anstett MC, Hossaert-McKey M, McKey D (1997) Modeling the persistence of small populations of strongly interdependent species: figs and fig wasps. Conserv Biol 11:204–213
Augspurger CK (1984) Seedling survival of tropical tree species: interactions of dispersal distance, light-gaps, and pathogens. Ecology 65:1705–1712
Beattie AJ (1985) The evolutionary ecology of ant-plant mutualisms. Cambridge University Press, New York
Borowicz VA (1997) A fungal root symbiont modifies plant resistance to an insect herbivore. Oecologia 112:534–542
Boucher DH, James S, Keeler KH (1982) The ecology of mutualisms. Annu Rev Ecol Syst 13:315–347
Bronstein JL (1994) Conditional outcomes in mutualistic interactions. Trends Ecol Evol 9:214–217
Bronstein JL, Gouyon P-H, Gliddon C, Kjellberg F, Michaloud G (1990) The ecological consequences of flowering asynchrony in monoecious figs: a simulation study. Ecology 71:2145–2156
Bronstein JL, Hossaert-McKey M (1995) Hurricane Andrew and a Florida fig pollination mutualism: resilience of an obligate interaction. Biotropica 27:373–381
Buchmann SL, Nabhan GP (1996) The forgotten pollinators. Island Press, Washington, DC
Carlsen TM, Menke JW, Pavlik PM (2000) Reducing competitive suppression of a rare annual forb by restoring native California grasslands. Restor Ecol 8:18–29
Caswell H (2000) Matrix population models, 2nd edn. Sinauer, Sunderland, MA
Christian CE (2001) Consequences of a biological invasion reveal the importance of mutualism for plant communities. Nature 413:635–638
Crawley MJ (1992) Seed predators and plant population dynamics. In: Fenner M (ed) Seeds: the ecology of regeneration in plant communities. CAB International, Wallingford, pp 157–192.
Cunningham SA (2000a) Depressed pollination in habitat fragments causes low fruit set. Proc R Soc Lond Ser B 267:1149–1152

Cunningham SA (2000b) Effects of habitat fragmentation on the reproductive ecology of four plant species in mallee woodland. Conserv Biol 14:758–768
Curran LM, Leighton M (2000) Vertebrate responses to spatiotemporal variation in seed production of mast-fruiting Dipterocarpaceae. Ecol Monogr 70:101–128
Cushman JH, Addicott JF (1991) Conditional interactions in ant-plant-herbivore mutualisms. In: Huxley CR, Cutler DF (eds) Ant-plant interactions. Oxford University Press, Oxford, pp 92–103
Damman H, Cain ML (1998) Population growth and viability analyses of the clonal woodland herb, *Asarum canadense*. J Ecol 86:13–26
Díaz I, Papic C, Armesto JJ (1999) An assessment of post-dispersal seed predation in temperate rain forest fragments in Chiloé Island, Chile. Oikos 87:228–238
Dick CW (2001) Genetic rescue of remnant tropical trees by an alien pollinator. Proc R Soc Lond B 1483:2391–2396
Fleming TH (1991) Fruiting plant-frugivore mutualism: the evolutionary theater and the ecological play. In: Price PW, Lewinsohn TM, Fernandez GW, Benson WW (eds) Plant-animal interactions. Evolutionary ecology in tropical and temperate regions. Wiley, New York, pp 119–144
Gange AC, Nice HE (1997) Performance of the thistle gall fly, *Urophora cardui*, in relation to host plant nitrogen and mycorrhizal colonization. New Phytol 137:335–343
Gautier-Hion A, Gautier J-P, Maisels F (1993) Seed dispersal versus seed predation: an inter-site comparison of two related African monkeys. In: Fleming TH, Estrada A (eds) Frugivory and seed dispersal: ecological and evolutionary aspects. Kluwer, Boston, pp 237–244
Gehring CA, Cobb NS, Whitham TG (1997) Three-way interactions among ectomycorrhizal mutualists, scale insects, and resistant and susceptible pinyon pines. Am Nat 149:824–841
Goodwin BJ, McAllister AJ, Fahrig L (1999) Predicting invasiveness of plant species based on biological information. Conserv Biol 13:422–426
Groom MJ (1998) Allee effects limit population viability of an annual plant. Am Nat 151:487–496
Hackney EE, McGraw JB (2001) Experimental demonstration of an Allee effect in American ginseng. Conserv Biol 1:129–136
Hambäck PA, Ågren J, Ericson L (2000) Associational resistance: insect damage to purple loosestrife reduced in thickets of sweet gale. Ecology 81:1784–1794
Hanzawa FM, Beattie AJ, Culver DC (1988) Directed dispersal: demographic analysis of an ant-seed mutualism. Am Nat 131:1–13
Harrison RD (2000) Repercussions of El Niño: drought causes extinction and the breakdown of mutualism in Borneo. Proc R Soc Lond Ser B 267:911–915
Heithaus ER (1981) Seed predation by rodents on three ant-dispersed plants. Ecology 62:136–145
Heithaus ER (1986) Seed dispersal mutualism and the population density of *Asarum canadense*, an ant-dispersed plant. In: Estrada A, Fleming EA (eds) Frugivores and seed dispersal. Junk, Dordrecht, pp 199–210
Herrera CM, Jordano P (1981) *Prunus mahaleb* and birds: the high efficiency seed dispersal system of a temperate fruiting tree. Ecol Monogr 51:203–218
Heywood VH (1993) Flowering plants of the world. Oxford University Press, New York
Hilborn R, Mangel M (1997) The ecological detective: confronting models with data. Princeton University Press, Princeton, NJ
Hochberg ME, Thomas JA, Elmes GW (1992) A modelling study of the population dynamics of a large blue butterfly, *MaculIn:ea rebeli*, a parasite of red ant nests. J Anim Ecol 61:397–410

Hochberg ME, Clarke RT, Elmes GW (1994) Population dynamic consequences of direct and indirect interactions involving a large blue butterfly and its plant and red ant hosts. J Anim Ecol 63:375–391

Holland JN, DeAngelis DL (2001) Population dynamics and the ecological stability of obligate pollination mutualisms. Oecologia 126:575–586

Holmes RD, Jepson-Innes K (1989) A neighborhood analysis of herbivory in *Bouteloua gracilis*. Ecology 70:971–976

Holsinger KE, Gottlieb LD (1991) Conservation of rare and endangered plants: principles and prospects. In: Falk DA, Holsinger KE (eds) Genetics and conservation of rare plants. Oxford University Press, New York, pp 195–208

Holthuijzen AMA, Sharik TL, Fraser JD (1993) Dispersal of eastern red cedar (*Juniperus virginiana*) into pastures: an overview. Can J Bot 65:1092–1095

Howe HF, Kerckove GAV (1979) Fecundity and seed dispersal of a tropical tree. Ecology 60:180–189

Howe HF, Primack RB (1975) Differential seed dispersal by birds of the tree *Casearia nitida* (Flacourtiaceae). Biotropica 7:278–283

Inouye DW (1982) The consequences of herbivory: a mixed blessing for *Jurinea mollis* (Asteraceae). Oikos 39:269–272

Inouye DW, Taylor OR (1979) A temperate region plant-ant-seed predator system: consequences of extrafloral nectar secretion by *Helianthella quinquenervis*. Ecology 60:1–7

Janzen DH (1979) How to be a fig. Annu Rev Ecol Syst 10:13–51

Janzen DH, Miller GA, Hackforth-Jones J, Pond CM, Hooper K, Janos DP (1976) Two Costa Rican bat-generated seed shadows of *Andira inermis* (Leguminosae). Ecology 57:1068–1075

Jennersten O, Nilsson SG (1993) Insect flower visitation frequency and seed production in relation to patch size of *Viscaria vulgaris* (Caryophyllaceae). Oikos 68:283–292

Johnson NC, Graham JH, Smith FA (1997) Functioning of mycorrhizal associations along the mutualism-parasitism continuum. New Phytol 135:575–586

Kaplan BA (1998) Variation in seed handling by two species of forest monkeys in Rwanda. Am J Primatol 45:83–101

Karban R (1997) Neighbourhood affects a plant's risk of herbivory and subsequent success. Ecol Entomol 22:433–439

Karban R, Kuc J (1999) Induced resistance against pathogens and herbivores: an overview. In: Agrawal AA, Tuzun S, Bent E (eds) Induced plant defenses against pathogens and herbivores. The American Phytopathological Society Press, St Paul, MN, pp 1–16

Karban R, Myers JH (1989) Induced plant responses to herbivory. Annu Rev Ecol Syst 20: 331–348

Kearns CA, Inouye DW (1994) Fly pollination of *Linum lewisii* (Linaceae). Am J Bot 81:1091–1095

Kearns CA, Inouye DW (1997) Pollinators, flowering plants, and conservation biology. BioScience 47:297–307

Kearns CA, Inouye DW, Waser NW (1998) Endangered mutualisms: the conservation of plant-pollinator interactions. Annu Rev Ecol Syst 29:83–112

Kelly D, Harrison AL, Lee WG, Payton IJ, Wilson PR, Schauber EM (2000) Predator satiation and extreme mast seeding in 11 species of *Chionochloa* (Poaceae). Oikos 90:477–488

Kretzer AM, Bidartondo MI, Grubisha LC, Spatafora JW, Szaro TM, Bruns TD (2000) Regional specialization of *Sarcodes sanguinea* (Ericaceae) on a single fungal symbiont from the *Rhizopogon ellenae* (Rhizopogonaceae) species complex. Am J Bot 87:1778–1782

Kunin WE, Gaston KJ (eds) (1997) The biology of rarity: causes and consequences of rare-common differences. Chapman and Hall, London

Lamont BB, Klinkhamer PGL, Witkowski ETF (1993) Population fragmentation may reduce fertility to zero in *Banksia goodii* – a demonstration of the Allee effect. Oecologia 94:446–450

Lapointe L, Molard J (1997) Costs and benefits of mycorrhizal infection in a spring ephemeral, *Erythronium americanum*. New Phytol 135:491–500

Larson DL (1996) Seed dispersal by specialist versus generalist foragers: the plant's perspective. Oikos 76:113–120

Laverty TM (1992) Plant interactions for pollinator visits: a test of the magnet species effect. Oecologia 89:502–508

Lesica P, Shelly JS (1996) Competitive effects of *Centaurea maculosa* on the population dynamics of *Arabis fecunda*. Bull Torrey Bot Club 123:111–121

Levey DJ, Byrne MM (1993) Complex ant-plant interactions: rain forest ants as secondary dispersers and post-dispersal seed predators. Ecology 74:1802–1812

Livingston RB (1972) Influence of birds, stones and soil on the establishment of pasture juniper, *Juniperus communis*, and red cedar, *J. virginiana* in New England pastures. Ecology 53:1141–1147

Louda SM, Potvin MA (1995) Effect of inflorescence-feeding insects on the demography and lifetime fitness of a native plant. Ecology 76:229–245

Maloof JE, Inouye DW (2000) Are nectar robbers cheaters or mutualists? Ecology 81:2651–2661

Maron JL, Simms EL (1997) Effect of seed predation on seed bank size and seedling recruitment of bush lupine (*Lupinus arboreus*). Oecologia 111:76–83

Mattson WJ, Addy ND (1975) Phytophagous insects as regulators of forest primary production. Science 190:515–520

Morales MA (1999) The role of space and behavior in an ant-membracid mutualism. PhD Thesis. University of Connecticut, Storrs, Connecticut

Morales MA (2000) Survivorship of an ant-tended membracid as a function of ant recruitment. Oikos 90:469–476

Morales MA, Heithaus ER (1998) Food from seed-dispersal mutualism shifts sex ratios in colonies of the ant *Aphaenogaster rudis*. Ecology 79:734–739

Murray KG (1988) Avian seed dispersal of three neotropical gap-dependent plants. Ecol Monogr 58:271–298

Nabhan GP, Buchmann SL (1997) Services provided by pollinators. In: Daily GC (editor) Nature's services. Societal dependence on natural ecosystems. Island Press, Washington, DC, pp 133–150

Nantel P, Gagnon D, Nault A (1996) Population viability analysis of American ginseng and wild leek harvested in stochastic environments. Conserv Biol 10:608–621

Nee S, May RM, Hassell MP (1996) Two-species metapopulation models. In: Hanski I, Gilpin ME (eds) Metapopulation biology: ecology, genetics, and evolution. Academic, San Diego, pp 123–148

Norconk MA, Grafton BW, Conklin-Brittain NL (1998) Seed dispersal by neotropical seed predators. Am J Primatol 45:103–126

Offenberg J (2001) Balancing between mutualism and exploitation: the symbiotic interaction between *Lasius* ants and aphids. Behav Ecol Sociobiol 49:303–310

Oostermeijer JGB (2000) Population viability of *Gentiana pneumonanthe*: the importance of genetics, demography, and reproductive biology. In: Young AG, Clarke GM (eds) Genetics, demography, and viability of fragmented populations. Cambridge University Press, Cambridge, pp 313–334

Paige KN, Whitham TG (1987) Overcompensation in response to mammalian herbivory: the advantage of being eaten. Am Nat 129:407–416

Parker IM (2000) Invasion dynamics of *Cytisus scoparius*: a matrix model approach. Ecol Appl 10:726–743
Pudlo RJ, Beattie AJ, Culver DC (1980) Population consequences of changes in an ant-seed mutualism in *Sanguinaria canadensis*. Oecologia 146:32–37
Rathcke BJ, Jules ES (1993) Habitat fragmentation and plant-pollinator interactions. Curr Sci 65:273–277
Reeves SA, Usher MB (1989) Application of a diffusion model to the spread of an invasive species: the coypu in Great Britain. Ecol Model 47:217–232
Sargent S (1990) Neighborhood effects on fruit removal by birds: a field experiment with *Viburnum dentatum* (Caprifoliaceae). Ecology 71:1289–1298
Schemske DW, Husband BC, Ruckelshaus MH, Goodwillie C, Parker IM, Bishop JG (1994) Evaluating approaches to the conservation of rare and endangered plants. Ecology 75:584–606
Schupp E, Howe H, Augsburger C, Levey D (1989) Arrival and survival in tropical treefall gaps. Ecology 70:562–564
Schupp EW (1990) Annual variation in seedfall, postdispersal predation, and recruitment of a neotropical tree. Ecology 71:504–515
Sih A, Baltus M-S (1987) Patch size, pollinator behavior, and pollinator limitation in catnip. Ecology 68:1679–1690
Simberloff D, Brown BJ, Lowrie S (1978) Isopod and insect root borers may benefit Florida mangroves. Science 201:630–632
Smith BH, deRivera CE, Bridgman CL, Woida JJ (1989) Frequency-dependent seed dispersal by ants of two deciduous forest herbs. Ecology 70:1645–1648
Snow AA, Spira TP, Simpson R, Klips RA (1996) The ecology of geitonogamous pollination. In: Lloyd DG, Barrett SCH (eds) Floral biology. Chapman and Hall, New York, pp 191–216
Snow DW, Snow BK (1986) Some aspects of avian frugivory in a north temperate area relevant to tropical forest. In: Estrada A, Fleming TH (eds) Frugivores and seed dispersal. Junk, Dordrecht, pp 159–164
Sousa WP, Mitchell BJ (1999) The effect of seed predators on plant distributions: is there a general pattern in mangroves? Oikos 86:55–66
Steffan-Dewenter I, Tscharntke T (1999) Effects of habitat isolation on pollinator communities and seed set. Oecologia 121:432–440
Stephens PA, Sutherland WJ, Freckleton RP (1999) What is the Allee effect? Oikos 87:185–190
Stowe KA, Marquis RJ, Hochwender CG, Simms EL (2000) The evolutionary ecology of tolerance to consumer damage. Annu Rev Ecol Syst 31:565–595
Terborgh J, Losos E, Riley MP, Bolanos Riley M (1993) Predation by vertebrates and invertebrates on the seeds of five canopy tree species of an Amazonian forest. In: Fleming TH, Estrada A (eds) Frugivory and seed dispersal: ecological and evolutionary aspects. Kluwer, Boston, pp 375–386
Thompson JN (1985) Postdispersal seed predation in *Lomatium* spp (Umbelliferae): variation among individuals and species. Ecology 66:1608–1616
Thompson JN (1988) Variation in interspecific interactions. Annu Rev Ecol Syst 19:65–87
Traveset A, Riera N, Mas RF (2001) Passage through bird guts causes interspecific differences in seed germination characteristics. Funct Ecol 15: 669–675
Turnbull LA, Crawley MJ, Rees M (2000) Are plant populations seed-limited? A review of seed sowing experiments. Oikos 88:225–238
Underwood N (1998) The timing of induced resistance and induced susceptibility in the soybean-Mexican bean beetle system. Oecologia 114:376–381
Underwood N (1999) The influence of induced plant resistance on herbivore population dynamics. In: Agrawal AA, Tuzun S, Bent E (eds) Induced plant defenses against

pathogens and herbivores. American Phytopathological Society Press, St. Paul, MN, pp 211–229

Wackers FL, Wunderlin R (1999) Induction of cotton extrafloral nectar production in response to herbivory does not require a herbivore-specific elicitor. Entomol Exp Appl 91:149–154

Walsburg GE (1975) Digestive adaptations of *Phainopepla nitens* with the eating of mistletoe berries. Condor 77:169–174

Whelan CJ, Willson MF (1991) Spatial and temporal patterns of postdispersal seed predation. Can J Bot 69:428–436

Whelan CJ, Schmidt KA, Steele BB, Quinn WJ, Dilger S (1998) Are bird-consumed fruits complementary resources? Oikos 83:195–205

White JA, Whitham TG (2000) Associational susceptibility of cottonwood to a box elder herbivore. Ecology 81:1795–1803

Widén B (1993) Demographic and genetic effects on reproduction as related to population size in a rare, perennial herb, *Senecio integrifolius* (Asteraceae). Biol J Linn Soc 50:179–195

Willson MF, Harmeson JC (1973) Seed preferences and digestive efficiency of cardinals and song sparrows. Condor 70:225–234

Wold EN, Marquis RJ (1997) Induced defense in white oak: effects on herbivores and consequences for the plant. Ecology 78:1356–1369

Wolin CL (1985) The population dynamics of mutualistic systems. In: Boucher DH (ed) The biology of mutualism. Oxford University Press, New York, pp 40–99

Young TP, Okello BD (1998) Relaxation of an induced defense after exclusion of herbivores: spines on *Acacia drepanolobium*. Oecologia 115: 508–513

Young A, Boyle T, Brown T (1996) The population genetic consequences of habitat fragmentation for plants. Trends Ecol Evol 11:413–418

Zalba SM, Sonaglioni MI, Compagnoni CA, Belenguer CJ (2000) Using a habitat model to assess the risk of invasion by an exotic plant. Biol Conserv 93:203–208

Zelmer CD, Currah RS (1995) Evidence for a fungal liaison between *Corallorhiza trifida* (Orchidaceae) and *Pinus contorta* (Pinaceae). Can J Bot 73:862–866

11 Modeling the Effects of Disturbance, Spatial Variation, and Environmental Heterogeneity on Population Viability of Plants

E.S. Menges and P.F. Quintana-Ascencio

11.1 Introduction

As daunting as it is to assemble data and estimate population viability for single populations, the real world is far more complicated. Ecological disturbances create spatial and temporal variation that will affect demography. Integration across many sites with different times since disturbance or disturbance regimes may be necessary to get a good idea of population persistence in a landscape. Species may also occur across environmental gradients or in patches that vary in quality. These complications challenge the simplicity implied by population viability analyses (PVAs). PVAs combine empirical data (usually on the entire life cycle) and modeling approaches to project the condition (and existence) of future populations (Menges 2000: Chaps. 1, 6, this Vol.). Most PVAs have considered populations as uniform and without explicit consideration of disturbance regimes. However, more recently, researchers are treating disturbances and spatially explicit demography with a range of innovative approaches (Chap. 6, this Vol.).

Ecological disturbances involve relatively discrete events that disrupt ecosystems, communities, or populations and create additional changes in structure and function (Pickett and White 1985). Most disturbances remove biomass and create an opportunity for species that may otherwise not be competitive. Typical ecological disturbances include fire, wind, ice, trampling, flooding, and other physical factors. Biotic disturbances such as herbivory and plant disease also reduce biomass and open up habitats, but there is not always a clear distinction between disturbances (which are defined as episodic) and chronic factors that may also reduce competition. For example, there is no clear line between chronic herbivory (a stress) and episodic herbivory (a disturbance).

Ecological Studies, Vol. 165
C.A. Brigham, M.W. Schwartz (Eds.)
Population Viability in Plants

Most human disturbances in ecological communities also reduce standing biomass and simplify community structure and composition. However, human disturbance regimes often deviate from historic ecological disturbance regimes and can lead to radical shifts, such as the introduction of exotic species (Chap. 2, this Vol.). Changes in frequency or intensity of disturbance can increase variance in population numbers and alter species demography beyond recovery capacity. Variation in the level and nature of the disturbances also leads to changes in population age structures. For example, older individuals are less likely to occur in areas disturbed deterministically, if large individuals are removed once they attain a certain age. In contrast, in areas disturbed stochastically, individuals of various ages may be removed and some older individuals are likely to remain (McCarthy and Burgman 1995).

In considering how disturbances may affect the demography and viability of plant populations, the episodic nature of disturbances determines how their effects are quantified. Many approaches consider these events as discrete, with typical return times and specific spatial patterns. This necessitates a departure from typical modeling of population viability, with its emphasis on equilibrium conditions. Transient analyses predicting analogs to equilibrium results such as elasticities (e.g., Fox and Gurevitch 2000) may be particularly appropriate for analyzing the effects of populations recovering from disturbances. Simulations may often be necessary, however, as complex disturbance patterns may not lead to analytical solutions.

Modeling of disturbance effects on plant population viability is a crucial area for species management and conservation. Not only are disturbance effects very important in many ecosystems and to many endangered species, but disturbances can often be controlled and manipulated by land managers (Chap. 2, this Vol.). The most-studied example is fire. Prescribed (controlled) fire is used to manage a range of forests, shrublands, and grasslands around the world. Several models have been developed to assess population viability under different fire regimes (e.g. Bradstock and O'Connell 1988; Silva et al. 1991; Burgman and Lamont 1992; Canales et al. 1994; Gross et al. 1998; Hoffmann 1999; Caswell and Kaye 2001; Quintana-Ascencio et al. 2002; Chap. 6, this Vol.).

In the following sections, we summarize and discuss examples of modeling approaches including the effects of disturbances, spatial variation, and environmental heterogeneity on population viability in plants. We emphasize the need to preserve observed variances among sites and years to avoid underestimates of extinction risks. We describe how comparative approaches contribute to our understanding of the factors creating demographic variation. Modeling approaches simulating disturbance regimes and integrating regional and local dynamics are also reviewed. Finally, we conclude by considering the strengths and limitations of current approaches to PVA.

11.2 The Issue of Variance and the Problems with Averaging

Point estimates of demographic parameters and the finite rate of increase (also known as lambda, λ, or population growth rate) give only one of a potential distribution of outcomes. Demographic rates vary across time and space (e.g., Horvitz and Schemske 1995). Elasticities vary widely among species and among populations within species, often in relation to environmental gradients and population growth (e.g., Silvertown et al. 1996; Golubov et al. 1999). Unfortunately, many researchers have created average matrices that mask true variation among populations and across years. This is a particular problem when the nature of demographic variation is attributable to disturbance, succession, environmental gradients, or other non-random phenomena. For example, successive years or nearby sites are likely to be similarly favorable (or unfavorable) and a few outstanding years or microsites may be counterbalanced by many poor years or microsites (e.g. Kalisz and McPeek 1993; Chap. 6, this Vol.). The mean matrix or mean finite rate of increase is a poor point estimate in these cases. Since variation generally increases extinction risk (Menges 1998), one problem with averaging will be a tendency to underestimate extinction risk. In addition, averaging will hinder our understanding of factors that may be driving demographic variation. Averaging may be a common response to problems with insufficient data for certain demographic parameters in certain years and populations, but there are alternatives to averaging matrices (Menges and Dolan 1998; Parker 2000; Chap. 7, this Vol.).

Bootstrap techniques have been frequently used to test hypotheses or to generate confidence intervals around demographic parameters (mostly finites rate of increase). Bootstrapping assumes random samples and statistically independent data (Manly 1991; Crowley 1992; Efron and Tibshirani 1998). The potential impact of data dependence on inference accuracy is not well understood and has been addressed case by case (Efron and Tibshirani 1998, and references therein). There are many sources of non-independence in matrix demographic models. Individual fates may be linked by exposure to common environmental conditions. Demographic parameters among life history stages may be positively correlated due to similar responses to weather, herbivores, and other environmental factors. Common historical effects may result in significant autocorrelations along time-series. Because non-independent data are so characteristic of demographic matrices, bootstrapped estimates of variance should be considered with caution. Bootstrapping of matrices can include unrealistic scenarios or combinations of demographic parameters that alter estimated variance in comparison to observed parameter correlations. Variance estimates generated for populations from different sites and years should be preserved whenever possible.

An alternative approach to bootstrapping with averaged matrices is to obtain variance estimates among a group of matrices from different sites and years. This clearly requires either many sites or many years of data, which are not available for the majority of plant PVA studies (Menges 2000b). However, there are advantages to dealing with matrices from multiple sites and years. Such groups of matrices often reveal strong patterns in finite rates of increase or elasticities that can be related to environmental gradients (e.g., Oostermeijer et al. 1996; Silvertown et al. 1996; Valverde and Silvertown 1997a; Menges and Dolan 1998; Quintana-Ascencio et al. 2003).

11.3 Comparing Populations Subject to Disparate Disturbance Regimes or Environmental Conditions

Perhaps the simplest and most common approach to understanding the effects of disturbance on population viability is to compare populations (e.g., Freckleton and Watkinson 1998; Reed et al. 2002; Chap. 6, this Vol.). For example, Eriksson and Eriksson (2000) found differences in time-to-extinction between two populations of *Plantago media* in semi-natural grasslands, and Mandujano et al. (2001) detected differences in the finite rates of increase and elasticities of *Opuntia rastrera* between nopaleras (cactus-dominated communities) and grasslands. However, neither study contrasted environmental conditions or disturbance regimes between sites to allow direct assessment of their effects on demographic variation.

Comparisons of disturbed vs. non-disturbed sites can be useful in evaluating population viability and suggesting management strategies (Table 11.1). For example, a simple comparison of burned and unburned plots showed that an annual grass of Venezuelan savannas benefited from frequent fires (Canales et al. 1994). Similar comparisons have been made to evaluate population viability as a function of community productivity (Werner and Caswell 1977), trampling (Maschinski et al. 1997), land-use history (Donohue et al. 2000), hurricane effects (Batista et al. 1998), and the interactive effects of fire, genetic variation, and geographic location (Menges and Dolan 1998).

Population comparisons have also been used to evaluate the demography of invading species, such as the exotic shrub *Cytisus scoparius* in Washington, USA (Parker 2000), and New South Wales, Australia (Downey and Smith 2000). The former study compared invading populations between pristine and human-disturbed urban habitats, and among different invasion stages. In contrast to expectations, the finite rate of increase was greater on relatively pristine prairie sites than on urban sites (Parker 2000). The latter compared long-term data (16 years) between established stands and stands at the edge of expanding populations. Recruitment occurred only where light levels were

Table 11.1. Selected recent (1995–2001) plant population viability analyses (PVAs) illustrating the approach of comparing populations to infer effects of disturbance regimes

Reference	Species	Life form	Disturbance or environmental gradient	Habitats	Key results from simulations
Donohue et al. (2000)	*Gaultheria procumbens*	Clonal shrub	Previous ploughing	Scrub oak and hardwood forest	Demography favorable on prior plowed sites, but colonization from undisturbed areas is limited.
Downey and Smith (2000)	*Cytisus scoparius*	Shrub	Stages of invasion	Eucalypt woodlands	Recruitment occurred only where light levels were high, either in recently invaded sites or after senescence produced canopy gaps in mature stands.
Ehrlén (1995)	*Lathryrus vernus*	Perennial herb	Herbivory intensity and type	Mixed deciduous forest/meadow	Finite rare of increase was more depressed by mollusk herbivory than vertebrate grazing or insect seed predation.
Lennartsson (2000); Lennartsson and Oostermeijer (2001)	*Gentianella campestris*	Biennial herb	Grazing intensity	Seminatural grasslands	Extinction risk varied depending on the month of mowing and whether grazing followed mowing An intermediate grazing intensity resulted in the lowest probability of extinction in mesic sites.
Nantel et al. (1996)	*Panax quinquefolium*, *Allium tricoccum*	Perennial herbs	Harvesting and environmental variation	Deciduous forest	Populations that are large enough to resist demographic stochasticity and harvesting are larger than most naturally occurring populations.
Parker (2000)	*Cytisus scoparium*	Shrub	Stages of invasion	Prairies-urban fields	In contrast to expectations, the finite rate of increase was greater on relatively pristine prairie sites than on urban sites.

high, either in recently invaded sites or within canopy gaps in mature stands (Downey and Smith 2000). Both studies showed higher finite rates of increase at invasion fronts, suggesting that the persistence of this exotic was influenced by changes created by the invasion itself.

Harvesting of plant parts can be considered as a disturbance or a chronic event. Designing sustainable harvesting strategies is essential for the long-term use of economically important species. Nantel et al. (1996) simulated various harvesting intensities on populations of two woodland herbs, examining how harvesting affected finite rates of increase, extinction probabilities, and minimum viable population sizes. Populations of ginseng and wild leek that would be large enough to resist demographic stochasticity and harvesting are larger than most naturally occurring populations (Nantel et al. 1996). Therefore, harvesting of wild populations has been banned in Quebec (Gagnon, personal communication).

In general, analyses of harvesting regimes suggest that species vary in their sensitivity to losses of different life stages, and removal frequency and intensity (e.g., Petters 1991; Soehartono and Newton 2001; Pinard 1993; Santos 1993; Olmsted and Alvarez-Buylla 1995; Menges et al. 2003). For example, frequent but small seed collections produced less extinction risk than infrequent but larger seed harvests, but species differed in their sensitivities (Menges et al. 2003). Most studies of harvesting effects concentrate on the impact of various, rather arbitrary levels of harvests. However, detailed information on harvest strategies can be incorporated into analyses of maximum sustainable harvest levels (Ticktin et al. 2002).

Grazing and mowing strongly affect population viability of many species and several comparative studies have evaluated their demographic consequences. Grazing intensity was positively associated with the finite rate of increase among populations of *Cirsium vulgare*, with seasonal changes in grazing intensity affecting different life stages (Bullock et al. 1994). While *Gentianella campestris* extinction risk decreased with grazing intensity in dry sites, intermediate grazing intensity resulted in the lowest probability of extinction in mesic sites (Lennartsson 2000). Extinction risk varied depending on the month of mowing and whether grazing followed mowing, with the historical management regime producing lower extinction risks than prevailing management strategies (Lennartsson and Oostermeijer 2001). Frequent mowing of a perennial herb increased initial finite rate of increase and equilibrium population size in density-dependent, phenologically explicit simulations (de Kroon et al. 1987). Herbivory experiments on a forest moss showed that it had reasonably good recovery of finite rate of increase due to regeneration of branch tips (Rydgren et al. 2001).

In addition to time and intensity of grazing, the type of grazer can also be important. For example, meristem damage by mollusks affected *Lathyrus vernus* finite rate of increase more than vertebrate grazing, even though the latter removed greater biomass (Ehrlén 1995). Sheep grazing (vs. cattle grazing) decreased mortality of vegetative and flowering *Ophrys sphegodes* plants, and increased the emergence of dormant plants, resulting in higher finite rates of increase (Waite and Hutchings 1991).

Although most PVAs have been concerned with perennial plants, annuals have been treated in several studies. Annual plant population dynamics can be modeled with difference equations when there is no overlap between generations. However, modeling of annual species with persistent seed banks may require the consideration of several life history stages (Kalisz and McPeek 1993). Cover of perennial vegetation and bare sand affected the population dynamics of the annual *Vulpia fasciculata* on fixed dunes (Watkinson 1990). Modeling suggested that persistence can only be expected where bare sand exceeds 50% of the total area. Perennial cover decreased persistence more through effects on soil crusts and on germination than from the effects of competition. Rainy season fires were predicted to reduce or eliminate populations of annual *Sorghum* spp. in tropical savannas, while dry season fires would only reduce densities (Watkinson et al. 1989; Lonsdale et al. 1998).

While comparisons of sites can hint at expected outcomes, disturbance cycles can have complex effects on population dynamics and persistence that go beyond simple contrasts between disturbed and undisturbed conditions. Explicit modeling of disturbance regimes has been accomplished using megamatrices as well as models of disturbance and recovery cycles.

11.4 Modeling Disturbance Explicitly with Megamatrices

When disturbance and recovery cycles affect demography, megamatrices can provide a compact summary of these complexities and afford analytical advantages (Table 11.2). The megamatrix approach summarizes demographic parameters in standard matrices that are then combined with disturbance/succession or patch dynamics transitions to produce a megamatrix. This megamatrix includes transitions showing the dynamics of both individual plants and stands.

The megamatrix approach has been particularly fruitful in analyzing the population dynamics of forest-gap-dependent species. The seminal work by Horvitz and Schemske (1986) analyzed the effects of seed dispersal and disturbance by treefalls on the persistence of *Calathea ovensis* in tropical rain forests. Forest dynamics were modeled as a linear Markovian process of succession. Growth, survival, and fertility were estimated using decreasing linear functions of canopy closure. Their matrix models predicted decreasing populations as the canopy closed, strong selection for local dispersal, and selection against long-distance dispersal. A study of canopy closure and its effects on *Cynoglossum virginianum* found that reproductive output was highest and

Table 11.2. Selected recent (1994–2001) plant PVAs illustrating the use of megamatrices to provide realistic simulations of complex disturbance regimes

Reference	Species	Life form	Disturbance or environmental gradient	Habitats	Key results from simulations
Abe et al. (1998)	*Styrax obassia*	Subcanopy tree	Forest disturbance	Deciduous forest	Populations were persistent in mature forest but could increase in gaps.
Alvarez-Buylla (1994)	*Cecropia obtusifolia*	Pioneer tree	Forest disturbance	Tropical rain forest	Recently dispersed seeds were key for seedling recruitment. Pioneer species lacking long-lasting seed banks in highly deforested areas may be vulnerable to extinction.
Cipollini et al. (1994)	*Lindera benzoin*	Understory shrub	Forest disturbance	Temperate forest	Increasing populations were predicted in all forest habitats, higher population growth rates for plants in gaps.
Pascarella and Horvitz (1998)	*Ardisia escallonioides*	Understory shrub	Hurricane	Subtropical forest	Open, post-hurricane gaps were characterized by higher finite rates of increase. However, regionally, these open patches were less common than closed patches due to rapid post-hurricane vegetation recovery.
Valverde and Silvertown (1997a,b)	*Primula vulgaris*	Perennial herb	Forest disturbance	Temperate forest	Seedling survival and fecundity interacted with forest disturbance and dispersal rates in affecting population dynamics.

seedling survival lowest in early successional environments (Cipollini et al. 1993).

The megamatrix approach was also used to investigate the effect of gap dynamics and seed dispersal on the patch-specific population dynamics of the dioecious shrub *Lindera benzoin* (Cipollini et al. 1994). Parameters were estimated by comparing growth of adults in gaps vs. understory and evaluating survival of seedlings and juveniles as a function of light levels. Individual populations increased most rapidly in gaps. At a regional level, an increasing rate of new gap formation was associated with higher metapopulation growth rates. Contrary to predictions for *Calathea ovensis* (Horvitz and Schemske 1986), models of *L. benzoin* indicated a benefit for long-distance seed dispersal. Another example of the megamatrix approach showed that a subcanopy tree, *Styrax obassia*, could maintain populations under shade, although populations had greater potential to increase in gaps (Abe et al. 1998).

Alvarez-Buylla and collaborators studied the effects of forest canopy dynamics on the demography of the pioneer tree *Cecropia obtusifolia* (Alvarez-Buylla and García-Barrios 1991; Martínez-Ramos and Alvarez-Buylla 1995). Their models included patch-specific demographic parameters from a broad range of patch ages. Recently dispersed seeds were critical for seedling recruitment, suggesting that the absence of a persistent seed bank predisposes pioneer species to high extinction risks in highly deforested areas. A detailed comparison of four matrix models confirmed the importance of gap dynamics and density dependence for *C. obtusifolia* population dynamics (Alvarez-Buylla 1994).

The temperate forest gap herb *Primula vulgaris* was studied with single and metapopulation modeling using megamatrices (Valverde and Silvertown 1997a,b). Gaps were classified based on canopy openness, and gap closure rates were estimated by comparisons of canopy photographs. Within-patch demographic rates varied with canopy openness and were summarized in stage-structured matrices. Fully closed gaps were unavailable for colonization. Colonization rates are often difficult to estimate in plant populations. In these studies, the authors explored the effects of different colonization rates for *P. vulgaris* using simulations. Relatively high seed dispersal among populations and higher forest disturbance rates each increased the metapopulation growth rate. However, frequent disturbances created conditions where local extinction did not occur, breaking down metapopulation structure. Fecundity in individual populations also affected metapopulation growth rate. Thus, for this species, individual population demography must be known to understand metapopulation dynamics.

To model hurricane effects on a tropical understory shrub, a megamatrix was formulated to include a transitions among seven forest patch types and individual stage-classified demography within each patch (Pascarella and Horvitz 1998). Open, post-hurricane environments were characterized by higher finite rates of increase than were closed, undisturbed sites. However, in the landscape, these open patches were (on average) less common than closed patches due to rapid post-hurricane vegetation recovery (Pascarella and Horvitz 1998). The stages with higher elasticities varied depending on the forest condition, and megamatrix analyses gave insights that would not be available from analyzing a series of single environments. This integration is one major advantage of the megamatrix approach.

Population and regional transition rates represented in megamatrices also vary temporally and spatially. Post-disturbance recovery rates may depend on the landscape context and regional dynamics of critical species. For example, in La Lacandona Tropical Rain Forest, Chiapas, Mexico, weed seed imports increased while arrival of native seeds decreased within forest gaps surrounded by field crops and grasslands, compared to gaps in a forest-dominated landscape (Quintana-Ascencio et al. 1996a). Megamatrix approaches may provide misleading estimates of population growth if they fail to consider the variable nature of landscape-level phenomena. Consideration of variation in disturbance regimes is one advantage of modeling episodic disturbances in an explicit manner.

11.5 Modeling Disturbance Cycles and Episodic Disturbances Explicitly

An alternative approach to megamatrices involves dynamic modeling using varying matrix elements: varying demographic parameters ("element selection") or alternating matrices ("matrix selection", Table 11.3). Variation of individual demographic parameters raises the question of correlation among parameters. Generally, given lack of data on correlation structure, individual demographic parameters are allowed to fluctuate independently (Chap. 6, this Vol.). This produces an underestimate of extinction risk since correlated fluctuations in demographic parameters produce more extreme results (Burgman et al. 1993; Greenlee and Kaye 1997). The alternative is to specify perfect correlations, over-estimating risk. In most cases, these two extreme assumptions bound the correct result. In a theoretical study comparing various correlation structures, estimates of stochastic population growth were compared (Fieberg and Ellner 2001; also Chap. 7, this Vol.).

Matrix selection chooses data-derived (not averaged) matrices to preserve the correlation structure implicit in the years and populations sampled. Simulation models choose among the possible matrices, with inclusion in any particular part of the simulation determined by disturbance intervals, time since disturbance, stochastic factors, or combinations. Commercially available software does not deal with this approach, so authors generally construct their own models to alternate matrices (but see Bearlin et al. 1999, Oostermeijer 2000). While this approach preserves empirical correlations, it limits out-

Table 11.3. Selected recent (mostly 1998–2001) plant PVAs illustrating explicit modeling of disturbance cycles to determine conditions promoting persistence of populations or metapopulations

Reference	Species	Life form	Disturbance or environmental gradient	Habitats	Key results from simulations
Caswell and Kaye (2001)	*Lomatium bradshawii*	Perennial herb	Fire (frequency)	Grasslands	Fire is important for persistence. Elasticities based on stochastic growth rate were highly correlated with deterministic elasticities.
Enright et al. (1998a,b)	*Banksia hookeriana* *B. attenuata*	Non-sprouting shrub Resprouting shrub	Fire and seasonal precipitation	Shrublands	Weather and fire frequency determine persistence. Longer fire return intervals and higher probabilities of inter-fire recruitment favor less canopy seed storage.
Gross et al. (1998)	*Hudsonia montana*	Shrub	Fire and trampling	Forested gorge	Trampling reduction and prescribed burning can promote persistence.
Hoffmann (1999)	*Perira mediterranea, Miconia albicans, Rourea induta, R. montana, Myrsine guianensis*	Shrubs and Trees	Fire (frequency)	Tropical savannas	Optimal fire regime can be conflicting among co-occurring species.
Pfab and Witkowski (2000)	*Euphorbia clivicola*	Perennial herb	Fire, herbivore exclusion	Succulent shrubland	Without intensive management, extinction of species is likely.
Oostermeijer (2000)	*Gentiana pneumonanthe*	Perennial herb	Succession	Abandoned meadows	Conditions along successional gradients, inbreeding, and reproductive failure with decreasing population size determine persistence.
Quintana-Ascencio et al. (2003)	*Hypericum cumulicola*	Perennial herb	Fire and winter precipitation	Florida scrub	Because optimal fire return interval is shorter than normal fire frequencies, metapopulation dynamics may occur.

comes to those observed in the data at hand, providing a conservative view of the extent of variation.

In an early approach to modeling disturbance effects on demography, Manders (1987) simulated the effect of fire cycles on the tree *Widdringtonia cedarbergensis*. He projected the growth of a single deterministic population until a fire, which reduced population size as a function of time-since-fire. The simulations suggested that *W. cedarbergensis* is able to increase in the absence of fire, and that fire regimes longer than 15–20 years are necessary to allow populations to include enough seed-producing trees.

The full approach with alternative matrices was first used for plant species to assess the effect of fire frequency on savanna grasses (Silva et al. 1991; Canales et al. 1994). Population matrices were built for burned and unburned populations. Deterministic models of populations under fixed fire frequencies were obtained using finite-state Markov chain models. The random occurrence of fires was introduced in a stochastic version of these models by changing the probability of occurrence of a burn year in a random fashion, with the probability of fire determined by a given fire frequency and an index of autocorrelation between burn and no burn years. Models predicted that fire probabilities higher than 0.89 and 0.29 were necessary for the persistence of *Andropogon semiberbis* and *A. brevifolius*, respectively.

One of the more thorough viability analyses concerning fire considered both deterministic and stochastic modeling, the latter using both matrix and element selection (Caswell and Kaye 2001; Kaye et al. 2001). Fire was beneficial to an endangered prairie perennial, *Lomatium bradshawii*, from Oregon and Washington, while unburned populations were likely to go extinct. The stochastic models indicated critical annual fire probabilities of 0.4–0.9, and predicted increasing population growth rates with more frequent fires and increasing negative autocorrelation between burn and no burn years. Although both matrix and element selection provided similar qualitative results on the effects of fire and site differences, matrix selection was more conservative in predicting higher extinction risks under intermediate scenarios.

The effects of ice rafting and scouring on the population dynamics of the seaweed *Ascophyllum nodosum* in the intertidal north Atlantic were evaluated with simulations from matrix sequences (Åberg 1992). The individual matrices described population dynamics of this species during a year with no ice, with moderate damage due to ice, and with large damage due to ice. Frequencies of ice years were used to determine the probability of selecting a given matrix during the simulation. Higher extinction risks and shorter times to extinction were found for the population with a higher frequency of ice years.

Stochastic computer simulations were developed to assess demographic effects of natural (e.g. El Niño) or anthropogenic alterations of irradiance and temperature on the giant kelp (*Macrocystis pyrifera;* Burgman and Gerard

1990). Modeled large seasonal fluctuations in giant kelp sporophytic recruitment and adult sporophytic density were in agreement with observations in natural populations. Simulations predicted reductions in adult density when El Niño preceded or coincided with major recruitment events, but rapid adult density increases in other circumstances, reflecting recovery variation in natural kelp forests in southern California. Similarly, a model was used to simulate increases in turbidity, such as that due to dredging operations or waste discharges, on the seagrass (*Zostera muelleri*; Bearlin et al. 1999). Higher chances of population decrease with this human impact were predicted.

To account for changes in population parameters with time-since-fire and weather, Quintana-Ascencio et al. (2003) extended this approach to evaluate *Hypericum cumulicola* persistence in Florida scrub. They used an algorithm that projected matrix sequences, randomly choosing different populations with data specific to the time-since fire interval, weighted by the relative frequency of the winter precipitation that occurred in each study year. Simulations predicted that populations with thousands of individuals after an initial fire might still become locally extinct within 300–400 years without additional fires. Fire regimes shorter than 50 years will be necessary for long-term persistence of most *H. cumulicola* populations. The authors suggested that, because optimal fire-return intervals for *H. cumulicola* are shorter than normal fire frequencies, metapopulation dynamics are likely in this Florida scrub endemic plant.

Combinations of two disturbances, prescribed fire and trampling, were dealt with in a study of a threatened shrub *Hudsonia montana* in North Carolina (Gross et al. 1998). Matrices representing unburned, year after burn, and two years post-burn were used in combination with matrices representing observed trampling and 50 or 100% reductions in trampling. Data were derived from a single 5-year field experiment. Only a combination of trampling reduction and prescribed burning created populations with positive finite rates of increase, although a variety of fire-return intervals were apparently effective in promoting population viability (Gross et al. 1998). Despite limited data, this study modeled 39 management strategies combining burning and trampling reduction.

Correlated variation among demographic rates was modeled using constrained values for each rate in a study evaluating management effects on the endangered *Euphorbia clivicola* in South Africa (Pfab and Witkowski 2000). Replicate simulations were run to predict extinction probability and the time to extinction. Results indicated that more intensive management, involving fire frequencies every 3 years and exclusion of herbivores, could reduce extinction risks.

Evolutionary and ecological effects of fire were explored in two papers by Enright and colleagues on sprouting and non-sprouting Australian *Banksia* shrubs (Enright et al. 1998a, b). They used dynamic models with recruitment, survival, and fecundity parameters determined by specific functional relationships with age, weather variables (especially summer rainfall), time-since-fire, and serotiny. Wet summers following fires were associated with increasing populations regardless of degree of serotiny, as long as fires were frequent. Dry summers were associated with increasing populations only under extreme serotiny and short fire return intervals. Longer fire-return intervals and higher probabilities of inter-fire recruitment favored lower rates of serotiny. Stochasticity in fire-return intervals predicted an intermediate level of serotiny as found in nature, interpreted as a bet-hedging strategy for the shrub (Enright et al. 1998a, b). This result also suggests that variation in fire return intervals has been likely in these Australian shrublands. Models predicted maximum finite rates of increase with a fire-return interval of 16 years for the non-sprouting shrub *Banksia hookeriana*, and 13 years for the resprouting *B. attenuata* (Enright et al. 1998a, b). Other studies also have predicted that single fire regimes will not be optimal across species. For example, simulations indicated that changes in fire frequency might cause shifts in the relative abundance of four woody species in the cerrado savanna of Brazil (Hoffman 1999).

Even for an individual species, there may not be a single optimal fire-return interval. In southwestern Australia, rains are often scarce and highly variable, not always providing enough moisture for seedling recruitment and survival of *Banksia* spp. Burgman and Lamont (1992) built a disturbance/recovery model involving demographic stochasticity, environmental stochasticity (related to rainfall), inbreeding depression (affecting recruitment) and fire effects. Interestingly, the fire frequency that maximizes population size of *B. cuneata* (15- to 25-year fire-return interval) does not minimize extinction risk. The post-fire environment, while potentially allowing increased population size, may also expose the population to higher extinction risk due to failure of entire seedling cohorts during drought years. Not burning is problematic because population sizes will slowly decrease. The authors suggested that seedling cohorts of this rare plant may have to be watered during years when drought follows fire (Burgman and Lamont 1992; also reviewed in Chap. 6, this Vol.). Another study included risks associated with fire-induced mortality of juveniles and post-fire germination failure in *B. ornata*. To increase the chances of persistence, the best strategy is to burn infrequently but before plant density declines (McCarthy et al. 2001).

Simultaneous use of a variety of modeling approaches can be informative. A study of the demography and population viability of the scrub buckwheat (*Eriogonum longifolium* var. *gnaphalifolium*) in relation with fire used deterministic analysis with single matrices and megamatrices, and stochastic analysis with alternating matrices (Satterthwaite et al. 2002). Deterministic models alone predicted long unburned populations to be nearly stable while stochastic simulations show a net decline over 250 years.

More complicated disturbance models can be built around functional demographic relationships with time-since-disturbance. Oostermeijer (2000) used a stochastic matrix-projection model to simulate population dynamics of *Gentiana pneumonanthe* during heathland succession. Using RAMAS/stage (Ferson 1991), regression equations relating matrix elements to time-since-sod-cutting were used instead of the standard transition probabilities. Average values and variances were used for those transitions not statistically related to time-since-sod-cutting. Three variants of this basic model incorporated equations summarizing relationships among reproductive success, density, population size, selfing rate, and inbreeding depression. Density-dependent fertility did not affect projected population performance, but inbreeding depression reduced peak population sizes and growth rates, and decreased time to extinction. Shorter intervals between disturbances decreased extinction risks, particularly among populations smaller than 100 individuals. This study highlights the importance of considering several factors during PVA.

None of the approaches that we have described consider how the spatial arrangement of populations, patches, or individual plants may influence population viability. Although spatially explicit models are more complex and demand additional data, they may give more precise answers and allow us to address important management questions (Chap. 6, this Vol.).

11.6 Spatially Explicit Demography and PVA

PVAs may incorporate spatial considerations in a number of ways, ranging from incidence-based metapopulation models that are not spatially explicit but do consider patch size and isolation (e.g. Quintana-Ascencio et al. 1996b; Chap. 6, this Vol.) to spatially explicit individual-based metapopulation models (Akçakaya 2000b; Chap. 6, this Vol.). To date, most metapopulation modeling has been applied to animals. Dispersal data for animals are often obtained when marked animals arrive at new patches (Akçakaya 2000a). Because direct evidence of arrival is difficult to ascertain for most seed plants, alternative approaches to modeling plant dispersal are necessary. Several approaches to metapopulation dynamics in plants have not been spatially explicit, and have used model simulations to estimate dispersal parameters or extinction probabilities (Valverde and Silvertown 1997a, b; discussed above).

Few plant PVAs have considered spatially explicit demography (Table 11.4), despite the fact that individual plant fate is closely tied to its neighborhood. One attempt utilizes "parameter fitting procedures" and "simulation experiments" to fill in poorly known parameters in a cellular automata (grid-based) modeling of the population dynamics of *Acacia* trees in Israeli deserts (Wiegand et al. 1999). Individual trees are considered, with rough demographic parameters and some information on mistletoe infection and water status.

Table 11.4. Selected recent (1996–2001) plant PVAs using spatially explicit approaches to population viability

Spatially Explicit					
Reference	Species	Life form	Disturbance or environmental gradient	Habitats	Key results from simulations
Bradstock et al. (1996, 1998)	*Banksia* spp.	Shrubs	Fire (size, frequency, and pattern)	Shrublands	Extinction risk is a function of complex interactions of fire size, fire frequency, and life history; but resprouters are less sensitive than obligate seeders.
Shimada and Ishihama (2000)	*Aster kantoensis*	Biennial herb	Flooding	River floodplains	Expected persistence of simulated pop ulations depended on flood frequency and time to perennial invasion.
Guàrdia et al. (2000)	*Achnatherum calamagrostis*	Perennial herb	Eroded areas	Eroded out-cropping clays	While sexual reproduction provides of nearly all colonization, tiller dynamics contribute mostly to maintenance of already occupied sites.
Watkinson et al. (2000)	*Vulpia ciliata*	Annual herb	Local disturbance	Infertile, dry soils	Spatial patterns found and modeled include density-dependent recruitment, positive spatial autocorrelations.
Wiegand et al. (1999)	*Acacia* species	Desert trees	Moisture, germination microsites	Israeli desert	Germination is crucial and episodic, tied to rainfall. Passage through mammal guts may stabilize populations.

Aggregated seed dispersal and safe sites for germination are considered in a spatially explicit manner. Sensitivity analyses showed that germination and precipitation are key factors affecting population dynamics (Wiegand et al. 1999).

Disturbance can also be dealt with in a spatially explicit manner. Bradstock et al. (1996) explored the effects of fire size and frequency on persistence of reseeding *Banksia* species. They used a cellular automaton model to treat both plant population and fire pattern. At each annual time step, fires, with parameters varying depending on the simulation, spread stochastically within the landscape. The population response included survival, dispersal, and population size. The life history of this species was sketched in broad strokes based on occurrence of major stages in each cell, time to fruiting, overall survival in fires, and assumptions about dispersal. Fire pattern was varied using contiguous spread algorithms and random fire ignitions. Extinctions tended to occur when fire frequencies were extremely high or low and with random (vs. contiguous) fire patterns. Interactions were complex. For example, seed dispersal mitigated extinctions when fire sizes were large but was unimportant when fire sizes were small. The initial hypothesis, that spatially patchy burns would prevent extinction, was not supported. In contrast, intermediate fire frequencies did promote population persistence. While this research is innovative and interesting, it is difficult to assess whether the specific results have generality to other systems or alternative model assumptions. For example, resprouters were not sensitive to fire parameters to the same extent as obligate seeders (Bradstock et al. 1998).

A cellular model was also used to simulate the effects of flooding and subsequent changes in herbaceous cover on populations of *Aster kantoensis* (Shimada and Ishihama 2000). Population processes in each cell were modeled using stage-structured matrices with elements changing with perennial herb cover, itself changing with time-since-flooding. Density-dependent fertility was assumed. *A. kantoensis* does not form a seed bank. Therefore, after disturbance, dispersed seeds are necessary to initiate new populations. As expected, persistence of simulated populations depended on flood frequency and the time until perennial invasion.

Smaller-scale spatially explicit population models have been used to examine details of population growth and persistence. For example, population spread was modeled for *Achnatherum calamagrostis*, modeling tussocks with different tiller numbers, in badlands of the Pyrenees, Spain (Guàrdia et al. 2001). Seedlings and colonizing vegetative tillers were distinguished from established tussocks. Matrix projection indicated higher population growth during non-drought years. Sexual reproduction was more important for spatial spread, while vegetative propagation provided the mechanism to persist at occupied sites. Local-scale population dynamics of the annual grass *Vulpia ciliata* were modeled using a spatially explicit, density-dependent, small-scale model to predict distributions of subplot densities (Watkinson et al. 2000).

Adding relatively large-scale disturbances was necessary to replicate the observed distributions, but these disturbances had only been observed in one population.

More complex treatments of disturbance regimes and spatial influences on demography may provide more realistic projections of population dynamics and estimates of population viability. However, many of the limitations of PVA cut across methodologies, and instead are functions of inherent limitations in most ecological data.

11.7 Conclusion: The Limitations and Uses of PVA

We believe that PVAs will not be able to predict exact population size distributions or discrete densities except in very exceptional circumstances. Predicting population sizes requires not only knowledge of species biology, but of initial population structure and future conditions. Of course, the details of future abiotic and biotic factors that will affect the population's demography cannot be perfectly known. The responses of the populations to the environment, as measured today, may also be different than the future responses under new environmental conditions or sequences. However, because PVAs can show what demographic processes seem to be critical to population persistence, they do provide some understanding that can help anticipate responses to future conditions.

These problems in prediction can be alleviated, to some extent, by choosing less sensitive, dependent variables than future population size. Extinction probability and time to extinction are commonly used. These also have several problems, including the fact that time to extinction has a skewed distribution (Akçakaya 2000a) and that extinction probability may be sensitive to small changes in parameters within certain regions of the simulated universe. Quasi-extinction thresholds, distributions of extinction times, and expected minimum population size may be more robust to moderate errors in demographic parameter estimation (Akçakaya 2000a; McCarthy and Thompson 2001). However, given uncertainty in future conditions, small differences in any projected outcomes must be treated cautiously.

We suggest that PVAs may be most useful in a relative sense, for comparing populations and management regimes or ranking alternatives. Increasingly, reviews of PVAs have been emphasizing their utility for these comparisons (Beissinger and Westphal 1998; Menges 2000a, b; Reed et al. 2002; Chaps. 6, 7, this Vol.). Explicit, realistic management regimes can be compared (Pfab and Witkowski 2000). Nonetheless, even comparative approaches may be limited by short-term data sets and the difficulties of knowing future conditions. A study comparing projections of population models of *Arisaema triphyllum* with actual population sizes 15 years later found that one population changed in the same direction as projected, while another changed in the opposite direction to model projections (Bierzychudek 1999).

PVAs can be used to show a range of outcomes within which most populations will tend (Chap. 7, this Vol.). This probabilistic approach explicitly acknowledges our uncertainty in the face of incomplete data and the unknowable future. Therefore, we suggest that PVAs should be used for projection, not prediction. PVAs also have an important role in making assumptions and data limitations transparent, and the logic of problem formulation internally consistent and unambiguous.

Acknowledgements. We thank Dorothy Mundell for helping us keep track of the ever-expanding literature on population viability analysis. Comments by Mark Burgman, Mario González-Espinosa, Mick McCarthy, David Matlaga, and Carl W. Weekley improved the manuscript. Our viewpoints on disturbance and demography have been influenced over the years by Orie Loucks, Lev Ginzburg, Jessica Gurevitch, Charles Janson, Jon Keeley, Jonathan Silvertown, Gerard Oostermeijer, Tom Kaye, Marc Abrams, Hal Caswell, Elena Alvarez-Buylla, Carol Horvitz, Phil Grime, Bill Platt, Teresa Valverde, Xavier Pico and many of the authors cited in this paper.

References

Abe S, Nakashizuka T, Tanaka H (1998) Effects of canopy gaps on the demography of the subcanopy tree *Styrax obassia*. J Veg Sci 9:787–796

Åberg P (1992) Size-based demography of the seaweed *Ascophyllum nodosum* in stochastic environments. Ecology 73:1488–1501

Akçakaya HR (2000a) Population viability analyses with demographically and spatially structured models. Ecol Bull 48:23–38

Akçakaya HR (2000b) Viability analyses with habitat-based metapopulation models. Popul Ecol 42:45–53

Alvarez-Buylla ER (1994) Density dependence and patch dynamics in tropical rain forests: matrix models applications to a tree species. Am Nat 143:155–191

Alvarez-Buylla ER, García-Barrios R (1991) Seed forest dynamics: a theoretical framework and an example from the neotropics. Am Nat 137:133–154

Alvarez-Buylla ER, García-Barrios R, Moreno CL, Martínez-Ramos M (1996) Demographic genetic models in conservation biology: applications and perspectives for tropical rain forest tree species. Annu Rev Ecol Syst 27:387–421

Batista WB, Platt WJ, Macchiavelli RE (1998) Demography of a shade-tolerant tree (*Fagus grandiflora*) in a hurricane-disturbed forest. Ecology 79:38–53

Bearlin AR, Burgman MA, Regan HM (1999) A stochastic model for seagrass (*Zostera muelleri*) in Port Phillip Bay, Victoria Australia. Ecol Model 118: 131–148

Beissinger SR, Westphal MI (1998) On the use of demographic models of population viability in endangered species management. J Wildl Manage 62:821–841

Bierzychudek P (1999) Looking backwards: assessing the projections of a transition matrix model. Ecol Appl 9:1278–1287

Bradstock RA, O'Connell MA (1988) Demography of woody plants in relation to fire: *Banksia ericifolia* L.f. and *Petrophile pulchella* (Schrad) R. Br. Aust J Ecol 13:505–518

Bradstock RA, Bedward M, Scott J, Keith DA (1996) Simulation of the effect of spatial and temporal variation of fire regimes on the population viability of a *Banksia* species. Conserv Biol 10:776–784

Bradstock RA, Bedward M, Kenny BJ, Scott J (1998) Spatially-explicit simulation of the effect of prescribed burning on fire regimes plant extinctions in shrublands typical of south-eastern Australia. Biol Conserv 86:83–95

Bullock JM, Clear Hill B, Silvertown J (1994) Demography of *Cirsium vulgare* in a grazing experiment. J Ecol 82:101–111

Burgman, MA, Gerard VA (1990) A stage-structured, stochastic population model for the giant kelp *Macrocystis pyrifera*. Mar Biol 105:15–23

Burgman MA, Lamont BB (1992) A stochastic model for the viability of *Banksia cuneata* populations: environmental, demographic, genetic effects. J Appl Ecol 29:719–727

Burgman MA, Ferson S, Akçakaya HR (1993) Risk assessment in conservation biology. Chapman and Hall, London

Canales J, Trevisan MC, Silva JF, Caswell H (1994) A demographic study of an annual grass (*Andropogon brevifolius* Schwarz) in burnt and unburnt savanna. Acta Oecol 15:261–273

Caswell H, Kaye TN (2001) Stochastic demography and conservation of an endangered perennial plant *(Lomatium bradshawii)* in a dynamic fire regime. Adv Ecol Res 32:1–49

Cipollini ML, Whigham DF, O'Neill J (1993) Population growth, structure, seed dispersal in the understory herb *Cynoglossum virginianum*: a population patch dynamics model. Plant Species Biol 8:117–129

Cipollini ML, Wallace-Senft DA, Whigham DF (1994) A model of patch dynamics, seed dispersal, and sex ratio in the dioecious shrub *Lindera benzoin* (Lauraceae). J Ecol 82:621–633

Crowley PH (1992) Resampling methods for computation-intensive data analysis in ecology and evolution. Annu Rev Ecol Syst 23:405–447

De Kroon H, Plaiser A, van Groenendael J (1987) Density-dependent simulation of the population dynamics of a perennial grassland species, *Hypochaeris radicata*. Oikos 50:3–12

Donohue K, Foster DR, Motzkin G (2000) Effects of the past the present on species distribution: land-use history demography of wintergreen. J Ecol 88:303–316

Downey PO, Smith JMB (2000) Demography of the invasive shrub Scotch broom (*Cytisus scoparius*) at Barrington Tops, New South Wales: insights for management. Aust Ecol 25:477–485

Efron B, Tibshirani RJ (1998) An introduction to the bootstrap. Chapman and Hall, New York

Ehrlén J (1995) Demography of the perennial herb *Lathyrus vernus*. II. Herbivory and population dynamics. J Ecol 83:297–308

Enright NJ, Marsula R, Lamont BB, Wissel C (1998a) The ecological significance of canopy seed storage in fire-prone environments: a model for resprouting shrubs. J Ecol 86:960–973

Enright NJ, Marsula R, Lamont BB, Wissel C (1998b) The ecological significance of canopy seed storage in fire-prone environments: a model for non-sprouting shrubs. J Ecol 86:946–959

Eriksson Å, Eriksson O (2000) Population dynamics of the perennial *Plantago media* in semi-natural grasslands. J Veg Sci 11:245–252

Ferson S (1991) RAMAS/stage 1.4. Manual. Applied Biomathematics, New York

Fieberg J, Ellner SP (2001) Stochastic matrix models for conservation and management: a comparative review of methods. Ecol Lett 4: 244–266

Fox GA, Gurevitch J (2000) Population numbers count: tools for near-term demographic analysis. Am Nat 156:242–256

Freckleton RP, Watkinson AR (1998) How temporal variability affects predictions of weed population numbers J Appl Ecol 35:340–344

Godínez-Alvarez H, Valiente-Banuet A, Valiente-Banuet L (1999) Biotic interactions and the population dynamics of the long-lived columnar cactus *Neobuxbaumia tetetzo* in the Tehuacán Valley, Mexico. Can J Bot 77:203–208

Golubov J, Mandujano MC, Franco M, Montaña C, Eguiarte LE, López-Portillo J (1999) Demography of the invasive woody perennial *Prosopis glulosa* (honey mesquite). J Ecol 87:955–962

Greenlee J, Kaye TN (1997) Stochastic matrix projection: a comparison of the effect of element and matrix selection methods on quasi-extinction risk for *Haplopappus radiatus* (Asteraceae). In: Kaye TN, Lister A, Love RM, Luoma DL, Meinke RJ, Wilson MV (eds) conservation management of native plants and fungi. Native Plant Society of Oregon, Corvalis, OR, pp 66–71

Gross K, Lockwood III JR, Frost CC, Morris WF (1998) Modeling controlled burning and trampling reduction for conservation of *Hudsonia montana*. Conserv Biol 12:1291–1301

Guàrdia R, Raventós J, Caswell H (2000) Spatial growth and population dynamics of a perennial tussock grass (*Achnatherum calamagrostis*) in a badland area. J Ecol 88:950–963

Hegazy AK (1992) Age-specific survival, mortality reproduction, and prospects for conservation of *Limonium delicatulum* J Appl Ecol 29:549–557

Hoffmann WA (1999) Fire and population dynamics of woody plants in a neotropical savanna: matrix model projections. Ecology 80:1354–1369

Horvitz CC, Schemske DW (1986) Seed dispersal and environmental heterogeneity in a neotropical herb: a model of population and path dynamics. In: Estrada A, Fleming EH (eds) Frugivores and seed dispersal. Junk, Dordrecht, pp 169–186

Horvitz CC, Schemske DW (1995) Spatiotemporal variation in demographic transitions of a tropical understory herb: projection matrix analysis. Ecol Monogr 65:155–192

Kalisz S, McPeek MA (1993) Extinction dynamics, population growth and seed banks. Oecologia 95:314–320

Kaye TN, Pendergrass KL, Finley K, Kauffman JB (2001) The effect of fire on the population viability of an endangered prairie plant. Ecol Appl 11:1366–1380

Lennartsson T (2000) Management and population viability of the pasture plant *Gentiabella campestris*: the role of interactions between habitat factors. Ecol Bull 48:111–121

Lennartsson T, Oostermeijer JGB (2001) Demographic variation and population viability in *Gentianella campetris*: effects of grassland management and environmental stochasticity. J Ecol 89:451–463

Lonsdale WM, Braithwaite RW, Lane AM, Farmer J (1998) Modelling the recovery of an annual savanna grass following a fire-induced crash. Aust J Ecol 23:509–513

Manders PT (1987) A transition matrix model of the population dynamics of the Clanwilliam cedar (*Widdringtonia cedarbergensis*) in natural stands subject to fire. For Ecol Manage 20:171–186

Mandujano MC, Montaña C, Franco M, Golubov J, Flores-Martínez A (2001) Integration of demographic annual variability in a clonal desert cactus. Ecology 82:344–359

Manly BFJ (1991) Randomization Monte Carlo methods in biology. Chapman and Hall, New York

Martínez-Ramos M, Alvarez-Buylla ER (1995) Seed dispersal and patch dynamics in tropical rain forests: a demographic approach. Ecoscience 2:223–229

Maschinski J, Frye R, Rutman S (1997) Demography and population viability of an endangered plant species before and after protection from trampling. Conserv Biol 11:990–999

McCarthy MA, Burgman MA (1995) Coping with uncertainty in forest wildlife planning. For Ecol Manage 74:23–36

McCarthy MA, Thompson C (2001) Expected minimum population size as a measure of threat. Anim Conserv 4:351–356

McCarthy MA, Possingham HP, Gill AM (2001) Using stochastic dynamic programming to determine optimal fire management for *Banksia ornata*. J Appl Ecol 38:585–592

Menges ES (1998) Evaluating extinction risks in plant populations. In: Fiedler PL, Kareiva PM (eds) Conservation biology for the coming decade. Chapman and Hall, New York, pp 59–65

Menges ES (2000a) Applications of population viability analysis in plant conservation. Ecol Bull 48:73–84

Menges ES (2000b) Population viability analysis in plants: challenges and opportunities. Trends Ecol Evol 15:51–62

Menges ES, Dolan RW (1998) Demographic viability of populations of *Silene regia* in midwestern prairies; relationships with fire management, genetic variation, geographic location, population size, and isolation. J Ecol 86:63–78

Menges ES, Guerrant E, Havens K, Maunder M (2002) What is the effect of seed collection on extinction risk? Saving the pieces: the value, limits and practice of off-site plant conservation in support of wild diversity. Island Press, Covello, CA (in press)

Nantel P, Gagnon D, Nault A (1996) Population viability analysis of American ginseng and wild leek harvested in stochastic environments. Conserv Biol 10:608–621

Olmsted I, Alvarez-Buylla ER (1995) Sustainable harvesting of tropical trees: demography and matrix models of two palm species in Mexico. Ecol Appl 5:484–500

Oostermeijer JGB (2000) Population viability analysis of the rare *Gentiana pneumonnthe*: importance of demography, genetics, and reproductive biology. In: Young A, Clarke G (eds) Genetics, demography, and viability of fragmented populations. Cambridge University Press, Cambridge, pp 313–334

Oostermeijer JGB, Burgman ML, de Boer ER, den Nijs HCM (1996) Temporal and spatial variation in the demography of *Gentiana pneumonanthe*, a rare perennial herb. J Ecol 84:153–166

Parker IM (2000) Invasion dynamics of *Cytisus scoparius*: a matrix model approach. Ecol Appl 10:726–743

Pascarella JB, Horvitz CC (1998) Hurricane disturbance and the population dynamics of a tropical, understory shrub: megamatrix elasticity analysis. Ecology 79:547–563

Peters CM (1991) Plant demography and the management of a tropical rain forest resource: a case study of *Brosimum alicastrum* in Mexico. In: Gómez-Pompa A, Whitmore TC, Hadley M (eds) Rain forest regeneration and management. UNESCO- man and biosphere series, vol 6. Parthenon, Paris, pp 265–272

Pfab MF, Witkowski ETF (2000) A simple population viability analysis of the critically endangered *Euphorbia clivicola* R.A. Dyer under four management scenarios. Biol Conserv 96:263–270

Pickett STA, White PS (1985) The ecology of natural disturbance and patch dynamics. Academic Press, Orlando, FL

Pinard M (1993) Impacts of stem harvesting on populations of *Iriartea deltoidea* (Palmae) in an extractive reserve in Acre, Brazil. Biotropica 25:2–14

Quintana-Ascencio PF, González Espinosa M, Ramírez Marcial N, Domínguez Vázquez G, Martínez Icó M (1996a) Soil seed banks and regeneration of tropical rain forest from milpa fields at the Selva Lacandona, Chiapas, Mexico. Biotropica 28:192–209

Quintana-Ascencio PF, Menges ES (1996b) Inferring metapopulation dynamics from patch-level incidence of Florida scrub plants. Conserv Biol 10:1210–1219

Quintana-Ascencio PF, Menges ES, Weekley CW (2003) A fire-explicit population viability analysis of *Hypericum cumulicola* in Florida rosemary scrub. Conserv Biol 17:1–18

Reed J M, Mills LS, Dunning Jr JB, Menges ES, McKelvey KS, Frye R, Beissinger S, Anstett MC, Miller P (2002) Use and emerging issues in population viability analysis. Conserv Biol 16:7–19

Rydgren K, de Kroon H, Okland RH, van Groenendael J (2001) Effects of fine-scale disturbances on the demography and population dynamics of the clonal moss *Hylocomium splendens*. J Ecol 89:395–405

Santos R (1993) Plucking or cutting *Gelidium sesquipedale*? A demographic simulation of harvest impact using a population projection matrix model. Hydrobiologia 260/261:269–276

Satterthwaite WH, Menges ES, Quintana-Ascencio PF (2002) Assessing scrub buckwheat population viability in relation to fire using multiple modeling techniques. Ecological Applications 12:1672–1687

Shimada M, Ishihama F (2000) Asynchronization of local population dynamics and the persistence of a metapopulation: a lesson from an endangered composite plant, *Aster kantoensis*. Popul Ecol 42:63–72

Silva JF, Raventos J, Caswell H, Trevisan MC (1991) Population responses to fire in a tropical savanna grass, *Andropogon semiberbis*: a matrix model approach. J Ecol 79:345–356

Silvertown J, Franco M, Menges ES (1996) Interpretation of elasticity matrices as an aid to the management of plant populations for conservation. Conserv Biol 10:591–597

Soehartono T, Newton AC (2001) Conservation and sustainable use of tropical trees in the genus *Aquilaria* II. The impact of gaharu harvesting in Indonesia. Biol Conserv 97:29–41

Ticktin T, Nantel P, Ramirez F, Johns T (2002) Effects of variation on harvest limits for non-timber forest species in Mexico. Conserv Biol 16:691–705

Valverde T, Silvertown J (1997a) An integrated model of demography, patch dynamics and seed dispersal in a woodland herb, *Primula vulgaris*. Oikos 80:67–77

Valverde T, Silvertown J (1997b) A metapopulation model for *Primula vulgaris*, a temperate forest understory herb. J Ecol 85:193–210

Waite S, Hutchings MJ (1991) The effects of different management regimes on the population dynamics of *Ophrys sphegodes*: analysis and description using matrix models. In: Wells TCE, Willems JH (eds) Population ecology of terrestrial orchids. SPB Academic, The Hague, pp 161–175

Watkinson AR (1990) The population dynamics of *Vulpia fasciculata*: a nine-year study. J Ecol 78:196–209

Watkinson AR, Lonsdale WM, Andrew MH (1989) Modeling the population dynamics of an annual plant *Sorghum intrans* in the wet-dry tropics. J Ecol 77:162–181

Watkinson AR, Freckleton RP, Forrester L (2000) Population dynamics of *Vulpia ciliata*: regional, patch, and local dynamics. J Ecol 88:1012–1029

Werner PA, Caswell H (1977) Population growth rates age versus stage-distribution models for teasel (*Dipsacus sylvestris* Huds.). Ecology 58:1103–1111

Wiegand K, Jeltsch F, Ward D (1999) Analysis of the population dynamics of *Acacia* trees in the Negev desert, Israel with a spatially-explicit computer simulation model. Ecol Model 117:203–224

12 Projecting the Success of Plant Population Restoration with Viability Analysis

T.J. Bell, M.L. Bowles, and A.K. McEachern

12.1 Introduction

Conserving viable populations of plant species requires that they have high probabilities of long-term persistence within natural habitats, such as a chance of extinction in 100 years of less than 5 % (Menges 1991, 1998; Brown 1994; Pavlik 1994; Chap. 1, this Vol.). For endangered and threatened species that have been severely reduced in range and whose habitats have been fragmented, important species conservation strategies may include augmenting existing populations or restoring new viable populations (Bowles and Whelan 1994; Chap. 2, this Vol.). Restoration objectives may include increasing population numbers to reduce extinction probability, deterministic manipulations to develop a staged cohort structure, or more complex restoration of a desired genetic structure to allow outcrossing or increase effective population size (DeMauro 1993, 1994; Bowles et al. 1993, 1998; Pavlik 1994; Knapp and Dyer 1998; Chap. 2, this Vol.). These efforts may require translocation of propagules from existing (in situ) populations, or from ex situ botanic gardens or seed storage facilities (Falk et al. 1996; Guerrant and Pavlik 1998; Chap. 2, this Vol.).

Population viability analysis (PVA) can provide a critical foundation for plant restoration, as it models demographic projections used to evaluate the probability of population persistence and links plant life history with restoration strategies. It is unknown how well artificially created populations will meet demographic modeling requirements (e.g., due to artificial cohort transitions) and few, if any, PVAs have been applied to restorations. To guide application of PVA to restored populations and to illustrate potential difficulties, we examine effects of planting different life stages, model initial population sizes needed to achieve population viability, and compare demographic characteristics between natural and restored populations. We develop and compare plant population restoration viability analysis (PRVA) case studies of two plant species listed in the USA for which federal recovery planning calls for

Ecological Studies, Vol. 165
C.A. Brigham, M.W. Schwartz (Eds.)
Population Viability in Plants

population restoration: *Cirsium pitcheri*, a short-lived semelparous herb, and *Asclepias meadii*, a long-lived iteroparous herb.

12.1.1 Developing PVA for Plants

Analysis of stage-structured transition matrix models of population growth based on demographic monitoring is the basic tool in plant PVA; it is commonly used to determine a finite rate of population growth (lambda, λ) and to project change in size and extinction risk of naturally occurring plant populations (e.g., Menges 1990, 1991, 1998, 2000; Fiedler et al. 1998; Chap. 6, this Vol.). This application is attractive for restorations that otherwise would have unknown potential for persistence. Small populations exposed to relatively high amounts of environmental stochasticity are more vulnerable to extinction because increased fluctuations in population growth rate can more frequently lead to years in which the population falls below a demographic extinction threshold (Gilpin and Soulé 1986; Goodman 1987; Chap. 2, this Vol.). Thus, for plants living in variable environments, modeling that incorporates environmental stochasticity provides a more realistic PVA than traditional deterministic approaches, but requires more data (Shaffer 1987; Lande 1993; Mangel and Tier 1994; Beissinger and Westphal 1998; Menges 2000; Chaps. 6, 11, this Vol.). Such applications may be useful for small restorations under variable environmental conditions with habitats of uncertain suitability, but may be limited by lack of information about specific environmental variability. Use of variance to mean ratios (V/M) as a measure of the relative amount of environmental stochasticity that populations are experiencing provides a solution, at least in part, to this problem (Menges 1998). Stochastic modeling can also be partitioned among different management treatments (e.g., experimental burning) to allow more useful comparative projections (Beissinger and Westphal 1998; Menges 2000; Kaye et al. 2001; Chaps. 6, 11, this Vol.). The effects of the three components of stochasticity (environmental, demographic, and genetic) can operate simultaneously on demographic variability but cannot be separated without experimental manipulations.

Elasticity analysis of contributions of different life stages to population growth can help guide management application and differentiate among approaches used for plants with different life history strategies (Crouse et al. 1987; Crowder et al. 1994; Doak et al. 1994; Heppell et al. 1994; Silvertown et al. 1996; Chap. 6, this Vol.). However, elasticities for some species may not follow expected models (e.g. *Calochortus*, Fiedler et al. 1998), and could focus management on inappropriate life stages of declining populations (Beissinger and Westphal 1998; Menges 1998, 2000; Chap. 9. this Vol.) or restorations. For this reason, modeling of multiple populations helps interpret elasticities.

12.1.2 PRVA Applications

Although PVA has been used to assess more than 100 naturally occurring plant populations, including many endangered and threatened plants (Fiedler 1987; Menges 2000; Chap. 1, this Vol.), its application in restoration ecology is in its infancy. Rarely have restorations been demographically modeled or placed in an experimental or theoretical framework, and no formal PRVAs using integrated modeling have been reported for North American endangered or threatened plant species (Pavlik 1994, 1996; Guerrant and Pavlik 1998).

Non-integrated approaches to PVA can provide information and a framework for restoration by developing short-term and long-term restoration objectives (Pavlik 1994, 1996). Short-term objectives for restored populations may include completion of life cycle, obtaining a particular life stage distribution, attaining a stable or positive growth rate, reaching a desired level of genetic diversity, having a seed/ovule ratio greater than one, increasing population area, achieving normal seed bank density, or utilizing native pollinators. Long-term restoration goals are the same as PVA goals: attaining and maintaining minimum viable population (MVP) size as defined by a 95% probability of survival after 100 years, as well as re-establishment of MVP size following perturbation (Pavlik 1996). In highly stochastic environments, comparatively larger initial population sizes are required to offset the increased risk of extinction.

Non-integrated demographic analyses (Pavlik 1994) have been used to examine cohort survivorship of restorations of the herbaceous perennials *Pediocactus knowltonii* (Olwell et al. 1990; Cully 1996) and *Isotria medeoloides* (Brumback and Fyler 1996) and the pine *Pinus torreyana* (Ledig 1996). Experimental treatments were used to assess effects of competition and burning on reintroduction of the shrub *Conradina glabra* (Gordon 1996), and herbs *Amsinckia grandiflora* (Pavlik et al. 1993; Pavlik 1994), *Asclepias meadii* (Bowles et al. 1998), and *Stephanomeria malheurensis* (Guerrant and Pavlik 1998). Performances of different seed sources were compared in a *Cirsium pitcheri* restoration (Bowles et al. 1993; Bowles and McBride 1996). For *Amsinckia grandiflora*, demographic monitoring was used in three phases of reintroduction, including an experimental phase to assess the response of the species to grass competition and to compare the performance of different propagule sources; a population enhancement phase using information from the first phase; and a third phase to evaluate the population's self sustainability in the absence of management, as well as its level of genetic diversity (Pavlik 1994; Guerrant and Pavlik 1998). These nonintegrated studies provide valuable but limited information about restorations and do not allow projection of MVP needed to assess whether long-term goals will be reached.

12.1.3 Theoretical Framework

Few theoretical models have been applied to plant population restoration. To examine the effect of size/stage class on founding success, Guerrant (1996) used published natural transition matrices to model extinction rates for three perennial plants with differing life histories: the woody iteroparous *Astrocaryum mexicanum*, the semelparous herb *Dipsacus sylvestris*, and the iteroparous herb *Calochortus pulchellus*. In these projections, all three species exhibited a decrease in extinction risk and accelerated population growth with initial establishment of larger or older plants relative to beginning with seedling establishment. Thus, Guerrant (1996) concluded that populations founded with seeds had a significantly higher extinction risk than those founded by larger life stages, and provided a testable theoretical basis for founding population restorations. Guerrant and Fiedler (2003) also modeled the "demographic cost" of reintroduction using studies of seven species with a range of life histories. They projected that substantial losses of individuals would be reduced by outplanting larger individuals, which reduces extinction probability. Therefore, in this chapter we investigate how planting different life stages affects the demographic cost of restoration and the initial population size needed to achieve population viability. Because so little is known about the viability of plant restorations, we compare demographic characteristics of a restored population against a viable natural population as a benchmark of success.

12.2 PRVA Case Studies

To address plant PRVA issues concerning MVP, initial planting size, life stage used for outplanting, and comparison with natural populations, we use case studies of two species in the USA with contrasting life history strategies that result in different restoration applications and PRVA outcomes (Table 12.1). Both species are classified by the federal authorities as threatened and have been the focus of long-term restoration projects coupled with demographic monitoring. Pitcher's thistle (*Cirsium pitcheri* [Torrey ex Eaton] T. and G.) is a short-lived monocarpic herb endemic to sand dunes of the western Great Lakes, where it colonizes successional habitats and requires frequent cohort replacement to maintain populations (McEachern et al. 1994). Mead's milkweed (*Asclepias meadii*) is a long-lived iteroparous herb of late-successional midwestern tallgrass prairie and glades (Bowles et al. 1998). Natural reproduction is infrequent in this species, and adult longevity is apparently a large component of population maintenance. Absence of complete life stage transition data prevented a complete PVA for this species, but we present preliminary analysis of population viability.

Table 12.1. Different habitats and biological characteristics result in contrasting restoration strategies for *Cirsium pitcheri* and *Asclepias meadii*

Species	*Cirsium pitcheri*	*Asclepias meadii*
Habitat	Early- successional shoreline dunes of the western Great Lakes	Late-successional prairies and glades of the Midwest
Life history	Short-lived monocarpic perennial with frequent cohort replacement	Long-lived perennial with infrequent cohort establishment
Genetic diversity	Comparatively low genetic diversity within and among populations	High genetic diversity within populations, low genetic differentiation among populations
Breeding system	Self compatible with mixed mating	Obligate outcrossing with different genotypes required for reproduction
Restoration strategies	Establish natural stage structure through repeated outplanting, manage for growth and reproduction	Establish adult cohort, maximize genetic diversity, manage for adult persistence, flowering and seed production

12.3 Pitcher's Thistle

12.3.1 Species Background

Pitcher's thistle (*Cirsium pitcheri*) is a threatened (Harrison 1988) species endemic to the western Great Lakes shoreline, and is extirpated from Illinois. *Cirsium pitcheri* inhabits shoreline sand dunes and beaches, where wind-generated disturbance processes maintain open sand and successional vegetation that allow successful seedling establishment (McEachern 1992). This semelparous species has no capacity for vegetative spread; plants reach a threshold size and flower after 3–8 years, disperse seeds, and then die (Loveless 1984). *Cirsium pitcheri* has a mixed mating system, with 35–88 % outcrossing through insect pollination, which produces higher seed set than with self-pollination (Keddy and Keddy 1984; Loveless 1984). Population structures of *Cirsium pitcheri* are temporally variable, depending upon cohort demographic histories and successional stages of vegetation, and this species may depend upon metapopulation processes for long-term persistence (McEachern et al. 1994). Although seed dormancy occurs in *Cirsium pitcheri* (Chen and Maun 1998), Rowland and Maun (2001) found little evidence for a seed bank in two Canadian populations. We observed consumption of flower heads of this species by eastern white tailed deer, as well as seed by small mammals, and assume that both could substantially reduce seedling numbers. Exclusion of insect seed predators from seed heads may increase fecundity (Louda 1994).

In accordance with recovery planning and to test whether *Cirsium pitcheri* could be successfully restored, a population restoration began in former habitat at Illinois Beach State Park in 1991(Bowles et al. 1993; McEachern et al. 1994; Bowles and McBride 1996; Pavlovic et al. 2003). The park is located along the west shoreline of Lake Michigan 70 km north of Chicago. It occupies a 1.5-km wide sand deposit with low dunes (up to 3 m), which extends for over 20 km. Secondary dunes were found to replicate appropriate habitat for this species and appeared to be free from problems of shoreline erosion and recreational impacts (Bowles et al. 1993). Two localities separated by <1 km were used to establish population units first south, and then north of the Dead River, which drains into Lake Michigan.

Cirsium pitcheri restoration propagules were grown from seeds collected from natural Indiana, southern Wisconsin, and southern Michigan thistle populations. Thistle cohorts were propagated for one season, over-wintered, and then translocated to the restoration site. Over 100 plants were established in the south unit by 1993, and the first two plants flowered in 1994. Naturally recruited seedlings from flowering plants are now replacing these artificial cohorts, and the first of these naturally recruited plants flowered in 1998.

Plant morphology, performance, and genetics differed between Indiana and Wisconsin seed sources when planted at Illinois Beach. Indiana plants had larger cotyledons and greater growth and survivorship than Wisconsin plants (Bowles et al. 1993; Bowles and McBride 1996). Random amplified polymorphic DNA analysis (RAPDs) also separated these seed sources (K. Havens, unpublished data). In addition, differences in population structure occurred north and south of the Dead River. The younger population north of the river had significantly lower relative abundance of flowering, large juvenile, and seedling plants, while plants south of the Dead River had lower mortality and lower abundance of small juveniles. In addition, naturally recruited plants had not yet flowered north of the Dead River. These differences suggest a greater rate of population growth south of the Dead River. The lack of representation by all demographic stages prevented us from developing an integrated PVA for the restoration north of the Dead River.

12.3.2 Restoration Viability Analysis

Our PRVA assessed the viability of the Illinois Beach restoration by comparing its population size to the MVP size required to achieve an extinction probability (P_e) of <5 % and by determining whether average, as well as annual, life stage transition matrices represented a growing (λ>1) or stable (λ=1) population. In addition, we determined the initial number of outplantings required to reach P_e<5% and how planting seeds vs. juveniles influences the initial numbers needed. We also determined whether demographic characteristics of the restoration were comparable to a natural population.

12.3.2.1 Demographic Modeling

Demographic monitoring data from 1994 through 2000 was used to develop a matrix model for the Illinois Beach State Park population restored south of the Dead River (Appendix 12.1; see Chap. 6, this Vol., for details on matrix modeling approaches). The matrix model was used to evaluate which stages to manipulate in order to increase the population growth rate, to evaluate the effect of environmental stochasticity on the MVP, and to simulate the effects of population size and initial transplant stage on the probability of extinction and MVP. Elasticities and λ for each matrix were determined using RAMAS/Stage (Ferson 1994). Popproj2 (Menges 1998) was used to calculate P_e using 1,000 runs for 100 years with a quasi-extinction population size of 1. Popproj2 introduces environmental stochasticity into the simulations by randomly choosing among transition matrices.

We analyzed five life stages for *Cirsium pitcheri* including seed, seedling, small juvenile, large juvenile and flowering plants, which may arise either from artificial transplant cohorts or as naturally recruited seedlings from flowering transplanted plants or their descendents (Fig. 12.1). Because greenhouse-grown transplants and naturally recruited individuals have different transition frequencies (Appendix 12.1), and transplants that flower earlier have fewer seed heads (Fig. 12.2), the restoration transition matrices incorporate transplants and naturally recruited individuals into the matrix separately (Fig. 12.1). Large juveniles differ from small juveniles by having attained a threshold flowering size measured by root crown size and a leaf area index. For *Cirsium pitcheri* plants that survived to the flowering stage, individuals remained in the small juvenile stage for 2 years (2.0±1.0 SD, range 1–6), were a large juvenile for only 1 year (1.0±0.7 SD, range 0–2), flowered in their fourth year (4.1±1.1 SD, range 2–7), and died. The transition from sown seed to seedling (G_{51}) was estimated by dividing the number of seedlings observed by the number of seeds produced the previous year. The number of seeds per flowering plant was estimated by multiplying the number of flower heads for a given year by the average number of seeds per flower head. We assumed that the restoration population, like natural populations (Rowland and Maun 2001), had no seed bank.

Although transplanting began in 1991 and naturally recruited plants began to appear in 1995, we could only develop complete matrices for 1998→1999 and 1999→2000 transitions because too few individuals occurred in each stage for previous years. In order to develop an average matrix, transition frequencies were calculated from transition numbers summed over the years

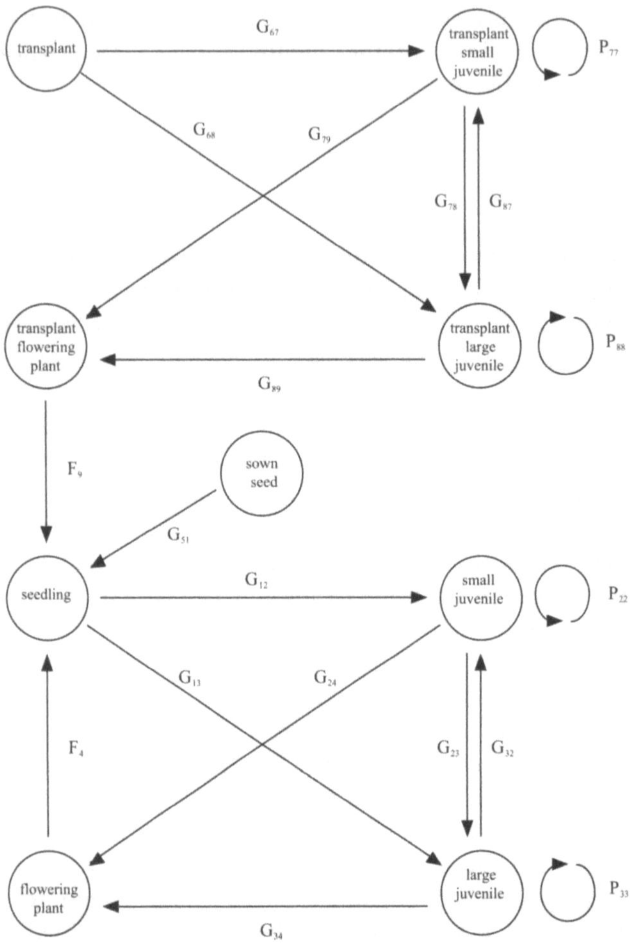

Fig. 12.1. Generalized stage-based population dynamics model for Pitcher's thistle (*Cirsium pitcheri*). *Circles* Stages, *arrows* possible transitions between stages. Transition labels indicate growth (G) from one stage to the next, the probability (P) of remaining in that stage and fecundity (F)

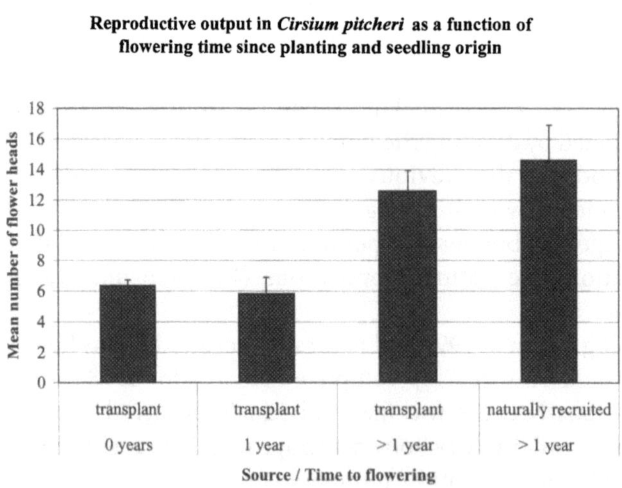

Fig. 12.2. Reproductive output is highest for flowering plants produced from naturally recruited seedlings or from transplants that require >1 year to flower (F=14.176, p<0.0001)

1991–2000. Thus the simulations for the Illinois Beach restoration included the three matrices in Appendix 12.1.

To determine whether the restored Illinois Beach population was demographically similar to a natural population, the demography of the restored Illinois Beach population was compared with that of a natural population at West Beach, Indiana Dune National Lakeshore (McEachern 1992; McEachern et al. 1994). The sample used for West Beach demographic monitoring included approximately 23% of a population of about 1,900 plants. Appendix 12.2 includes projection matrices of transition frequencies for the natural *Cirsium pitcheri* population at Indiana Dunes West Beach from 1988 to 1993.

Annual transitions were generally similar but variable for both the Illinois Beach restoration and natural West Beach population. A small percentage of seedlings at Illinois Beach became large juveniles the next year but not at the

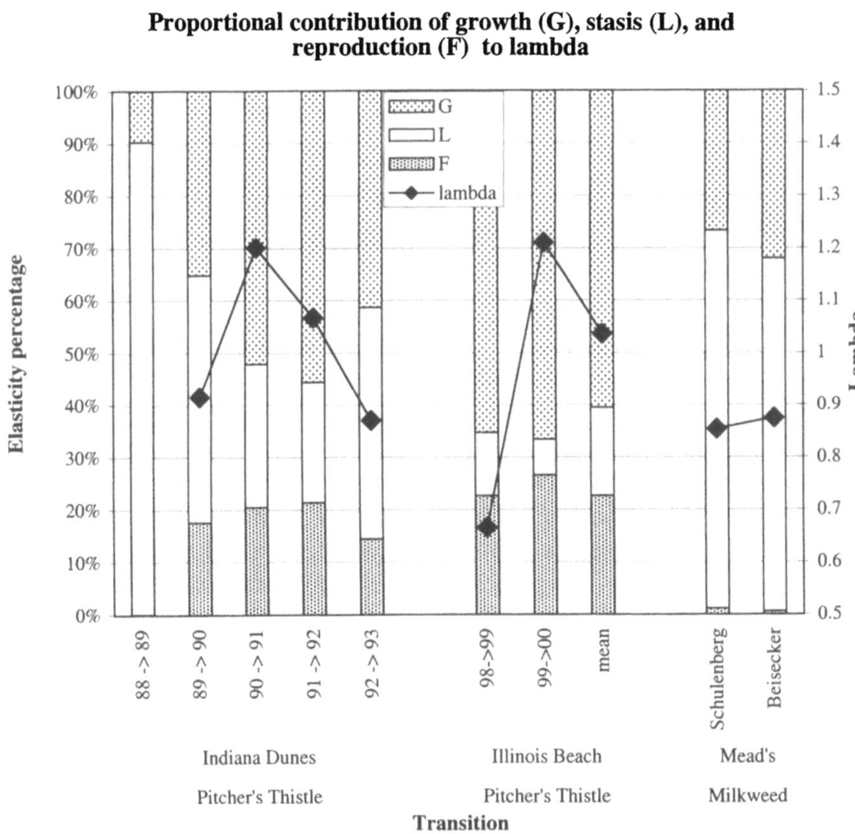

Fig. 12.3. Proportional contribution of growth (*G*), stasis (*L*) and reproduction (*F*) to λ for *Cirsium pitcheri* natural (Indiana Dunes) and restoration (Illinois Beach) populations, and *Asclepias meadii* restoration populations

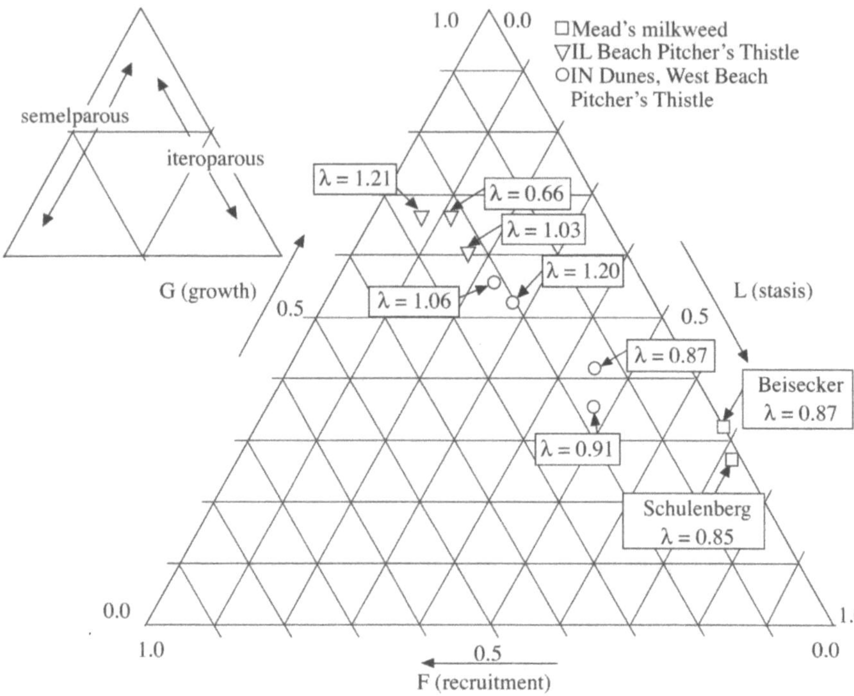

Fig. 12.4. Ordination of growth (*G*), stasis (*L*) and reproduction (*F*) elasticities for *Cirsium pitcheri* restoration (Illinois Beach, *triangles*) and natural (Indiana Dunes, *circles*) populations, and *Asclepias meadii* restoration populations (*squares*)

Indiana Dunes population. Some large juveniles at Indiana Dunes shrank to small juveniles the next year but not at Illinois Beach.

There were variations in λ from year to year in both the restoration and natural *Cirsium pitcheri* populations (Fig. 12.3). Restoration population growth rates ranged from 0.66 to 1.21, with λ=1.03 for the average matrix (Figs. 12.3, 12.4). These are similar to the values of λ calculated for the natural Indiana Dunes population, which ranged from 0.87 to 1.21 (Figs. 12.3, 12.4). One Indiana Dunes matrix representing the transitions from 1988 to 1989 had a λ=0 because no seedlings survived (Appendix 12.2) apparently because 1988 had the lowest rainfall record in 50 years.

12.3.2.2 Elasticity Analysis

Following Silvertown et al. (1996), ordination of elasticities was used to evaluate the effect that changes to groups of stage transitions have on λ (see Chaps. 6, 11, this Vol., for an in-depth discussion of elasticities). We grouped elasticities into the *G* region representing the combined effects of changes to

Table 12.3. **a** *Cirsium pitcheri* variance-to-mean ratios for Illinois Beach State Park restoration population; median (all elements)=0.1695, median (recruited only)=0.2069. **b** Variance-to-mean ratios for Indiana Dunes National Lakeshore West Beach population; median=0.1044. *R* Naturally recruited individuals, *T* transplanted individuals

a Illinois Beach restored population

	State (t) → Naturally recruited					Transplants			
Fate (t+1)	R seedling	R small juvenile	R large juvenile	R flowering	Sown seed	T "seedling"	T small juvenile	T large juvenile	T flowering
R seedling				0.4939	0.0029				1.4161
R small juvenile	0.3795	0.3493							
R large juvenile	0.1687	0.2081	0.0313						
R flowering		0.0474	0.2057						
Sown seed									
T "seedling"									
T small juvenile						0.1661	0.1181	0.1376	
T large juvenile						1.0728	0.1086	0.1703	
T flowering							0.0510	0.2063	

Median (all elements) = 0.1695
Median (recruited only) = 0.2069

b Indiana Dunes natural population

Fate (t+1)	State (t)→ Seedling	Small juvenile	Large juvenile	Flowering
Seedling				6.4924
Small juvenile	0.4393	0.0266	0.0564	
Large juvenile		0.1852	0.0740	
Flowering		0.1349	0.0665	

Median = 0.1044

growth on λ, the L region representing the combined effects of changes to stasis and retrogression on λ, and the F region representing the combined effects of changes to recruitment on λ (Table 12.2).

Although most *Cirsium pitcheri* G/L/F elasticity ratios occur in the semelparous region (Fig. 12.4), several were positioned toward the region where Silvertown et al. (1996) tended to find iteroparous herbs. For *Cirsium pitcheri*, λ tends to increase with increasing F in both the natural Indiana Dunes and restored Illinois Beach populations (Figs. 12.3, 12.4). When Silvertown et al. (1996) ordinated 16 populations of *Cirsium vulgare* (semelparous) and 15 *Pedicularis furbishiae* populations (iteroparous), λ also increased with increasing contribution of F. They interpreted this to mean that the best management strategy would focus on increasing recruitment. This suggests we should be able to increase *Cirsium pitcheri* λ by increasing recruitment, which makes sense for a semelparous perennial because increasing longevity only (i.e. without growth) does not increase recruitment as it does in iteroparous plants (Silvertown et al. 1996). However, managing for both survivorship and growth may also increase fecundity.

12.3.2.3 Variance/Mean Ratios

High levels of environmental stochasticity are expected for *Cirsium pitcheri* because it inhabits a highly dynamic habitat. The median Illinois Beach V/M was 0.17 for all matrix elements (0.03–0.38 for recruited non-fecundity elements and 0.49 for recruited fecundity) and the West Beach V/M median was 0.10 (0.03–0.44 for non-fecundity elements and 6.49 for fecundity) (Table 12.3).

The variance-to-mean ratio for the *Cirsium pitcheri* restoration (0.17) was higher than for the natural population (0.10) and both corresponded to very strong environmental stochasticity. In comparison, only one of eleven species

Table 12.2. Generalized matrix for *Cirsium pitcheri*, grouping elasticities into growth (G), stasis (L) and reproduction (F) regions

	State (year t)→			
Fate (year t+1)	Seedling	Small juvenile (crown <1 cm)	Large juvenile (crown >1 cm)	Flowering
Seedling		F=fecundity, recruitment		
Small juvenile		L=stasis		Shrinking
Large juvenile	G=growth			
Flowering				

analyzed by Menges (1998) had a median *V/M* greater than 0.17. Although seven of these species had populations with *V/M* greater than 0.10, all but one species had at least one population with median *V/M* below 0.10. The high variation, especially in non-fecundity elements, requires a relatively high MVP size to reduce extinction probability.

The effect of environmental stochasticity on MVP decreases as λ increases (Menges 1998). Populations with $\lambda<1$ were sensitive to $V/M<0.0025$, while populations with $\lambda\approx1$ were sensitive to $V/M>0.005$ and populations with $\lambda>1.15$ were sensitive to $V/M>0.14$. Three of the five transition matrices for the natural *Cirsium pitcheri* population had $\lambda<1$ and average $\lambda=0.96$, indicating that this population, as well as the restoration (average $\lambda=1.03$), are very sensitive to environmental stochasticity.

12.3.2.4 Minimum Viable Population Estimates

We used the matrix model to simulate the effects of population size on P_e and MVP. Extinction probability varied as a function of initial simulated population size for both populations of *Cirsium pitcheri* (Fig. 12.5). The minimum

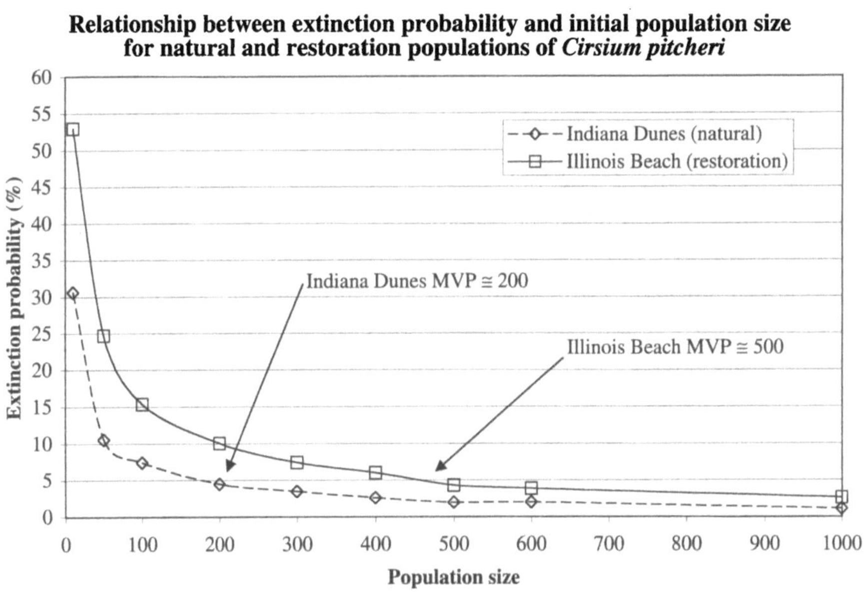

Fig. 12.5. Extinction probabilities are lower for the Indiana Dunes population than for Illinois Beach restoration of *Cirsium pitcheri.* Three empirically derived transitions (Appendix 12.1) from the Illinois Beach State Park restoration were alternatively chosen with equal probability each year of each simulation. For the five Indiana Dunes matrices (Appendix 12.2), the drought transition matrix, 1988->1989, had a probability of 0.132 and $p=0.217$ for all other matrices

Table 12.4. Minimum viable population (*MVP*) size projections (at P_e=5 %) is greater for the Illinois Beach State Park restoration of *Cirsium pitcheri* than for the Indiana Dunes West Beach natural population. Observed stage distribution and population size are smaller than MVP for the Illinois Beach restoration

	Observed stage distribution		Projected MVP and distribution ($P_e \leq 5$ %)	
Stages	Indiana Dunes West Beach average	Illinois Beach 2000	Indiana Dunes West Beach	Illinois Beach
Seedlings	222 (50 %)	58 (60 %)	100	250
Small juveniles	141 (32 %)	13 (14 %)	60	150
Large juveniles	34 (8 %)	11 (11 %)	20	50
Flowering plants	43 (10 %)	14 (15 %)	20	50
Total	440	96	200	500

population size needed for the restored population to be viable for *Cirsium pitcheri*, as indicated by P_e<5 %, was ~500 for the Illinois Beach State Park restoration, and ~200 for the Indiana Dunes West Beach natural population (Table 12.4).

Simulation methods used to estimate MVP size differed for the restoration and natural population. Each of the three Illinois Beach matrices had an equal probability (0.333) of being chosen in the simulation. The Indiana Dunes drought transition matrix, 1988→1989, had a probability of 0.132, based on the proportion of years the Palmer drought severity index was equal to or lower than the index value that occurred in the drought year of 1988 (e.g., Karl and Knight 1985). For all other Indiana Dunes matrices p=0.217. As the frequency of the drought year matrix increases in the simulation, extinction probability increases exponentially. For example, estimated MVP size for the Indiana Dunes population was approximately 10,000 when drought year matrix frequency was equal to all other matrix frequencies (p=0.20) compared to MVP size ~200 for our simulation (Table 12.4). The difference between these two MVPs reflects the increased year-to-year population size variability as a result of stochastic events, such as drought (see discussion in Chap. 11, this Vol.).

Estimation of MVP agrees with *V*/*M*. Since the natural Indiana Dunes population has a lower *V*/*M* than the Illinois Beach restoration, it is not surprising that the MVP for Indiana Dunes (200) is lower than for Illinois Beach (500) (Fig. 12.5).

We suspect that the Illinois Beach MVP is higher than Indiana Dunes MVP because the 1998→1999 transition matrix, for which λ=0.66, has a high frequency in the Illinois Beach simulation. We had no rationale for reducing the frequency of the 1998→1999 matrix in the simulation as we did for the Indiana

Dunes drought matrix because both 1998 and 1999 had Palmer drought severity indices that indicated normal rainfall. We expect estimated restoration MVPs to become more similar to the natural population as monitoring continues and the number of transition matrices in the simulation increases.

Because the restoration population size of 96 is below the estimated MVP, more than 400 transplants are required to increase population numbers to a viable level (Table 12.4). Matrix models for the Illinois Beach State Park restoration of *Cirsium pitcheri* were used to compare the effects of initial transplant number and seed number on P_e. Approximately 1,600 seedlings would need to be planted to establish a viable population compared to approximately 250,000 seeds, illustrating the increased demographic cost (Guerrant and Fiedler 2002) of planting seeds (Fig. 12.6). Using seeds to establish a population of *Cirsium pitcheri* is inefficient because high seed mortality in the restoration results in frequencies for transition from seed to seedling ranging from 0.0032 to 0.0086 (see sown seed transitions, Appendix 12.1).

The number of transplants used to create a viable restoration at Illinois Beach (Fig. 12.6) differs from the MVP for the Illinois Beach restoration (Table 12.4). This is because in the transplant simulation all individuals entering the restoration were in the same transplant "seedling" stage and needed to mature through the larger transplant stages before seed production began

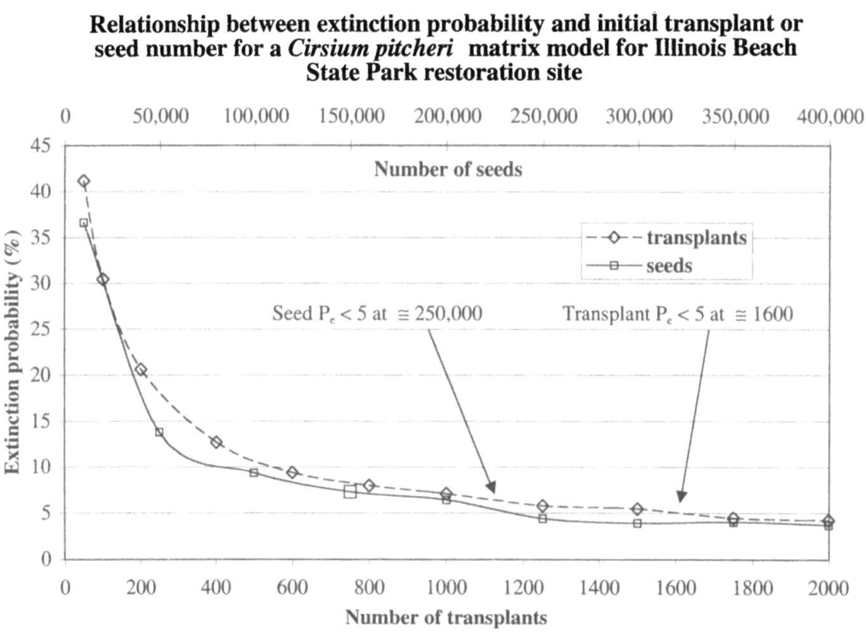

Fig. 12.6. Extinction probability decreases with increasing initial transplant and seed number for a *Cirsium pitcheri* matrix model for Illinois Beach State Park restoration site. Empirically derived transition matrices were alternatively chosen with equal probability each year of each simulation

and naturally recruited individuals were present. In the MVP simulation, individuals in the simulation were all naturally recruited and were divided among the stages in a ratio approximating the natural stage distribution (5:3:1:1, see Table 12.4). The later stages have greater reproductive value than transplant seedlings and the number of plants needed for $P_e<5\%$ is smaller when divided among all naturally recruited stages compared to transplant "seedlings".

As with Guerrant's (1996) simulations, the minimum number of transplants required to establish a viable restoration population of *Cirsium pitcheri*, as well as the extinction risk at the same initial transplant number, were greater for seed compared to non-seed transplants (Fig. 12.6). However, caution must be used when generalizing transplant performance from matrices based on natural populations, since we observed that transplants flower earlier and have lower fecundity (Fig. 12.2) and different transition frequencies (Appendix 12.1) than naturally recruited plants. Our data for *Cirsium pitcheri* suggest that projections based on naturally recruited seedlings would actually have a lower extinction probability than transplants.

12.4 Mead's Milkweed

12.4.1 Species Background

Mead's milkweed (*Asclepias meadii*), an iteroparous herb of late-successional tallgrass prairies of the midwestern USA, is classified by federal authorities as a threatened species because of conversion of its habitat to agriculture (Harrison 1988; Betz 1989). As with many other milkweeds, this genetically diverse species expresses late-acting self-incompatibility and has high sensitivity to inbreeding depression and low rates of reproduction (Betz 1989; Betz et al. 1994; Wyatt and Broyles 1994; Bowles et al. 1998; Tecic et al. 1998; Kettle et al. 2000). Large clones may develop by rhizomes when sexual reproduction is prevented (Hayworth et al. 2002). Genets of this species are apparently very long-lived, although ramet numbers vary annually (Betz 1989), and individual plants have persisted for up to 30 years. Most populations in the western part of its range occur in haymeadows, where mowing removes seed pods and promotes clonal spread, and fragmented eastern populations apparently comprise single or few clones that persist vegetatively (Bowles et al. 1998; Tecic et al. 1998; Hayworth et al. 2002).

Recovery planning for *Asclepias meadii* includes restoration of viable populations in the eastern part of its range, which will require large numbers of different genotypes to facilitate outcrossing (Tecic et al. 1998; Hayworth et al. 2002). To help achieve this goal, a genetically diverse ex situ garden population of this species was developed from multiple seed sources, and serves as a propagule source for restoration (Bowles et al. 1998, 2001). Since 1991, nine restorations have been initiated in Illinois and adjacent Indiana, in which plantings of seeds and 1-year old juveniles have been repeated over time to simulate recruitment (Bowles et al. 1998, 2001). This work has allowed comparison of different establishment methods, as well as in situ vs. ex situ growth rates and environmental and management effects on demographic processes. We analyze viability of restored populations at the Schulenberg Prairie of the Morton Arboretum, DuPage County, Illinois, and the Biesecker Prairie Nature Preserve, Lake County, Indiana.

Field establishment of *Asclepias meadii* is constrained by abiotic factors (Fig. 12.7). Dormant-season prescribed burns enhance seedling establishment and growth, especially in years with greater than average rainfall, while planted juveniles show a greater response to fire during years of normal rainfall (Bowles et al. 1998). There is little evidence for a seed bank, as <1 % of field-planted seeds have germinated the second year after planting. Seedling establishment rates in the field are much lower than greenhouse germination, and establishing cohorts by transplanting greenhouse propagated juveniles is therefore more efficient than planting of seeds. Nevertheless, there is little difference in survivorship once plants become established either by planting of seeds or juveniles (Bowles et al. 2001). Cohort mortality has been high over

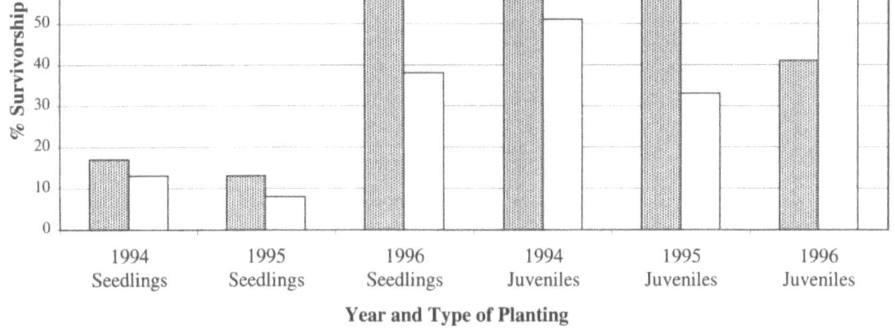

Fig. 12.7. Planted seedling and juvenile Mead's milkweeds respond differently to burning and rainfall. Greater than normal rainfall and higher seedling survivorship (P =0.001) occurred in 1996, and seedling survivorship was higher (P=0.057) in burned habitat in 1996. Juvenile survivorship was higher in burned habitat in 1994 (P<0.001) and in 1995 (P=0.013). (Data from Bowles et al. 1998, with permission of the Annals of the Missouri Botanical Garden)

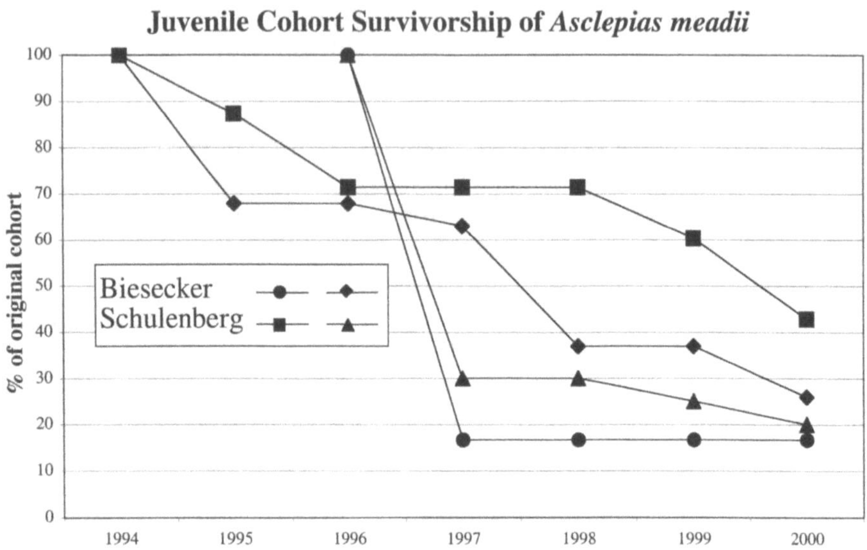

Fig. 12.8. Cohort survivorship of spring-planted 1-year-old Mead's milkweeds

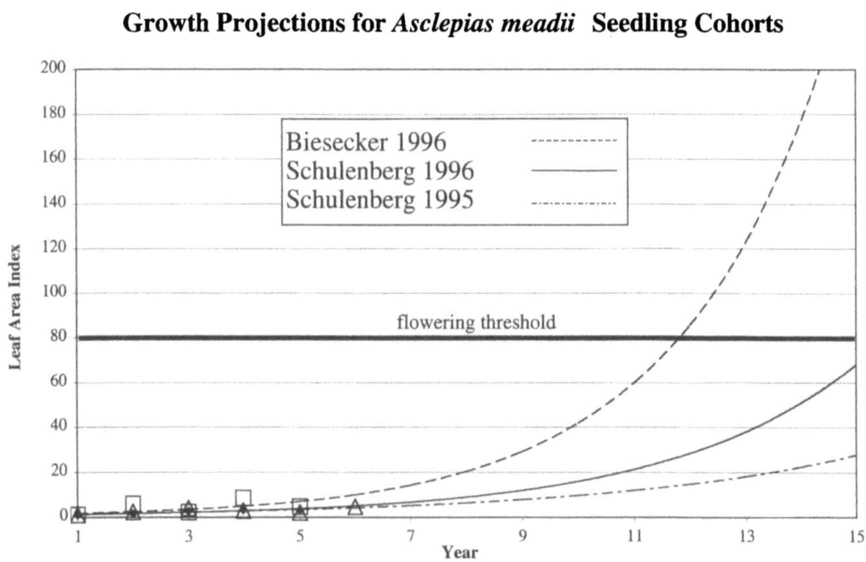

Fig. 12.9. Exponential growth projections of Mead's milkweed seedling cohorts predict 12 years or more to attain flowering threshold

the first winter after spring planting, and survivorship has then declined to about 20–40 % (Fig 12.8). Despite similar survivorship, plants established as juveniles have greater growth than seedlings and have attained flowering size thresholds after as few as 2 years. Seedling growth is suppressed by competition (Bowles et al. 1998), and 12 or more years may be required for seedlings to reach reproductive size in the field (Bowles et al. 2001; Fig. 12.9). Reproduction may be further constrained by other factors, as natural seed production has occurred only once after five consecutive years of flowering at the Schulenberg restoration. This may have been due to lack of successful crossing between compatible genotypes, or because plants were too small to develop fruits. Dormancy has occurred among plants established from seed or planted juveniles, as well as in cultivated plants.

12.4.2 Restoration Viability Analysis

Projections of population growth and size required to achieve a 95 % probability of persistence over time are not currently possible for the Mead's milkweed restorations because of missing data. Except for a single event of seed production, all transitions have been the result of artificial planting, and natural transitions are therefore not yet available. Likewise, despite extensive monitoring of several natural populations (Betz 1989; Kettle et al. 2000), data are also unavailable on complete stage transitions because of the great longevity and low fecundity of this species. As a result, no complete demographic data were available from a natural population with which to compare the stage distribution of the restoration. Also, because natural recruitment has not been observed in the restoration populations, reliable matrix models are not available with which to estimate a MVP.

12.4.2.1 Demographic Modeling

To develop transition frequencies for the restored Schulenberg and Biesecker populations, we used six stages, including seedlings, three juvenile size classes, dormant plants, and flowering plants (Fig. 12.10). The juvenile life stages are size classes based on leaf-area indices, with the smallest including juveniles that retain the linear leaves characteristic of seedlings (Bowles et al. 1998). All juveniles or flowering plants may revert to the smallest stage, or to dormancy. Seedling transitions were based on planted seeds, because naturally recruited seedlings have not been observed, while planted juveniles were entered into the transitions at their appropriate size class. Transition frequencies for the restored populations (Appendix 12.3) were calculated using cumulative annual transitions from 1995 to 2000. At Schulenberg, flowering

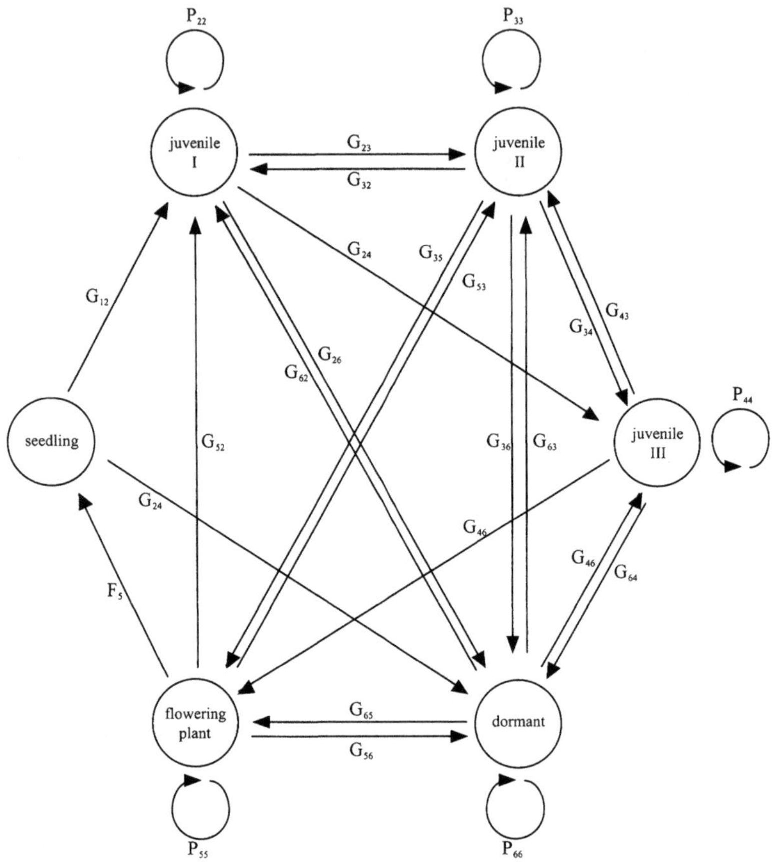

Fig. 12.10. Generalized stage-based population dynamics model for Mead's milkweed (*Asclepias meadii*). *Circles* Stages, *arrows* possible transitions between stages. Transition labels indicate growth (*G*) from one stage to the next, the probability (*P*) of remaining in that stage and fecundity (*F*)

occurred in all 5 years, but successful seed production occurred only in 2000, and λ=0.85. To compensate for lack of seed production at Biesecker, the fecundity transition frequency (F_{51}) was calculated using seed production from Schulenberg. This change in the transition frequency matrix for Biesecker had a relatively small effect on λ ($\lambda_{(F51=0)}$=0.869, $\lambda_{(F51=0.4424)}$=0.874).

Modeling was also hampered by the slow growth and low fecundity of this species. Although our reproduction transitions based on artificial seedling establishment could reflect expected survivorship under natural environmental conditions, it is unknown how well our seed planting technique replicated the natural process of seed dispersal and germination.

12.4.2.2 Variance/Mean Ratios

The *V/M* ratios for Schulenberg matrix elements ranged from 0.01–0.76 for non-fecundity elements, with 0.33 for fecundity and a median value of 0.10. For Biesecker, *V/M* ratios ranged from 0.006–0.65 for non-fecundity elements, with 25.81 for fecundity and a median value of 0.18 (Table 12.5). The high fecundity *V/M* ratio occurred for Biesecker because this site had much lower flowering frequency than Schulenberg, even though seed production was considered identical to Schulenberg.

The comparatively high median *V/M* ratios for the Schulenberg and Biesecker populations (Table 12.5) may indicate strong environmental stochasticity (Menges 1998), possibly due to effects of fire and rainfall on survivorship and growth of this species (Bowles et al. 1998). Also, because λ was <1 for both restorations, they appear to be vulnerable to environmental stochasticity (Menges 1998). However, variability in growth and survivorship in restorations may have a genetic origin if ex situ propagation promotes establishment of juvenile plants that might not have survived germination or seedling stages under more stressful natural conditions. Therefore, our λ values of <1 for could reflect loss of less fit individuals, while survivorship and growth of more fit plants could lead to eventual population stability. Management to maintain these plants, and restoration to increase numbers of differ-

Table 12.5. Variance-to-mean ratios for Mead's milkweed (*Asclepias meadii*): Schulenberg and Biesecker restorations

	Seedling	Juvenile-I	Juvenile-II	Juvenile-III	Flowering	Dormant
Schulenberg	State (t)→					
Fate (t+1)						
Seedling					0.3311	
Juvenile-I	0.0154	0.1377	0.0159			0.1931
Juvenile-II		0.0884	0.0269	0.12	0.6917	0.0512
Juvenile-III			0.0266	0.45		0.0216
Flowering				0.1	0.1771	0.0773
Dormant	0.0581	0.7638	0.0941	0.1438		0.1166
Median = 0.1000						
Biesecker	State (t)→					
Fate (t+1)						
Seedling					25.807	
Juvenile-I	0.0954	0.0383	0.106			0.1276
Juvenile-II		0.0733	0.131	0.3556	0.6	0.1647
Juvenile-III			0.6488	0.2		0.0361
Flowering				2.4	0.6	0.0059
Dormant	0.2125	0.1165	0.2263	0.4533		0.1801
Median = 0.1801						

ent genotypes that can contribute to successful outcrossing and seed production, are important objectives.

12.4.2.3 Elasticities

Elasticities for growth (*G*), stasis (*L*) and reproduction (*F*) categories ordinated in positions expected for late-successional plants, primarily trees and shrubs (Fig. 12.4). Elasticity values were highest for stasis, slightly lower for growth, and lowest for fecundity (Figs. 12.3, 12.4). The ordination also corresponded to findings of Fiedler et al. (1998), that for species of *Calochortus*, which are also iteroparous herbs, most *G*/*L*/*F* ratios positioned toward the lower right of the triangle. Although Fiedler considered this problematic given the more central ordination of iteroparous herbs found by Silvertown et al. (1996), it corresponds to a successional sequence. The ordination positions for *Asclepias meadii* and *Calochortus* may be appropriate for long-lived herbaceous plants, as well as for trees and shrubs as proposed by Silvertown et al. (1996), fitting a late successional position proposed based on Grimes triangle (Silvertown et al. 1992).

Some management applications can be suggested based on ordination position. The similar ordinations of Schulenberg and Biesecker indicate that stasis and growth had much stronger effects on λ than fecundity (Fig. 12.4). This suggests that restoration management should focus on growth and stasis rather than increasing reproduction, which seems applicable because of the time required for flowering, the expected longevity, and the apparently low fecundity of this species. Continuing to introduce plants will have little to do with increasing λ until large numbers of plants can attain threshold sizes for reproduction and then overcome further constraints on seed production. However, continued introductions of additional genotypes would result in larger populations that will have greater reproductive potential, and, presumably, less vulnerability to environmental stochasticity. These population characteristics could veil long-term population change for late-successional iteroparous herbs, such as *Calochortus* (Fiedler et al. 1998) and *Asclepias meadii* (Kettle et al. 2000), and many years of monitoring may be required before realistic PVA projections can be attained for such species (Bowles et al. 2001).

Caution in interpreting *G*/*L*/*F* ratios is advised, however, because *L*, G and *F* change when the number of stages change and the effects of stasis, growth and reproduction are not clearly separated among *L*, *G* and *F* (Caswell 2001). A greater number of restoration transition years and natural populations will help illuminate the usefulness of *G*/*L*/*F* ratio ordination to conservation biology. Despite these potential problems with *G*/*L*/*F* ratio ordination, the process of developing the ordination is valuable because it encourages careful thought about the life history of *Asclepias meadii* and implications to management decisions.

12.5 Conclusions

12.5.1 Application of PRVA

Although it remains difficult to project how well PRVA can be applied to endangered plant restorations, our analyses provide some potentially useful conclusions about projecting plant restoration viability.

Encouragingly, our data show that demographic life stage transitions can be used to project population growth rates of restored populations and to identify relative contributions of different life stages to population growth. This may allow projecting the numbers of transplants or seeds needed to achieve population growth - an important objective in population restoration. The most meaningful interpretation of PRVA modeling requires transitions for all life stages, comparative data from multiple natural populations as well as from replicated restorations, and an understanding of how environmental stochasticity (such as drought for *Cirsium pitcheri* or rainfall for *Asclepias meadii*) may affect viability.

Our comparisons indicate that PRVA will be most useful for short-lived species (see discussion of difficulties of PVA with long-lived species in Chap. 9, this Vol.). For example, because of its short life span and rapid cohort development, *Cirsium pitcheri* transplants provided transitions for each stage within 5 years, and demographic data from a natural population strengthened interpretation of elasticities. The Illinois Beach State Park restoration of *Cirsium pitcheri* south of the Dead River has successfully reached short-term goals of completion of its life cycle, a natural stage distribution, overall stable λ, and variation in λ comparable to a natural population. Although natural recruitment is taking place, population expansion into available habitat does not yet appear to be occurring. The long-term goal of reaching MVP size has not occurred, and we do not know whether the population can recover from environmental or demographic stochasticity. However, the 1998→1999 transition matrix had a very low λ, and λ the following year was relatively high, suggesting that the restoration has sufficient size to recover.

As with natural populations (Fiedler et al. 1998), PRVA of long-lived species with low reproductive rates appears problematic without long-term data sets. For *Asclepias meadii*, slow growth and low fecundity limited our interpretation of elasticities and population growth and prevented estimation of a MVP. Further difficulty in interpretation arose because stages were missing and matrices were composites of 6 years of monitoring data rather than separate matrices for each pair of years. Finally, we had no data from natural populations of *Asclepias meadii* with which to compare demographic characteristics. Nevertheless, interpretation of *Asclepias meadii* elasticities supported a logical management need to increase growth. Because Mead's milkweed requires many years to reach maturity, and because natural recruitment has not occurred in the restoration, reliable matrix models are not available with which to estimate MVP let alone determine whether the restorations have reached short-term or long-terms goals. In addition, we do not have a viable natural population with which to compare the stage distribution of the restoration. On the positive side, after initial transplant mortality, survivorship is high. In addition, the single event of seed production indicates that genetic diversity may be sufficient to overcome self-incompatibility in a restoration, which has apparently prevented seed production in many eastern populations (Tecic et al. 1998; Hayworth et al. 2002).

A theoretical basis for plant restoration remains largely undeveloped and our results with *Cirsium pitcheri* suggest that restoration may not match models based on natural populations. For example, although Guerrant's (1996) model indicated that larger initial propagule sizes would reduce extinction probability and demographic cost of restoration plantings, he used published transition data from natural populations assuming similar transition rates (Guerrant and Fiedler 2002). Transitions from natural populations may differ from those initiated by artificial planting. For *Asclepias meadii*, although larger propagules will accelerate time to flowering, seedlings established from seed may result in similar survivorship to transplanted juveniles and will not increase the probability of extinction. Although outplanting larger *Cirsium pitcheri* plants could increase transplant success, naturally recruited flowering plants may have greater fecundity (Fig. 12.2). Thus, projections of λ using transplant-derived transitions could be comparatively low, although earlier flowering of transplants could counter this effect if it increases recruitment over time. Outplanting of propagated plants also may allow establishment of less fit plants that might not have survived natural environmental rigors in the seedling stage and might have lower fecundity. Nevertheless, based on extinction risk analysis, transplanting greenhouse propagated plants that require more than one year to reach flowering threshold is more efficient than planting seeds, thereby reducing the "demographic cost" of reintroduction (Guerrant and Fiedler, 2002). Reintroduction using adults may be more efficient than for juveniles in animals as well (Sarrazin and Legendre 2000). Repeated smaller outplantings to represent natural cohort replacement may also spread over time an otherwise large drain on seeds or propagules provided from ex situ or wild collections.

12.5.2 Differences Between PVA of Restorations vs. Natural Populations

Applications of PVA to restored populations differ in a number of important ways from PVA of natural populations (Table 12.6). Initially, many restored populations are nonviable, or below "quasi-extinction" levels. Here, PRVA can be viewed as projecting how experimental planting and management treatments contribute toward population growth to achieve a viability threshold,

Table 12.6. How PVA factors may differ between natural and restored populations

Factor	Natural population	Restoration
Population viability	Maintain viability by managing to sustain $\lambda>1$ or to prevent $\lambda<1$	Manage to increase $\lambda>1$ and reach viability
Demographic stochasticity	Minor importance in large population	Major importance in small population
Genetic stochasticity	Minor importance in large population	Major importance in small population
Increasing population growth ($\lambda>1$)	More likely to indicate viability for large population	Small restoration may be non-viable due stochastic effects
Matrix model stages	Based only on natural recruitment	Natural recruitment and transplants are separate elements
Life stage transitions	May be easily obtainable in natural populations	May be difficult to obtain in non-functional restorations
Monitoring cycle required to produce a transition matrix of naturally recruited individuals	Two or more years if all stages are present	Continue until all stages are present from natural reproduction, which may require many years (e.g., 8 for *Cirsium pitcheri*)
Incorporation of multiple years into matrix model	Separate matrices for each pair of years monitored	Composite matrices for initial restoration years
Elasticity analysis	More likely to identify important life stage	May emphasize or de-emphasize incorrect life stages if non-functional
Elasticities and λ	Affected by natural birth and death	Manipulated by out-planting

while PVA of many natural populations can be most often viewed as determining how to prevent growth trends from falling below thresholds of viability. Small restored populations with increasing growth may remain highly vulnerable to environmental stochasticity, as well as genetic and demographic stochasticity, which usually are less important in larger populations (Lande 1988; Menges 1991). For example, small numbers of flowering plants may not attract pollinators, or may enforce selfing or inbreeding among related plants (Chaps. 2, 3, 10, this Vol.).

If transition frequencies differ between transplants and naturally recruited individuals, matrix models for restorations will require more complex matrices which have separate elements for stages derived from natural recruitment and transplants.

All transition stages are more likely to be present in natural populations than in restorations, especially for early restoration years. As a result, important variables, such as fecundity, may be difficult to obtain or may differ from natural populations. Also, developing a transition matrix for a natural population may require only 2 years of demographic monitoring if all stages are present each year. Many more years of monitoring will be necessary for restorations in which transplants must complete their life cycle before naturally occurring cohorts can be established and develop all stages. For *Cirsium pitcheri*, all stages were present in the restoration 5 years after transplanting began. *Asclepias meadii* had not completed its life cycle 7 years after initial seed planting and translocation efforts. In addition, since transplant transition frequencies may differ from those of naturally recruited individuals, reliable transition matrices cannot be constructed for a restoration until all stages are present among naturally recruited individuals. For *Cirsium pitcheri*, a transition matrix for naturally recruited individuals was first constructed only after the eighth year of restoration. Another consequence of missing stages at the beginning of restoration for both *Cirsium pitcheri* and *Asclepias meadii* was the necessity to construct composite matrices rather than using separate matrices for each pair of years as can be done for natural populations.

In restorations, as well as in stressed natural populations, elasticities could de-emphasize important but poorly functioning life history stages, leading to improper management conclusions unless compared with natural populations. Demographic characteristics, such as elasticities and λ are affected by birth and death in natural populations in contrast to restorations in which these characteristics may be more strongly affected by the number of transplants introduced.

These differences between restorations and natural populations suggest that traditional PVA approaches may be difficult for restored populations. For example, small short-term data sets may be most interpretable using nonintegrated approaches, and experimental approaches will be critical, especially if data are missing for mean vital rates (Pavlik 1994). A greater emphasis could also be placed on absolute growth rate and cohort depletion rates, as opposed to modeling extinction probabilities. Comparisons with natural populations may be important for developing and interpreting models, especially elasticities. We recommend use of both integrated and nonintegrated methods to evaluate plant restoration viability. If both methods agree the validity of conclusions is increased, if they do not agree it suggests that there may be problems with the one or both methods of analysis.

12.5.3 Future Directions

PVAs for natural populations have usually ignored genetic and demographic stochasticity because these factors are considered important only when population size is small, originally suggested as less than 50 (for demographic stochasticity and inbreeding depression) or 500 (for genetic stochasticity) individuals (Franklin 1980; Frankel and Soulé 1981; Shaffer 1981; Lande 1988, 1993; Chap. 11, this Vol.). However, because population restorations may initially be small, demographic and genetic stochasticity could significantly impact their initial stages and may need to be incorporated into PVAs of small populations. As yet, little information is available on interactions between these factors in extinction of small populations (Lande 1988; Beissinger and Westphal 1998).

Genetic applications to restoration often face a tradeoff between maintaining local genetic adaptation, which commonly occurs in plant species, vs. avoiding inbreeding and maximizing genetic diversity and adaptive evolutionary potential (Huenneke 1991; Knapp and Dyer 1998; Barrett and Kohn 1991; Fenster and Galloway 2000). Measures of genetic variation and how it is allocated within and among natural populations, such as the G_{ST} statistic, can help guide species restoration (Hamrick et al. 1991; Weller 1994; Pavlik 1996; Steinberg and Jordon 1998). For example, Pavlik et al. (1993) compared levels of genetic variability between natural and restored populations of *Amsinckia grandiflora*. Genetic applications are strengthened when linked with plant breeding systems, especially effects of self-incompatibility and inbreeding depression. DeMauro (1993, 1994) used multiple genotypes to maximize genetic diversity and reproductive potential in restored populations of the self-incompatible *Hymenoxys acaulis* var. *glabra*. Likewise, Bowles (Bowles et al. 1998, 2001) maximized numbers of genotypes within restorations of *Asclepias meadii*, a species apparently highly susceptible to inbreeding depression.

Metapopulation and spatially explicit models may be more appropriate for plants, but they have not yet received much application (Menges 2000; Chaps. 6, 11, this Vol.). For example, *Cirsium pitcheri* may depend upon metapopulation processes for long-term persistence (McEachern et al. 1994). Metapopulation level analysis of the number, spatial patterns, and dispersal of populations needed for a high probability of persistence (Morris et al. 1999) may be most realistic for this species. There will also be a need for understanding how demographic-genetic linkage could enhance viability among small isolated restorations, and how carrying capacity of small habitat remnants affects species restoration potential. For example, pollen transfer among small populations could increase effective population sizes of self-incompatible species such *Asclepias meadii*, and would help avoid high density packing of individuals that could lead to increased herbivory or attack by pathogens, with cascading effects on population viability (Bowles et al. 1998; Chaps. 10, 3, this Vol.).

Applying PRVA modeling to help understand whether recovery targets are being met by achieving minimum numbers of viable populations could become an important application for agencies charged with listing and delisting endangered species (Bowles and Bell 1999). Coupling this work with experimental procedures (Menges 2000; Kaye et al 2001) would be useful for learning how management treatments, such as prescribed burning, can help accelerate growth of restored populations. However, Coulson et al. (2001) warned that PVA projections will be problematic without good data and unless distributions of vital rates between individuals and years can be predictable over time, a tall order given the impact of humans on environmental change and stress in small habitat remnants.

Acknowledgements. We thank Kay Havens, Pati Vitt, and Susanne Masi at the Chicago Botanic Garden for collaboration on *Cirsium pitcheri* restoration, Jenny McBride and Bob Betz for assistance on restoration of *Cirsium pitcheri* and *Asclepias meadii*, Eric Menges and Pedro Quintana-Ascencio for assistance with analyses, and Ed Guerrant for sharing unpublished information. We also thank the University of Wisconsin Press and Annals of the Missouri Botanical Garden for permission to use data and figures, and the Shawnee National Forest, Illinois Endangered Species Protection Board, US Fish and Wildlife Service, Illinois Department Of Natural Resources, Morton Arboretum and Indiana Division of Nature Preserves for funding and permission to conduct research. Finally, we thank Christy Brigham and Mark Schwartz for the opportunity to participate in this volume.

Appendix 12.1. Stage-based projection matrices of transition frequencies for a restored *Cirsium pitcheri* population at Illinois Beach State Park. *R* Naturally recruited individuals, *T* transplanted individuals. The average matrix represents 1991 through 2000

	Naturally recruited					Transplants			
	R seedling	R small juvenile	R large juvenile	R flowering	Sown seed	T "seedling"	T small juvenile	T large juvenile	T flowering
	State (1999)→								
Fate (2000)									
R seedling	0	0	0	3.2060	0.0086	0	0	0	5.1588
R small juvenile	0.5556	0.0000	0	0	0	0	0	0	0
R large juvenile	0.3333	0.8000	0.2500	0	0	0	0	0	0
R flowering	0	0.0000	0.6250	0	0	0	0	0	0
Sown seed	0	0	0	0	0	0	0	0	0
T "seedling"	0	0	0	0	0	0	0	0	0
T small juvenile	0	0	0	0	0	0.5556	0.7273	0.0000	0
T large juvenile	0	0	0	0	0	0.3333	0.0000	0.1000	0
T flowering	0	0	0	0	0	0	0.0909	0.8000	0
	State (1998)→								
Fate (1999)									
R seedling	0	0	0	1.8659	0.0032	0	0	0	1.4268
R small juvenile	0.1333	0.2308	0	0	0	0	0	0	0
R large juvenile	0.0000	0.4615	0.0000	0	0	0	0	0	0
R flowering	0	0.0769	1.0000	0	0	0	0	0	0
Sown seed	0	0	0	0	0	0	0	0	0
T "seedling"	0	0	0	0	0	0	0	0	0
T small juvenile	0	0	0	0	0	0.1333	0.2692	0.3333	0
T large juvenile	0	0	0	0	0	0.0000	0.3846	0.0000	0
T flowering	0	0	0	0	0	0	0.0769	0.6667	0

Appendix 12.1. (*Continued*)

	Naturally recruited					Transplants			
	R seedling	R small juvenile	R large juvenile	R flowering	Sown seed	T "seedling"	T small juvenile	T large juvenile	T flowering
	Average state (t) →								
Fate (t+1)									
R seedling	0	0	0	2.7593	0.0052	0	0	0	2.1600
R small juvenile	0.5366	0.3929	0	0	0	0	0	0	0
R large juvenile	0.0732	0.3929	0.2222	0	0	0	0	0	0
R flowering	0	0.0714	0.6667	0	0	0	0	0	0
Sown seed	0	0	0	0	0	0	0	0	0
T "seedling"	0	0	0	0	0	0	0	0	0
T small juvenile	0	0	0	0	0	0.4545	0.5878	0.0959	0
T large juvenile	0	0	0	0	0	0.0136	0.2041	0.2740	0
T flowering	0	0	0	0	0	0	0.0449	0.5479	0

Appendix 12.2. Stage-based projection matrices of transition frequencies for a naturally occurring *Cirsium pitcheri* population at Indiana Dunes West Beach

	Seedling	Small juvenile	Large juvenile	Flowering
	State (1992)→			
Fate (1993)				
Seedling	0.0000	0.0000	0.0000	2.2000
Small juvenile	0.2786	0.6364	0.0000	0.0000
Large juvenile	0.0000	0.2121	0.2500	0.0000
Flowering	0.0000	0.0303	0.7500	0.0000
	State (1991)→			
Fate (1992)				
Seedling	0.0000	0.0000	0.0000	12.7273
Small juvenile	0.4545	0.5217	0.1111	0.0000
Large juvenile	0.0000	0.1304	0.1111	0.0000
Flowering	0.0000	0.0435	0.4444	0.0000
	State (1990)→			
Fate (1991)				
Seedling	0.0000	0.0000	0.0000	4.0000
Small juvenile	0.7857	0.6250	0.1053	0.0000
Large juvenile	0.0000	0.2500	0.2632	0.0000
Flowering	0.0000	0.1250	0.4737	0.0000
	State (1989)→			
Fate (1990)				
Seedling	0.0000	0.0000	0.0000	0.9333
Small juvenile	1.0000	0.6667	0.2051	0.0000
Large juvenile	0.0000	0.0000	0.2821	0.0000
Flowering	0.0000	0.2222	0.4359	0.0000
	State (1988)→			
Fate (1989)				
Seedling	0.0000	0.0000	0.0000	0.1818
Small juvenile	0.0000	0.3684	0.0408	0.0000
Large juvenile	0.0000	0.5789	0.5714	0.0000
Flowering	0.0000	0.0000	0.3061	0.0000

Appendix 12.3. Stage-based projection matrices of transition frequencies for two restored Mead's milkweed (*Asclepias meadii*) populations

	Seedling	Juvenile-I	Juvenile-II	Juvenile-III	Flowering	Dormant
Schulenberg Prairie						
	State (t)→					
Fate (t+1)						
Seedling	0	0	0	0	0.3575	0
Juvenile-I	0.7451	0.5484	0.12	0	0.0667	0.4272
Juvenile-II	0	0.0753	0.5133	0.2778	0.3333	0.1845
Juvenile-III	0	0.0054	0.0467	0.1111	0	0.0583
Flowering	0	0	0.0333	0.0556	0.2667	0.068
Dormant	0.0588	0.0588	0.2473	0.4444	0.1333	0.2621
Biesecker Prairie						
	State (t)→					
Fate (t+1)						
Seedling	0	0	0	0	0.4424	0
Juvenile-I	0.3088	0.4406	0.1667	0	0	0.5167
Juvenile-II	0	0.104	0.3667	0.25	0.3333	0.175
Juvenile-III	0	0	0.0667	0.1667	0	0.0333
Flowering	0	0	0	0.0833	0.3333	0.0083
Dormant	0.6176	0.2178	0.3444	0.4167	0	0.2417

References

Barrett SCH, Kohn JR (1991) Genetic and evolutionary consequences of small population size in plants: implications for conservation. In: Falk DA, Holsinger KE (eds) Genetics and conservation of rare plants. Oxford University Press, New York, pp 3–30

Beissinger SR, Westphal MI (1998) On the use of demographic models of population viability in endangered species management. J Wildl Manage 62:821–841

Betz RF (1989) Ecology of Mead's milkweed (*Asclepias meadii* Torrey). In: Bragg TB, Stubbendieck J (eds) Proceedings of the eleventh North American Prairie Conference. University of Nebraska, Lincoln, pp 187–191

Betz RF, Struven RD, Wall JE, Heitler FB (1994) Insect pollinators of 12 milkweed (*Asclepias*) species. In: Wickett RG, Lewis PD, Woodliffe A, Pratt P (eds) Proceedings of the thirteenth North American Prairie Conference. Department of Parks and Recreation, Windsor, Ontario, Canada, pp 45–60

Bowles ML, Bell TJ (1999) Establishing recovery targets for Illinois plants. Report to the Illinois Endangered Species Protection Board. The Morton Arboretum, Lisle, Illinois

Bowles ML, McBride JL (1996) Status and structure of a Pitcher's thistle (*Cirsium pitcheri*) population reintroduced to Illinois Beach Nature Preserve. In: Falk D, Olwell P, Millar C (eds) Restoring diversity: ecological restoration and endangered plants. Island Press, New York, pp 194–218

Bowles ML, Whelan CJ (eds) (1994) Restoration of endangered species: conceptual issues, planning, and implementation. Cambridge University Press, Cambridge

Bowles ML, Flakne R, McEachern K, Pavlovic N (1993) Recovery planning and reintroduction of the federally threatened Pitcher's thistle (*Cirsium pitcheri*) in Illinois. Nat Area J 13:164–176

Bowles ML, McBride JL, Betz RF (1998) Management and restoration ecology of the federal threatened Mead's milkweed, *Asclepias meadii* Torrey (Asclepiadaceae). Ann Mo Bot Gar 85:110–125

Bowles ML, McBride JL, Bell TJ (2001) Restoration of the federally threatened Mead's milkweed (*Asclepias meadii*). Ecol Restor 19:235–241

Brown JS (1994) Restoration ecology: living with the prime directive. In: Bowles ML, Whelan CJ (eds) Restoration of endangered species: conceptual issues, planning, and implementation. Cambridge University Press, Cambridge, pp 355–380

Brumback WE, Fyler CW (1996) Small whorled pogonia (*Isotria medeoloides*) transplant project. In: Falk D, Olwell P, Millar C (eds) Restoring diversity: ecological restoration and endangered plants. Island Press, New York, pp 445–451

Caswell H (2001) Matrix population models: construction, analysis and interpretation. 2nd edn. Sinauer, Sunderland, MA

Chen H, Maun MA (1998) Population ecology of *Cirsium pitcheri* on Lake Huron sand dunes: III. Mechanisms of seed dormancy. Can J Bot 76:575–586

Coulson T, Mace GM, Hudson E, Possingham H (2001) The use and abuse of population viability analysis. Trends Ecol Evol 16:219–221

Crouse DT, Crowder LB, Caswell H (1987) A stage-based population model for loggerhead sea turtles and implications for conservation. Ecology 68:1412–1423

Crowder LB, Couse DT, Heppel SS, Martin TH (1994) Predicting the impact of turtle excluder devices on loggerhead sea turtle populations. Ecol Appl 4:437–445

Cully A (1996) Knowlton's cactus (*Pediocactus knowltonii*) reintroduction. In: Falk D, Olwell P, Millar C (eds) Restoring diversity: ecological restoration and endangered plants. Island Press, New York, pp 403–410

DeMauro MM (1993) Relationship of breeding system to rarity in the Lakeside daisy (*Hymenoxys acaulis* var. *glabra*). Conserv Biol 7:542–550

DeMauro MM (1994) Development and implementation of a recovery program for the federal threatened Lakeside daisy (*Hymenoxys acaulis* var. *glabra*). In: Bowles ML, Whelan CJ (eds) Restoration of endangered species: conceptual issues, planning, and implementation. Cambridge University Press, Cambridge, pp 298–321

Doak D, Kareiva P, Klepteka B (1994) Modeling population viability for the desert tortoise in the western Mohave Desert. Ecol Appl 4:446–460

Falk DA, Millar CI, Olwell M (eds) (1996) Restoring diversity: strategies for reintroduction of endangered plants. Island Press, Washington, DC

Fenster CB, Galloway LF (2000) Inbreeding and outbreeding depression in natural populations of *Chamaecristata fasciculate* (Fabaceae). Conserv Biol 14:1406–1412

Ferson S (1994) RAMAS/Stage: Generalized stage-based modeling for population dynamics. Applied Biomathematics, Seatuket, New York

Fiedler PL (1987) Life history and population dynamics of rare and common mariposa lilies. J Ecol 75:977–995

Fiedler PL, Knapp FE, Fredricks N (1998) Rare plant demography: lessons from the mariposa lilies (*Calochortus*: Liliaceae). In: Fiedler PL, Kareiva PM (eds) Conservation biology for the coming decade, 2nd edn. Chapman and Hall, New York, pp 28–48

Frankel OH, Soulé ME (1981) Conservation and evolution. Cambridge University Press, Cambridge

Franklin IA (1980) Evolutionary change in small populations. In: Soulé ME, Wilcox BA (eds) Conservation biology: an evolutionary-ecological perspective. Sinauer, Sunderland, MA, pp 135–149

Gilpin ME, Soulé ME (1986) Minimum viable populations: processes of species extinction. In: Soulé ME (ed) Conservation biology: the science of scarcity and diversity. Sinauer, Sunderland, MA, pp 19–34

Goodman D (1987) The demography of chance extinction. In: Soulé ME (ed) Viable populations for conservation. Cambridge University Press, New York, pp 2–34

Gordon DR (1996) Apalachicola rosemary (*Conradina glabra*) reintroduction. In: Falk D, Olwell P, Millar C (eds) Restoring diversity: ecological restoration and endangered plants. Island Press, New York, pp 417–422

Guerrant EO Jr (1996) Designing populations: demographic, genetic, and horticultural dimensions. In: Falk D, Olwell P, Millar C (eds) Restoring diversity: ecological restoration and endangered plants. Island Press, New York, pp 171–207

Guerrant EO Jr, Fiedler PL (2003) Attrition during ex situ storage and reintroduction. In: Guerrant EO Jr, Havens K, Maunder M (eds) Saving the pieces: the role, value and limit of ex situ plant conservation. Island Press, Washington, DC

Guerrant EO Jr, Pavlik BM (1998) Reintroduction of rare plants: genetics, demography, and the role of *ex situ* conservation methods. In: Fiedler PL, Kareiva PM (eds) Conservation biology for the coming decade, 2nd edn. Chapman and Hall, New York, pp 80–108

Hamrick JL, Godt MJW, Murawski DA, Loveless MD (1991) Correlations between species traits and allozyme diversity: implications for conservation biology. In: Falk DA, Holsinger KE (eds) Genetics and conservation of rare plants. Oxford University Press, New York, pp 75–86

Harrison WF (1988) Endangered and threatened wildlife and plants; determination of threatened status for *Asclepias meadii* (Mead's milkweed). Fed Regis 53:33982–33994

Hayworth D, Bowles ML, Schaal B, Williamson K (2002) Clonal population structure of the federal threatened Mead's Milkweed, as determined by RAPD analysis, and its conservation implications. In: Bernstein N, Ostrander LJ (eds) Proceedings of the Seventeenth North American Prairie Conference: seeds for the future; roots of the past. North Iowa Area Community College, Mason City, Iowa

Heppell SS, Walters JR, Crowder LB (1994) Evaluating management alternatives for red-cockaded wood-peckers: a modeling approach. J Wildl Manage 58:479–487

Huenneke LF (1991) Ecological implications of genetic variation in plant populations. In: Falk DA, Holsinger KE (eds) Genetics and conservation of rare plants. Oxford University Press, New York, pp 31–44

Karl TR, Knight RW (1985) Atlas of monthly Palmer drought severity index (1931–1983) for the contiguous United States. National Climatic Data Center, Asheville, NC; see also www.ncdc.noaa.gov/onlineprod/drought/main.html

Kaye TN, Pendergrass KL, Finley K, Kauffman JB (2001) The effect of fire on the population viability of an endangered prairie plant. Ecol Appl 11:1366–1380

Keddy CJ, Keddy PA (1984) Reproductive biology and habitat of *Cirsium pitcheri*. Mich Bot 23:57–67

Kettle WD, Alexander HM, Pittman GL (2000) An 11-year ecological study of a rare perennial (*Aslcepias meadii*): implications for monitoring and management. Am Midl Nat 144:66–79

Knapp RE, Dyer AR (1998) When do genetic considerations require special approaches to ecological restoration? In: Fiedler PL, Kareiva PM (eds) Conservation biology for the coming decade, 2nd edn. Chapman and Hall, New York, pp 345–363

Lande R (1988) Genetics and demography in biological conservation. Science 241:1455–1460

Lande R (1993) Risks of population extinction from demographic and environmental stochasticity and random catastrophes. Am Nat 142:911–927

Ledig FT (1996) *Pinus torreyana* at Torrey Pines State Reserve, California. In: Falk D, Olwell P, Millar C (eds) Restoring diversity: ecological restoration and endangered plants. Island Press, New York, pp 265–271

Louda SM (1994) Experimental evidence for insect impact on populations of short-lived, perennial plants, and it's application in restoration ecology. In: Bowles ML, Whelan CJ (eds) Restoration of endangered species: conceptual issues, planning, and implementation. Cambridge University Press, Cambridge, pp 118–138

Loveless MD (1984) Population biology and genetic organization in *Cirsium pitcheri*, an endemic thistle. PhD Diss, University of Kansas, Lawrence

Mangel M, Tier C (1994) Four facts every conservation biologist should know about persistence. Ecology 75:607–614

McEachern K (1992) Disturbance dynamics of pitcher's thistle (*Cirsium pitcheri*) in Great Lake sand dune landscapes. PhD Diss, University of Wisconsin, Madison

McEachern K, Bowles ML, Pavlovic N (1994) A metapopulation approach to recovery of the federally threatened Pitcher's thistle (*Cirsium pitcheri*) in southern Lake Michigan dunes. In: Bowles ML, Whelan CJ (eds) Restoration of endangered species: conceptual issues, planning, and implementation. Cambridge University Press, Cambridge, pp 194–218

Menges ES (1990) Population viability analysis for an endangered plant. Conserv Biol 4:52–62

Menges ES (1991) The application of minimum viable population theory to plants. In: Falk DA, Holsinger KE (eds) Genetics and conservation of rare plants. Oxford University Press, New York, pp 45–61

Menges ES (1998) Evaluating extinction risks in plant populations. In: Fiedler PL, Kareiva PM (eds) Conservation biology for the coming decade, 2nd edn. Chapman and Hall, New York, pp 49–65

Menges ES (2000) Population viability analyses in plants: challenges and opportunities. Trends Ecol Evol 15:51–56

Morris W, Doak D, Groom M, Kareiva P, Fieberg J, Gerber L, Murphy P, Thomson D (1999) A practical handbook for population viability analysis. The Nature Conservancy, Arlington, VA

Olwell P, Cully A, Knight P (1990) The establishment of a new population of *Pediocactus knowltonii*: third year assessment. In: Mitchell RS, Sheviak CJ, Leopold DJ (eds)

Ecosystem management: rare species and significant habitats. NY State Mus Bull 471:189–193
Pavlik BM (1994) Demographic monitoring and the recovery of endangered plants. In: Bowles ML, Whelan CJ (eds) Restoration of endangered species: conceptual issues, planning, and implementation. Cambridge University Press, Cambridge, pp 322–350
Pavlik BM (1996) Defining and measuring success in rare plant reintroductions. In: Falk D, Olwell P, Millar C (eds) Restoring diversity: ecological restoration and endangered plants. Island Press, New York, pp 127–155
Pavlik BM, Nickrent DL, Howald AM (1993) The recovery of an endangered plant. I. Creating a new population of *Amsinckia grandiflora*. Conserv Biol 7:510–526
Pavlovic NB, Bowles ML, Crispin S, Gibson T, Kavetsky R, McEachern KA, Penskar M (2003) Pitchers's thistle (*Cirsium pitcheri*) recovery plan. US Department of the Interior, Fish and Wildlife Service, Minneapolis
Rowland J, Maun MA (2001) Restoration ecology of an endangered plant species: establishment of new populations of *Cirsium pitcheri*. Restor Ecol 9:60–70
Sarrazin F, Legendre S (2000) Demographic approach to releasing adults versus young in reintroductions. Conserv Biol 14:488–500
Shaffer ML (1981) Minimum population size for species conservation. BioScience 3:131–134
Shaffer ML (1987) Minimum viable populations: coping with uncertainty. In: Soulé ME (ed) Viable populations for conservation. Cambridge University Press, New York, pp 69–86
Silvertown J, Franco M, McConway K (1992) A demographic interpretation of Grimes's triangle. Funct Ecol 6:130–136
Silvertown J, Franco M, Menges E (1996) Interpretation of elasticity matrices as an aid to the management of plant populations for conservation. Conserv Biol 10:591–597
Steinberg EK, Jordon CE (1998) Using molecular genetics to learn about the ecology of threatened species: the allure and illusion of measuring genetic structure in natural populations. In: Fiedler PL, Kareiva PM (eds) Conservation biology for the coming decade, 2nd edn. Chapman and Hall, New York, pp 440–460
Tecic D, McBride JL, Bowles ML, Nickrent DL (1998) Genetic variability in the federal threatened Mead's milkweed, *Asclepias meadii* Torrey (Asclepiadaceae) as determined by allozyme electrophoresis. Ann Mo Bot Gar 85:97–109
Weller SG (1994) The relationship of rarity to plant reproductive biology. In: Bowles ML, Whelan CJ (eds) Restoration of endangered species: conceptual issues, planning, and implementation. Cambridge University Press, Cambridge, pp 90–117
Wyatt R, Broyles SB (1994) Ecology and evolution of reproduction in milkweeds. Annu Rev Ecol Syst 25:423–441

IV. Conclusions

13 Plant Population Viability: Where to from Here?

C.A. BRIGHAM

13.1 Introduction

Over the past 20 years, threats to plant populations have increased worldwide. Habitat destruction, invasive species, climate change, and disrupted disturbance regimes are just a few of the factors negatively affecting plant populations globally. As threats to plants have increased, so has interest and research in population viability analysis (PVA). Although there are relatively few PVAs for plant species, the number is increasing yearly (Menges 2000). PVA can be used to assess future status of populations or management strategies, compare populations, and identify sensitive life stages. In this volume, Bell and Bowles (Chap. 12, this Vol.) presented a novel use of PVA to explore the status and future of restoration projects. However, as many of the authors in this volume have pointed out, PVAs are very data-intensive, requiring years of monitoring data. The future of PVA and plant conservation is uncertain. In the balance, will land managers and conservation biologists use PVAs successfully to aid in rare plant management, or will this tool simply be too data-intensive? The future of PVAs and plants seems to lie in several areas outlined below.

13.2 Threats to Population Viability

Our understanding of the threats facing plant populations has increased over the past decades (see Chap. 2, this Vol.). While we currently understand many of the direct impacts of factors such as habitat destruction and invasive species, indirect effects are only beginning to be quantified. Ouborg and Biere (Chap. 4, this Vol.) discussed some of the possible indirect effects of habitat fragmentation on plant populations. In order to manage threats to plant population viability correctly, it is imperative that we learn more about the indirect effects of these factors. Work on the potential impact of exotic species on biotic interactions such as pollination or herbivory has only just begun. If all impacts of a given factor are negative, then enumerating each impact is

Ecological Studies, Vol. 165
C.A. Brigham, M.W. Schwartz (Eds.)
Population Viability in Plants

unnecessary for a correct management decision (i.e., remove the factor if at all possible). However, it is possible that some factors (such as habitat fragmentation, see Chap. 4, this Vol.) may have both positive and negative effects. Further, synergistic interactions may exist and may not be clear from field data that is not integrated into a modeling framework. As a result, designing management strategies becomes quite difficult. Additionally, many management strategies for rare plants involve attempts to ameliorate negative effects of a threat since the factor cannot be completely removed (for instance, the impossibility of getting rid of all exotic species in California grasslands). In such cases, we must identify all the negative impacts of a factor before we can work to ameliorate them. Thus, the first stage in PVA may be identification of threats and their effects. PVA provides a standardized method to quantify the effects of threats on population persistence.

13.3 Quantifying the Limits of Applicability of PVA Models

Over the past several years, a number of authors have examined the accuracy of different types of PVA models under different circumstances (see discussion in Chaps. 6, 7, this Vol.). These efforts indicate that the reliability of PVAs for prediction of the future status of populations based on limited data is suspect; however, some analyses suggest that PVAs may be quite good at predicting the general trend of populations (i.e., growing or declining) even with limited data sets and data uncertainties (e.g., unmeasured seed bank, see Chap. 7, this Vol.). More work needs to be done to identify under what circumstances PVAs make robust predictions. For instance, how many years of monitoring data are required to make predictions 50 years into the future? How does the number of years needed change depending on the life history of the species involved (e.g., plants with seed banks vs. long-lived perennials vs. annuals with no seed bank)? Specific analyses of both the uncertainty and reliability of estimates are required before conservation biologists and managers will be able to determine whether it is worthwhile to conduct a PVA in the first place, and also how to interpret PVA results.

13.4 PVAs and Relative Rankings

Several authors in this volume have suggested that perhaps the best use of PVA is to rank different populations or sites with regard to viability or extinction risk (see Chaps. 6, 7, this Vol.). This use of PVAs has been increasing (see discussion in Chaps. 6, 7, this Vol.); however, the reliability of rankings made through PVA has seldom been evaluated (although see Chap. 7, this Vol.). Specific analyses need to be conducted to determine the accuracy of PVAs when used to rank populations based on viability or relative risk.

13.5 Increasing the Complexity of PVAs

Several chapters in this volume addressed the inclusion of complicating factors such as genetics, spatial variability, biotic interactions, and disturbance regimes in PVA models. Although methods are now available to include such factors in PVA models, the models become more complicated and require more data to parameterize. Due to the increase in complexity and data needs for such models, such PVAs likely will not be conducted on a regular basis. Now is the time to ask under what circumstances we expect a given complicating factor to be important to population viability and thus required for inclusion in order for the PVA to be accurate. Oostermeijer (2000) has begun such an effort with his PVA for *Gentiana pneumonanthe*, which examines the importance of different small population effects in population decline. Such analyses may be conducted using species for which we have a large amount of data on the factors involved or may be simulated using artificial life histories (see example of simulated life histories in Chaps. 7, 9, this Vol.). Because complicating factors are likely to be left out of PVAs unless compelling evidence exists to the demonstrate their necessary inclusion, we must identify the conditions under which factors such as genetics, biotic interactions, and disturbance have significant impacts on estimates of viability.

13.6 Role of PVA in Plant Conservation and Management

Although PVA is a tool that has been used in managing several high-profile animal species (e.g., spotted owl, loggerhead sea turtles), its use in plant management is still relatively uncommon. One of the reasons for this is that the most common type of PVA model used is a stage-based matrix model (see description in Chap. 6). This type of model requires data on transitions between life stages for many years in order to generate extinction times and population projections. Linking PVAs to conservation will require more work assessing exactly what information is necessary for a PVA, on the reliability of PVA results, and the consequences of insufficient data.

Several approaches presented in this volume have great promise for increasing the use of PVA in plant management and conservation. First, Eldred et al. (Chap. 7, this Vol.) described the use of the diffusion approximation approach for modeling population viability. This approach requires only yearly counts of data, by far the most common type of data collected for plants by managers (see Chap. 7, this Vol.; Morris et al. 1999). Although this approach has been criticized as needing large numbers of years worth of data to predict persistence over a short time period (see discussion in Chap. 7, this Vol.), Eldred et al. (Chap. 7, this Vol.) actually found that the method is robust, generating predictions based on 10 years of data. They also found that the method tolerates measurement error and unknown life stages (e.g., seed banks) (for a full discussion of this method, see Chap. 7, this Vol.).

Assuming managers have the information required for some type of PVA, what can such analyses be used for? Several chapters highlighted the use of PVA for assessing relative risks of different populations. Managers are frequently in the position of having to choose among populations or species in terms of allocation of sparse resources. PVA may be a quantitative tool to assess the best directions of such efforts (although the accuracy of PVA for this use has not been well tested, see discussion above). Additionally, PVAs, or similar model types, have been used to simulate the effects of different management regimes on population persistence. Although greatest accuracy of these methods requires information from experimental approaches, PVA modeling can extend or broaden the information received from experiments to make predictions of the overall impact of different management strategies (see discussion in Chap. 6).

Several alternatives to strict PVAs are also described in this volume for use by managers who do not have the data necessary for a full PVA. Schwartz (Chap. 9) presented a decision-tree-based model for determining the level of threat faced by a plant population for which data are sparse. Data requirements for this type of analysis are limited (basically all that is required is sufficient knowledge of the species to make accurate estimations of the types of threats facing it), and it provides a structured framework for thinking about threats to species and overall trends in viability. Such a decision-tree-based model could be used to rank populations or species with regard to protection, management, or conservation actions. Another alternative was presented by Bell and Bowles for use in modeling populations when information on some transitions are missing. They used averages and data from other populations to make an initial estimate of population viability for restored populations with missing transitions.

Finally, Bell and Bowles (Chap. 12) also presented an innovative use of PVA that may be valuable to many restoration ecologists. One of the most difficult aspects of restoration is determining an endpoint or matrix of success for a project (see discussion in Chap. 12, this Vol.). Bell and Bowles used a PVA approach to examine viability of restored populations and compare transition rates and population structure of restored populations to natural populations. This approach provides a quantitative method for assessing restoration project success and also highlights ways in which restored populations may differ from natural populations.

13.7 Conclusions: Future of Plant Conservation and Population Viability

Sufficient research now exists on modeling plant population viability to examine the future of PVA and plant conservation. Although PVA will be useful to conservation biologists for all the reasons outlined above, there will still be many cases in which a PVA is unwarranted or impractical. For many plants, causes of decline and potential impacts of management are clear (see discussion in Chap 9, this Vol.). For such species, construction of a PVA may be a misdirection of limited time and resources due to the large input of funds and research hours necessary to obtain the monitoring data required for a formal PVA. However, research indicates that many endangered plants, at least in the United States, are in close proximity to urban centers (see Schwartz et al. 2002). The possibility of increased volunteer monitoring efforts for such species suggests that in the future there may be more data available for PVAs of plants.

Overall, PVAs will be most useful for species that have complex interactions, alternative management schemes, or indirect effects that are difficult to examine without a formal model. For species with simple biology and known population trends, a more simple approach may be preferable. Additionally, PVAs may be useful in designing management strategies or choosing targets for conservation (see discussion above). While this volume has highlighted the many possibilities for plant PVAs, it is clear that PVA is not the end all be all of plant conservation.

References

Menges ES (2000) Population viability analyses in plants: challenges and opportunities. Trends Ecol Evol 15:51–56

Morris WF, Doak D, Groom M, Kareiva P, Fieberg J, Gerber L, Murphy P, Thomson D (1999) A practical handbook for population viability analysis. The Nature Conservancy Press, New York

Oostermeijer J (2000) Population viability of the rare *Gentiana pneumonanthe*: the importance of genetics, demography, and reproductive biology. In: Young A, Clarke G (eds) Genetics, demography and viability of fragmented populations. Cambridge University Press, Cambridge, pp 313–333

Schwartz MW, Jurjavcic N, O'Brien J (2002) Conservation's disenfranchised urban poor. Bioscience 52:601–606

Subject Index

Ecological Studies

Volumes published since 1997

Volume 126
The Central Amazon Floodplain: Ecology of a Pulsing System (1997)
W.J. Junk (Ed.)

Volume 127
Forest Decline and Ozone: A Comparison of Controlled Chamber and Field Experiments (1997)
H. Sandermann, A.R. Wellburn, and R.L. Heath (Eds.)

Volume 128
The Productivity and Sustainability of Southern Forest Ecosystems in a Changing Environment (1998)
R.A. Mickler and S. Fox (Eds.)

Volume 129
Pelagic Nutrient Cycles: Herbivores as Sources and Sinks (1997)
T. Andersen

Volume 130
Vertical Food Web Interactions: Evolutionary Patterns and Driving Forces (1997)
K. Dettner, G. Bauer, and W. Völkl (Eds.)

Volume 131
The Structuring Role of Submerged Macrophytes in Lakes (1998)
E. Jeppesen et al. (Eds.)

Volume 132
Vegetation of the Tropical Pacific Islands (1998)
D. Mueller-Dombois and F.R. Fosberg

Volume 133
Aquatic Humic Substances: Ecology and Biogeochemistry (1998)
D.O. Hessen and L.J. Tranvik (Eds.)

Volume 134
Oxidant Air Pollution Impacts in the Montane Forests of Southern California (1999)
P.R. Miller and J.R. McBride (Eds.)

Volume 135
Predation in Vertebrate Communities: The Białowieża Primeval Forest as a Case Study (1998)
B. Jędrzejewska and W. Jędrzejewski

Volume 136
Landscape Disturbance and Biodiversity in Mediterranean-Type Ecosystems (1998)
P.W. Rundel, G. Montenegro, and F.M. Jaksic (Eds.)

Volume 137
Ecology of Mediterranean Evergreen Oak Forests (1999)
F. Rodà et al. (Eds.)

Volume 138
Fire, Climate Change and Carbon Cycling in the North American Boreal Forest (2000)
E.S. Kasischke and B. Stocks (Eds.)

Volume 139
Responses of Northern U.S. Forests to Environmental Change (2000)
R. Mickler, R.A. Birdsey, and J. Hom (Eds.)

Volume 140
Rainforest Ecosystems of East Kalimantan: El Niño, Drought, Fire and Human Impacts (2000)
E. Guhardja et al. (Eds.)

Volume 141
Activity Patterns in Small Mammals: An Ecological Approach (2000)
S. Halle and N.C. Stenseth (Eds.)

Volume 142
Carbon and Nitrogen Cycling in European Forest Ecosystems (2000)
E.-D. Schulze (Ed.)

Volume 143
Global Climate Change and Human Impacts on Forest Ecosystems: Postglacial Development, Present Situation and Future Trends in Central Europe (2001)
J. Puhe and B. Ulrich

Volume 144
Coastal Marine Ecosystems of Latin America (2001)
U. Seeliger and B. Kjerfve (Eds.)

Volume 145
Ecology and Evolution of the Freshwater Mussels Unionoida (2001)
G. Bauer and K. Wächtler (Eds.)

Volume 146
Inselbergs: Biotic Diversity of Isolated Rock Outcrops in Tropical and Temperate Regions (2000)
S. Porembski and W. Barthlott (Eds.)

Volume 147
Ecosystem Approaches to Landscape Management in Central Europe (2001)
J.D. Tenhunen, R. Lenz, and R. Hantschel (Eds.)

Volume 148
A Systems Analysis of the Baltic Sea (2001)
F.V. Wulff, L.A. Rahm, and P. Larsson (Eds.)

Volume 149
Banded Vegetation Patterning in Arid and Semiarid Environments (2001)
D. Tongway and J. Seghieri (Eds.)

Volume 150
Biological Soil Crusts: Structure, Function, and Management (2001)
J. Belnap and O.L. Lange (Eds.)

Volume 151
Ecological Comparisons of Sedimentary Shores (2001)
K. Reise (Ed.)

Volume 152
Global Biodiversity in a Changing Environment: Scenarios for the 21st Century (2001)
F.S. Chapin, O. Sala, and E. Huber-Sannwald (Eds.)

Volume 153
UV Radiation and Arctic Ecosystems (2002)
D.O. Hessen (Ed.)

Volume 154
Geoecology of Antarctic Ice-Free Coastal Landscapes (2002)
L. Beyer and M. Bölter (Eds.)

Volume 155
Conserving Biological Diversity in East African Forests: A Study of the Eastern Arc Mountains (2002)
W.D. Newmark

Volume 156
Urban Air Pollution and Forests: Resources at Risk in the Mexico City Air Basin (2002)
M.E. Fenn, L. I. de Bauer, and T. Hernández-Tejeda (Eds.)

Volume 157
Mycorrhizal Ecology (2002)
M.G.A. van der Heijden and I.R. Sanders (Eds.)

Volume 158
Diversity and Interaction in a Temperate Forest Community: Ogawa Forest Reserve of Japan (2002)
T. Nakashizuka and Y. Matsumoto (Eds.)

Volume 159
Big-Leaf Mahogany: Genetic Resources, Ecology and Management (2003)
A. E. Lugo, J. C. Figueroa Colón, and M. Alayón (Eds.)

Volume 160
Fire and Climatic Change in Temperate Ecosystems of the Western Americas (2003)
T. T. Veblen et al. (Eds.)

Volume 161
Competition and Coexistence (2002)
U. Sommer and B. Worm (Eds.)

Volume 162
How Landscapes Change: Human Disturbance and Ecosystem Fragmentation in the Americas (2003)
G.A. Bradshaw and P.A. Marquet (Eds.)

Volume 163
Fluxes of Carbon, Water and Energy of European Forests (2003)
R. Valentini (Ed.)

Volume 164
Herbivory of Leaf-Cutting Ants: A Case Study on *Atta colombica* in the Tropical Rainforest of Panama (2003)
R. Wirth, H. Herz, R.J. Ryel, W. Beyschlag, B. Hölldobler

Volume 165
Population Viability in Plants: Conservation, Management, and Modeling of Rare Plants (2003)
C.A Brigham, M.W. Schwartz (Eds.)

FSC
www.fsc.org
MIX
Papier aus verantwortungsvollen Quellen
Paper from responsible sources
FSC® C105338

If you have any concerns about our products,
you can contact us on
ProductSafety@springernature.com

In case Publisher is established outside the EU,
the EU authorized representative is:
Springer Nature Customer Service Center GmbH
Europaplatz 3, 69115 Heidelberg, Germany

Printed by Libri Plureos GmbH
in Hamburg, Germany